EXPLORING THE REAL NUMBERS

EXPLORING THE REAL NUMBERS

Frederick W. Stevenson
University of Arizona

PRENTICE Hall, Upper Saddle River, New Jersey 07458

Library of Congress Cataloging-in-Publication Data

Stevenson, Frederick W.
    Exploring the real numbers / Frederick W. Stevenson.
    p. cm.
    Includes index.
    ISBN: 0-13-040261-3
    1. Numbers, Real. I. Title
QA241.S78 2000
512'.7—dc21

99-055704

Acquisition Editor: *George Lobell*
Production Editor: *Bayani Mendoza de Leon*
Assistant Vice President of Production and Manufacturing: *David W. Riccardi*
Executive Managing Editor: *Kathleen Schiaparelli*
Senior Managing Editor: *Linda Mihatov Behrens*
Manufacturing Buyer: *Alan Fischer*
Manufacturing Manager: *Trudy Pisciotti*
Marketing Manager: *Melody Marcus*
Marketing Assistant: *Vince Jansen*
Director of Marketing: *John Tweeddale*
Editorial Assistant: *Gale Epps*
Art Director: *Jayne Conte*
Cover Designer: *Bruce Kenselaar*
Cover Photo: *"Trackwork"* ©*Vikki Chenette. Cottons with New York City Transit
    Authority subway token, hand and machine pieced, hand quilted, 36 x 36 in. Collection
    of the artist. Design by Tom Spence. Photo by Karen Bell.*

Printed in the United States of America

10 9 8 7 6 5 4 3 2 1

ISBN 0-13-040261-3

Prentice-Hall International (UK) Limited, *London*
Prentice-Hall of Australia Pty. Limited, *Sydney*
Prentice-Hall Canada Inc., *Toronto*
Prentice-Hall Hispanoamericana, S.A., *Mexico*
Prentice-Hall of India Private Limited, *New Delhi*
Prentice-Hall of Japan, Inc., *Tokyo*
Pearson Education Asia Pte. Ltd.
Editora Prentice-Hall do Brasil, Ltda., *Rio de Janeiro*

Contents

Preface

This book is about the real numbers. These are the numbers that can be used to count things and measure lengths of objects. They lie on the mathematical ruler − the real number line. The real number line includes the counting numbers $1, 2, 3, \ldots$, zero, the negative integers, the fractions, and all the in-between numbers that aren't fractions. It is our intent to bring the reader up to date with the study of the nature of real numbers and, at the same time, give a flavor of the historical journey that got us where we are. It is a journey of tens of thousands of years.

There is evidence that as long ago as 30,000 years people were counting by tallying marks in stone. For discrete quantities, like people and objects, notches in stone would be a way to do this. Over the centuries various notations were invented to accommodate more easily large numbers of notches. Useful notations for nonintegral quantities were slow in coming. Since not everything around us is whole, surely a need for denoting parts of a whole was recognized long ago, but it wasn't until around 3000 B.C. that the Egyptians developed a notation that was useful. The Egyptians used unit fractions (that is, fractions with numerator 1) to express most of their fractions. The existence of quantities different from whole numbers or fractions of whole numbers was recognized by the Greek mathematicians. This came from their study of geometry. Questions of measurement in geometry, such as ratios of lengths, were the object of inquiry of the Pythagorean school of mathematicians, beginning in about 500 B.C. They understood that, regardless of the size of a unit that might be chosen, there were contrasting lengths that could not be measured exactly with the same unit. With this came the realization that irrational numbers existed. Thus, from whole numbers to irrational numbers was a journey through history of perhaps $30,000$ years. In the last $2,000$ years, mathematics has become a discipline

in its own right and questions about the nature of numbers, as well as
the use of numbers, have occupied mathematicians. During the past
200 years, mathematical abstraction has flourished. Many abstract
systems have been created to address problems related to number
theory. During the past 20 years, computers have provided us with
incredible powers of calculation.

With all the great thinkers that have come before and all the
newfound power to compute that exists today, it is ironic that there
are even more questions about numbers now than ever before. One
reason for the seemingly inexhaustible questions that continue to
challenge the minds of mathematicians is the inexhaustible supply
of numbers. There are infinitely many whole numbers. And the rationals and the irrationals are tucked in infinite quantity between
any pair of real numbers. Computers are powerful and will become
more powerful, but they can never reach really large numbers because
there is no limit to the size of numbers. Only the mind can approach
infinity. For example, it can be shown that there are infinitely many
prime numbers, but it is generally impossible to determine if a particular number is prime or not. And it has not yet been determined
it there are infinitely many twin primes; that is, primes that are two
units apart. Also, it is known that the decimal expansion of π is nonrepeating, but little else is known about it. For example, it is not
known if the digits in the expansion are equally represented. Some
would argue that since the expansion is not known, the number itself
is not known. This is a philosophical question, but it is like many
of the mathematical questions we explore here; it will remain forever
an intriguing challenge for the human mind yet will remain forever
outside the reach of supercomputers.

This book is written with two purposes in mind; first, to present
the many interesting topics that arise during the study of the real
numbers; second, to provide the readers an opportunity to further
study these topics on their own.

The first purpose is addressed in Chapters 1 through 4. The topics are many, varied, and fascinating. They include prime numbers,
perfect numbers, Carmichael numbers, constructible numbers, arithmetic numbers, algebraic numbers, polygonal numbers, transcendental numbers, uncountability, the golden ratio, π, cube duplication,
angle trisection, unit fractions, continued fractions, Pell's equations,
Diophantine equations, Farey sequences, cryptography, Pythagorean
triples, Fermat's Last Theorem, and the formulas for general cubics

and quartics.

The second purpose is addressed in Chapter 5. This chapter contains 21 research projects. Increasingly, students are being encouraged to pursue research at the undergraduate level. It is believed that the frontiers of science are no longer the private domain of professionals. The computer has been a chief instrument in allowing young students to participate in scientific research. And even if the student does not discover new and significant results the process of exploration is very much worth the effort. Doing mathematical research is especially risky, however, because there is no guarantee that anything will be discovered. These projects are meant to lessen that risk by providing paths that lead in promising directions. The hope is that the journey will remain challenging but will also be fruitful.

The book is written as a textbook, but it should not be viewed as a book that needs to be studied in the classroom. With more than 250 carefully worked examples, this material is quite accessible to the layman as well as the student. A calculus background is helpful for mathematical maturity and a calculator is helpful when large numbers or lengthy decimal expansions arise. The book has 350 exercises along with the 21 mathematical projects; the exercises are meant to keep the reader current with the text, and the projects are meant to stretch beyond.

As a university course, this book is intended to be covered in one semester, but teaching every section of every chapter would be an ambitious goal indeed. There are natural compromises; for example, one could go directly from Section 3.2 to 4.3 and introduce irrational numbers through continued fractions. Generally, if real numbers are the preferred focus, selected sections of Chapters 2 and 3 can be omitted so that Chapter 4 is afforded plenty of time. If number theory is the focus, then it is easy to settle on the first three chapters and topics of the teacher's choosing from Chapter 4. Each section of Chapter 4 studies a different aspect of irrational numbers: algebraic, geometric, trigonometric, and analytic. The last section, 4.4, deals with transcendental numbers, and perhaps there would be time only to view the beautiful expressions for π. It should be kept in mind that the degree of emphasis placed on the projects can greatly affect the pace. It is expected that a project will be an integral part of the course. In one semester one project, done completely, would probably suffice. Mathematical research can take dozens of hours and it is important that the research be carefully written up and the results be communicated to colleagues. It would be natural to have the students present their research orally to the class and this could easily take two weeks. In any case, the book should not be seen as a work that has to be studied from cover to cover. Rather it should conform to the interests and the tempo of the instructor and the reader.

This book can be used in a variety of settings. At the University of Arizona it has been used as a course for future high school teachers. This is a good fit. Not only should teachers be conversant with the ideas presented in this book, they should do a research project so that they can gain experience in exploratory mathematics. In its 1991 publication, *Standards for Teaching Mathematics*, The National Council of Teachers of Mathematics urged that the next generation of teachers be taught the importance of student discovery. Routine computation should be buttressed, if not entirely replaced, by inquiry, conjecture and exploration of nonroutine problems.

The book can also serve as an introduction to mathematical thought, a short course in number theory, an honors course at either the high school and the college level, or an introduction to mathematical research. But no matter how this book is used it must be remembered that mathematics is more than a subject with procedures to solve problems and an appendix with answers. It is a human endeavor, complete with false approaches, unanswered questions, multitudes of guesses, and plenty of curiosity and passion. It is more than a subject to be learned, it is a journey to be traveled.

I want to thank the students in my classes who have been forced to pore over this book in its many forms. There is nothing more helpful to an author of a textbook than to teach the material to students. While the students had no choice in the matter the same cannot be said of my colleagues. Olga Yiparaki, John Brillhart, and Bill Martin actually volunteered to read the manuscript. Thanks go to them for their valuable suggestions. And then there's Larry Grove. He not only read the manuscript but gave a remarkable amount of his time to help me learn TeX and put the book in final form. Surely there are few, if any, people more generous than Larry. And, finally, there are few if any people who could put up with me during the countless hours I spent thinking about this project. On numerous occasions my wife, Cheryl, would be discussing something with me and I would be absolutely oblivious to what she was saying. She is not expecting thanks but she is getting it anyway.

The Real Numbers

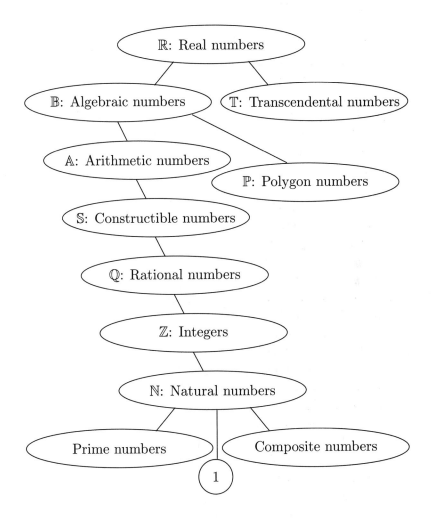

$\mathbb{R}$: Real numbers

$\mathbb{B}$: Algebraic numbers

$\mathbb{T}$: Transcendental numbers

$\mathbb{A}$: Arithmetic numbers

$\mathbb{P}$: Polygon numbers

$\mathbb{S}$: Constructible numbers

$\mathbb{Q}$: Rational numbers

$\mathbb{Z}$: Integers

$\mathbb{N}$: Natural numbers

Prime numbers

Composite numbers

1

EXPLORING THE REAL NUMBERS

Chapter 1

The Natural Numbers

The numbers 1, 2, 3, 4, ... go on forever. They are often called the "natural numbers" probably because they are used to record the most elemental of processes: counting. And while these numbers used for counting are the simplest of all numbers, they are the most fascinating. This was true in ancient times, when the knowledge of the universe of numbers was limited, and it remains true today after extensive study and significant discovery. In ancient civilizations, certain numbers were accorded special significance. For example, the number 5 was a mystical number to the ancient Babylonians, Greeks, and Egyptians. The mystery was tied in with the pentagram, or five-pointed star, which in turn is related to the magical Golden Section. The number 6 was considered "perfect" by Euclid, the great Greek mathematician who lived around 300 B.C. While his name is synonymous with geometry, he was fascinated with numbers as well. A perfect number, as defined by Euclid, is a number whose proper divisors, including 1, add up to itself. For the number 6 those divisors are 1, 2, and 3, and it's true that $1 + 2 + 3 = 6$. The number 10 has had special significance for centuries. It is of obvious importance to us because it is the base of our number system, the decimal system. It also has nice arithmetic and geometric properties. As the bowling pin arrangement tells us, we can arrange 10 dots in a equilateral triangle. Numbers of dots that have this property are called triangular numbers. The first few triangular numbers are 1, 3, 6, and 10. It is also possible, with 10 balls, to build a tetrahedron, a triangular pyramid. Such numbers of balls for which this is true are called tetrahedral numbers. The first three tetrahedral

1

numbers are 1, 4, and 10. Larger numbers have special properties too. In fact, there are mathematicians who see fascinating properties in almost every number. G. H. Hardy, the great English mathematician (1877−1947), recounted a visit he had with the Indian genius Ramanujan (1887−1920). Hardy writes in his book *Ramanujan*,

> I remember going to see him once when he was lying ill in Putney. I had ridden in taxi-cab No. 1729, and remarked that the number seemed to me to be a rather dull one, and I hoped it was not an unfavorable omen. "No", he reflected, "it is a very interesting number; it is the smallest number expressible as the sum of two cubes in two different ways."

Those two ways are $1729 = 12^3 + 1^3 = 10^3 + 9^3$.

In this chapter we will study this fascinating and remarkably complex set of numbers. Our main focus will be on the prime numbers. Perhaps more than any other set of numbers, mathematicians have studied the prime numbers. They command such attention because of their theoretical importance to number theory. Primes are the atoms of the multiplicative structure of that universe; every natural number can be uniquely built out of a product of primes. This is known as the fundamental theorem of arithmetic. In practice it is not easy to analyze the prime makeup of numbers. Of course, if the numbers are small enough (for example, less than 1 million), the table of primes given in Section 1.3 can be used to analyze the prime makeup. If numbers have fewer than 100 digits, computers can effectively do the job. But most numbers are much larger. Super computers are helping humans see deeper into the number universe, but analyzing the makeup of numbers will always present one of the ultimate challenges for mathematicians.

1.1   The Basics

We begin the book with a potpourri of fundamental topics that should be familiar to the reader.

Let us start with a most basic study: how we write numbers. It was probably very early in the cultural development of humankind that it became necessary to record the results of counting objects. And the number of objects counted likely became so large that a simple tallying became too tedious. To handle large amounts a system needed to be invented. Many of the systems used the concepts of grouping; the most organized of such systems is called the base system. Here's how it works: Fix a number greater than 1, call it n, and begin counting. Arrange the objects in groups of n; then arrange the groups themselves in groups of n, and then do the same with those groups and continue in this fashion until all the objects are accounted for. Let's see how this works.

Example 1.1.1

(a) Suppose you were sentenced to serve six months in jail and you want to count off the days as you serve them. As we know, prisoners tally up the days in groups of five by marking four vertical lines and a cutting a fifth line through them. Suppose also that the groups of five are tallied in an analagous way as well; that is, five groups of five are lumped together in a single super group. And, continuing in this fashion, five super groups are lumped together into an ultra super group and so on. If your sentence is one hundred and eighty three days (we write the numbers out here so as not to favor a particular base system) here is how the grouping works.

Your first grouping yields thirty six groups of five days with three days left over. In mathematical notation it reads $186 = 36(5) + 3$. The thirty six groups are then organized into seven super groups of five fives with one group of five left over. That is, $36 = 7(5) + 1$. The seven super groups are organized into one ultra super group of five super groups with two super groups left over. That is, $7 = 1(5) + 2$. And we are finished. Our sentence is served. The length of the sentence is three days plus one group of five days plus two super groups of days plus one ultra super group of days. This is $3 + 1(5) + 2(5 \times 5) + 1(5 \times 5 \times 5)$. We can write it this way: $183 = 1(5^3) + 2(5^2) + 1(5^1) + 3$. In shorthand notation we write 183 as 1213_5.

(b) In base 10 this same example loses its mystery. The number 183 gets parsed as follows: There are eighteen groups of ten days with three days left over. That is, $183 = 18(10) + 3$. In grouping the groups of ten together we have one super group of ten groups with eight left over. That is, $18 = 1(10) + 8$. Thus we have one super group of ten tens plus eight groups of ten plus three. So one hundred and eighty three is $3 + 8(10) + 1(10 \times 10)$. Rewriting we get

$1(10^2) + 8(10^1) + 3$. So 183, in base 10, looks like 183_{10} which is no surprise.   ∎

Formally, we may put it this way:

Definition 1.1.2 *The number N is written in* **base** *n if*

$$N = a_k n^k + a_{k-1} n^{k-1} + \cdots + a_2 n^2 + a_1 n + a_0,$$

where n is a given number > 1, $k \geq 0$, and $0 \leq a_j < n$.

Notation: In base n, the number N of Definition 1.1.2 is denoted by $a_k a_{k-1} \ldots a_2 a_1 a_0$.

Historically several different base systems have been used. A vigesimal system, based on groupings by 20, was used by American Indians. Base 20 was also used, in a more developed form, by the Mayans. The largest known base system was base 60, the sexagesimal system. It was used by the Babylonians and still persists in our measurement of time and of angles. For example, there are 60 seconds in a minute, 60 minutes in an hour, 360 degrees in a circle. There are traces of a base 12 system that we see today in our use of dozens and grosses. There is also evidence of rudimentary base 3, base 4, and base 5 systems. Base 2 is especially attractive because of its simplicity and it is used today in the fields of electrical engineering and computer science. The most well known of the nonbase systems is the system of Roman numerals. Numbers in this system follow groupings, but not by a fixed size. There are groupings of 5 and 10 and 50 and 100 and beyond. And place value is used in an elementary way.

Of course, the most widely used system today is the base 10 system of notation. This dates back to Hindu mathematicians around 600 A.D. Decimal notation was introduced into Europe in 1202 by Leonardo of Pisa, also known as Fibonacci. Decimal notation for numbers between 0 and 1 dates back to about 1600 in Europe, but it was known 600 years before by Arab mathematicians. The strength of base 10, aside from its being the number of fingers on both hands, is that it seems to strike a happy medium between using few symbols for digits yet enough digits so that really large numbers can easily be accommodated.

Historically number systems have had difficulty dealing with the enormity of the number universe. The Romans had no symbol for numbers beyond 100,000, and the Greeks' largest number was "myriad", which was 10,000. For the ancient Chinese 10,000 years was eternity. Interest in the size of the universe of numbers dates back to antiquity. In the third century B.C. Archimedes wanted to estimate the number of grains of sand in the universe. He devoted a whole treatise to this purpose, *The Sand Reckoning*. This work is addressed to King Gelon of Syracuse. It begins like this:

> There are some, King Gelon, who think that the number of grains of sand is infinite in multitude; and I mean by the sand not only that which exists about Syracuse and the rest of Sicily, but also that which is found in every region, whether inhabited or uninhabited. Again there are some who, without regarding it as infinite, nevertheless think that no number has been named which is great enough to exceed its size.

In addressing this problem Archimedes devised a way of expressing large quantities. He did not use a base system but his estimation, expressed in base 10, is somewhere on the order 10^{50}; that is, 1 followed by 50 zeros. The early Hindus developed a number system that could name even larger numbers, numbers as large as 153 digits in base 10. We have a whimsical name for certain convenient large numbers. The number 1 followed by 100 zeros (that is, 10^{100}) is called a **googol**. To get an idea of the size of the googol you can compare it with the estimated number of atoms in the universe. That number is something like 10^{80}. Then there is the **googolplex**. This number is 1 followed by a googol of zeros (that is, $10^{10^{100}}$). You can see how we can build up super large numbers using exponents piled one on top of another.

For the mathematician the fact that base n notation can express quantities of any size is due to the division theorem. We shall give it in its general form later in this section (Theorem 1.1.9). Here we give a specialized version of the theorem that applies to base 10.

Theorem 1.1.3 *Given the natural number N, there exist unique numbers q and r such that $N = 10q + r$ where $0 \leq r < 10$.*

The process of numbering an amount, N, is a repetitive use of Theorem 1.1.3. After finding r from Theorem 1.1.3, we may relabel it a_0 and apply the theorem to the number q. Using the labeling

$q = 10q_1 + a_1$, we have

$$N = 10(10q_1 + a_1) + a_0 = 10^2q_1 + 10a_1 + a_0.$$

Repeating this a second time, we get $q_1 = 10q_2 + a_2$. So

$$N = 10^3q_2 + 10^2a_2 + 10a_1 + a_0.$$

And so it goes until we reach a quotient that is smaller than 10. Thus the number

$$N = 10^k a_k + 10^{k-1}a_{k-1} + \ldots + 10a_1 + a_0$$

is written in shorthand as $a_k a_{k-1} \ldots a_0$.

It is important that this notation be unique; that is, there is only one base 10 representation for N. If there were ambiguity in the way numbers are written it would lead to chaos, both in practice and in mathematics. Let us examine the uniqueness of representation of natural numbers in base 10 a bit more carefully. Suppose that $a_k a_{k-1} \ldots a_i \ldots a_1 a_0$ and $b_k b_{k-1} \ldots b_i \ldots b_1 b_0$ represent the same number. Suppose, for the sake of argument, that the representations are different. Thus there is a place, i, where $a_i \neq b_i$. Let i mark the largest occurrence of inequality (that is, $a_k = b_k$ for $k > i$) and suppose that $a_i > b_i$. We can look at the difference; since the numbers are the same it must by 0. So

$$\begin{aligned}(a_k a_{k-1} \ldots a_i \ldots a_1 a_0) &- (b_k b_{k-1} \ldots b_i \ldots b_1 b_0) \\ &= (a_i \ldots a_1 a_0) - (b_i \ldots b_1 b_0) \\ &= 10^i(a_i - b_i) + (a_{i-1} \ldots a_1 a_0) - (b_{i-1} \ldots b_1 b_0) \\ &= 0.\end{aligned}$$

It follows that

$$10^i(a_i - b_i) = (b_{i-1} \ldots b_1 b_0) - (a_{i-1} \ldots a_1 a_0).$$

But the largest the righthand side can be is $99 \ldots 9$ (i nines) while the smallest the lefthand side can be is $100 \ldots 0$ (that is, 1 followed by i 0s) a larger number (by 1). Thus they cannot be the same.

Numbers can be added, subtracted, multiplied, and divided. In the set of natural numbers, addition and multiplication can be carried out without restriction, while subtraction and division have their limitations. What we mean is that the sum of two natural numbers is always a natural number. The same goes for the product of two natural numbers. But this is not true in general for the difference and the quotient of two natural numbers. For example, $5 - 13$ is not a natural number; nor is $11 \div 7$. This makes life interesting for us. In particular, it is the lack of ready divisibility that drives our study in this chapter. Let us recall some familiar definitions. The word "number" here refers to "natural number", although the concepts of factoring and divisibility transfer to many other systems of numbers.

Definition 1.1.4

*(a) Given numbers a and b, we say that a **divides** b if there exists a number c such that $ac = b$. If a divides b then we say that a is a **factor** of b or a is a **divisor** of b. Also, we say that b is a **multiple** of a.*

*(b) p is a **prime number** if the only divisors of p are itself and 1.*

*(c) A number, other than 1, that is not a prime number is a **composite number**.*

(d) 1 itself is neither prime nor composite.

*(e) The **greatest common divisor** of a set of numbers is the largest of the divisors common to all of the numbers. This is also called the **greatest common factor**.*

*(f) The **least common multiple** of a set of numbers is the smallest of the multiples common to all of the numbers.*

*(g) Two numbers are **relatively prime** if they share no common prime factor; that is, their greatest common factor is 1.*

Notation: If a divides b we shall write $a \mid b$. The greatest common divisor of a and b is denoted by $\gcd(a, b)$, the least common multiple of a and b is denoted by $\mathrm{lcm}(a, b)$.

Here are some elementary theorems that follow directly from the definitions.

Theorem 1.1.5 *If a, b, and k are natural numbers, then $a \mid b$ if and only if $ak \mid bk$.*

Proof Suppose that $a \mid b$. Then there is a c such that $ac = b$, Therefore, $(ac)k = bk$. Hence $(ak)c = bk$ and so $ak \mid bk$. Conversely, if $ak \mid bk$, then there is a c such that $(ak)c = bk$. Thus $(ac)k = bk$.

Since $k \neq 0$, we may divide it evenly into both sides, thus giving us $ac = b$. So $a \mid b$. □

Theorem 1.1.6 *Let a, b, d, and r be natural numbers. Then*

1. *if $d \mid a$ then $d \mid ra$*

2. *if $d \mid a$ and $d \mid b$ then $d \mid (a + b)$*

3. *if $d \mid (a + b)$ and $d \mid a$ then $d \mid b$*

The proof is left as an exercise.

Mixing our understanding of base 10 with our understanding of divisibility, we can make some statements about when numbers are divisible by other numbers just by staring at the representation. For example, we know that the number $a_k a_{k-1} \ldots a_i \ldots a_1 a_0$ is divisible by 2 if a_0 is divisible by 2. We know that $a_k a_{k-1} \ldots a_i \ldots a_1 a_0$ is divisible by 3 when the sum of the digits is divisible by 3. Here is a theorem, most of which is familiar to you.

Theorem 1.1.7 *Let $N = a_n a_{n-1} \ldots a_2 a_1 a_0$. Then*

1. $2 \mid N$ *if and only if* $2 \mid a_0$

2. $4 \mid N$ *if and only if* $4 \mid a_1 a_0$

3. $5 \mid N$ *if and only if* $5 \mid a_0$

4. $3 \mid N$ *if and only if* $3 \mid (a_n + a_{n-1} + \cdots + a_2 + a_1 + a_0)$

5. $9 \mid N$ *if and only if* $9 \mid (a_n + a_{n-1} + \cdots + a_2 + a_1 + a_0)$

6. $11 \mid N$ *if and only if*
$$11 \mid ((-1)^n a_n + (-1)^{n-1} a_{n-1} + \cdots + (-1)^1 a_1 + (-1)^0 a_0)$$

Proof (1) We may write N as

$$(10^n a_n + 10^{n-1} a_{n-1} + \cdots + 10^2 a_2 + 10 a_1) + a_0.$$

We know from Theorem 1.1.6 that since $2 \mid 10^k$ for $k \geq 1$,

$$2 \mid (10^n a_n + 10^{n-1} a_{n-1} + \cdots + 10^2 a_2 + 10 a_1).$$

So, by 1.1.6 (2) if $2 \mid a_0$, then $2 \mid N$. By 1.1.6 (3), if $2 \mid N$, then $2 \mid a_0$.

(4) We may write N as

$$[(10^n - 1)a_n + a_n] + \cdots + [(10^2 - 1)a_2 + a_2] + [(10^1 - 1)a_1 + a_1] + a_0$$
$$= [(10^n - 1)a_n + \cdots + (10^2 - 1)a_2 + (10^1 - 1)a_1] + [a_n + \cdots + a_2 + a_1 + a_0].$$

Since $3 \mid (10^k - 1)$ for any k, it follows from Theorem 1.1.6 that 3 divides the first summand; that is,

$$3 \mid [(10^n - 1)a_n + \cdots + (10a_1 - 1)a_1].$$

It then follows by Theorem 1.1.6 (2) that if $3 \mid (a_n + a_{n-1} + \cdots + a_1 + a_0)$, then $3 \mid N$ and by Theorem 1.1.6 (3) that if $3 \mid N$, then $3 \mid (a_n + a_{n-1} + \cdots + a_1 + a_0)$.

We leave the proofs of the other statements for the exercises. $\square$

We shall finish this section with some axioms, or principles, and some basic theorems. The set of natural numbers can be formalized into an abstract system, and while it is not our intention to prove everything rigorously in this book it is important to set a context for our study. At the end of the twentieth century we are comparatively sophisticated about mathematics and we must address the question of "why is this true?" if and when we can. So let us set the scene. The natural numbers, as stated previously, begin with the number 1 and continue on forever. Because of the immensity of this number universe —it has infinitely many members— it can be difficult to establish facts that hold true for all numbers. The most familiar method of establishing facts is the method of proof by induction. Let us review this.

The First Induction Principle *Let $\mathcal{P}(n)$ be a statement that is either true or false for any number $n \geq 1$.*

Suppose that

(a) $\mathcal{P}(1)$ is true

(b) For each number $n \geq 1$, if we know that $\mathcal{P}(n)$ is true, then $\mathcal{P}(n+1)$ is true as well

Then we conclude that $\mathcal{P}(n)$ is true for all numbers, n.

The first induction principle says simply that if a statement can be shown to be correct for the number 1 and furthermore if the truth of the statement for some given number is good enough to ensure the truth of the statement for the following number, then the statement is true for all numbers. This makes sense because the numbers themselves are arranged in an orderly progression beginning with the number 1 and proceeding one after another. Here is an example showing how we can use induction to prove facts about the number system.

Example 1.1.8

(a) Pick a number, any number. Now cube it; add 5 times your number to it; finally, add 6 more to your total. I'll bet your answer is divisible by 3. Putting it into the language of algebra, we are saying that any number of the form $n^3 + 5n + 6$ is divisible by 3. Try some examples: If $n = 1$ we get 12, $n = 7$ gives 384, and $n = 103$ obtains 1093248. Clearly 3 divides 12, it is easily figured that 3 divides 384 and with a calculator we find that $1093248 \div 3 = 364416$. Of course, we could also add the digits $1 + 0 + 9 + 3 + 2 + 4 + 8 = 27$ and notice $3 \mid 27$. So a spot check gives the statement some credibility. Here's an induction proof:

Let $\mathcal{P}(n)$ be the statement $n^3 + 5n + 6$ is divisible by 3.

We have already seen that $\mathcal{P}(1)$ is true; 12 is divisible by 3.

Next we suppose that $\mathcal{P}(n)$ is true and, given that assumption, check out the truth of $\mathcal{P}(n+1)$. $\mathcal{P}(n+1)$ says that $(n+1)^3 + 5(n+1) + 6$ is divisible by 3. Here is the algebra.

$$
\begin{aligned}
(n+1)^3 &+ 5(n+1) + 6 \\
&= n^3 + 3n^2 + 3n + 1 + 5n + 5 + 6 \\
&= (n^3 + 5n + 6) + (3n^2 + 3n + 6)
\end{aligned}
$$

Both of these summands are divisible by 3, the first because of the induction hypothesis, $\mathcal{P}(n)$, the second because every term is a multiple of 3. So $\mathcal{P}(n+1)$ is true given that $\mathcal{P}(n)$ is true. By induction we have proved that the statement is true for all numbers n.

Notice that we can play with this statement and create other facts. For example, we can change the 5 to a 2 or an 8, or any number of the form $3k + 2$, where k stands for a natural number. And we can change the 6 to be any number that is a multiple of 3. The exercises offer the opportunity for you to be creative.

(b) It's a fact that $10^n + (3 \times 4^n) + 5$ is always divisible by 9. This is a weird sort of statement but it is perfect for an induction proof.

Let $\mathcal{P}(n)$ be the statement $10^n + (3 \times 4^n) + 5$ is divisible by 9.

$\mathcal{P}(1)$ is true because $10^1 + (3 \times 4^1) + 5 = 27$ and this is divisible by 9.

Suppose that $\mathcal{P}(n)$ is true and check out the truth of $\mathcal{P}(n+1)$. $\mathcal{P}(n+1)$ says that $10^{n+1} + (3 \times 4^{n+1}) + 5$ is divisible by 9.

Using some clever rearrangement of terms, we get

$$10^{n+1} + (3 \times 4^{n+1}) + 5$$
$$= 10^n(10-1) + 10^n + (3 \times 4^n)(4-1) + (3 \times 4^n) + 5$$
$$= (10^n + (3 \times 4^n) + 5) + (10^n(10-1) + (3 \times 4^n)(4-1)).$$

Both of these summands are divisible by 9, the first because of the induction hypothesis, $\mathcal{P}(n)$, the second because each term is a multiple of 9. So $\mathcal{P}(n+1)$ is true given that $\mathcal{P}(n)$ is true. By induction we have proved that the statement is true for all numbers n.

As in (a), we can play with this and create new theorems. For example, we can change 4^n to 4^{n+1} or 4^{n+2} or 4^{n+k} for any positive k. We can change the 4 to 7 or to any number of the form $3l+1$. And we can change the 5 to anything of the form $9l+5$ for any integer l.

(c) Add the first n numbers and then square the result. Now add the first n perfect cubes. Compare your two sums. For example, $(1+2+3+4)^2 = 10^2 = 100$. And $1^3 + 2^3 + 3^3 + 4^3 = 1+8+27+64 = 100$. It turns out they are the same; we wonder if they will always be. Here is how we can phrase this for an induction inquiry.

Let $\mathcal{P}(n)$ be the statement $1^3 + 2^3 + \cdots + n^3 = (1+2+\cdots+n)^2$.

$\mathcal{P}(1)$ is true because $1^3 = 1^2 = 1$.

Suppose $\mathcal{P}(n)$ is true and consider $\mathcal{P}(n+1)$. That is, consider the truth of the equality:

$$1^3 + 2^3 + \cdots + n^3 + (n+1)^3 = (1+2+\cdots+n+(n+1))^2.$$

Starting with the righthand side, we have

$$(1+2+\cdots+n+(n+1))^2$$
$$= [(1+2+\cdots+n)^2] + [2(1+2+\cdots+n)(n+1)] + [(n+1)^2]$$
$$= [1^3 + 2^3 + \cdots + n^3] + [(n)(n+1)(n+1)] + [(n+1)^2]$$
$$= 1^3 + 2^3 + \cdots + n^3 + (n+1)^3.$$

The first equality follows from algebra; the second equality follows from the induction hypothesis. The third equality follows from algebra.

It follows by induction that the statement is true for all numbers n.

(d) The statement $2^n < n!$ is not true for all natural numbers. It fails for $n = 1, 2$, and 3. But for numbers larger than 3 it is true. While technically the first induction principle, as stated, does not apply, we can alter it and proceed by beginning with the number $n = 4$ rather than $n = 1$. Here is how it goes.

Let $\mathcal{P}(n)$ be the statement: $2^n < n!$ for all numbers $n > 3$.

Clearly $\mathcal{P}(4)$ is true because $2^4 = 16$ and $4! = 24$.

Suppose $\mathcal{P}(n)$ is true and consider $\mathcal{P}(n+1)$; that is $2^{n+1} < (n+1)!$. The following series of equalities and inequalities show what we want.

$$2^{n+1} = 2 \times 2^n < 2 \times n! < (n+1) \times n! = (n+1)!.$$

The first inequality is true because of the induction hypothesis, and the second follows from the fact that $n + 1 > 2$.

This completes the induction proof that $2^n < n!$ is true for numbers $n \geq 4$.   ■

Here are two other useful principles that are true of the natural numbers:

The Second Induction Principle   *Let $\mathcal{P}(n)$ be a statement that is either true or false for any number $n \geq 1$.*

Suppose that

(a) $\mathcal{P}(1)$ is true.

(b) For each number $n \geq 1$, if we know that $\mathcal{P}(m)$ is true for all $m \leq n$, then $\mathcal{P}(n+1)$ is true as well

Then we can conclude that $\mathcal{P}(n)$ is true for all numbers, n.

The Well-Ordering Principle   *Every nonempty set of natural numbers contains a smallest member.*

Like the first induction principle, these two principles agree with common sense. The second principle appears to be stronger than the first, but they are logically equivalent. The well-ordering principle simply says what we already believe; that any subset of natural numbers has a first element. This statement would not be true, for example, for the set of integers or for the set of positive fractions.

Let us now use the well-ordering principle to prove a more general theorem about the natural numbers. It is the general form of the division theorem that we cited earlier in Theorem 1.1.3. This theorem, like the principles upon which it depends, makes good sense. In a practical sense, if we seek to reach a destination that is b units away and we walk toward it with steps of length a units, we will eventually reach it. Either we reach it exactly after q steps, or we fall just short by the length $r < a$ and we can go beyond our destination with the next step. To put it concisely, b lies between two consecutive multiples of a.

Theorem 1.1.9 (The Division Theorem) *If a and b are natural numbers such that $a < b$, and a is not a divisor of b, then there exist unique numbers q and r such that $b = aq + r$ where $1 \leq r < a$.*

Proof   We prove the existence of q and r using the well-ordering principle.

Let S be the set $\{b - ax : x \geq 1 \text{ and } b - ax \geq 1\}$.

Since $b > a$, it follows that $b - (a \cdot 1)$ is in S, so S is a nonempty set of natural numbers. Therefore, by the well-ordering principle, S has a least element, call it r. Since r is in S, it follows that $r \geq 1$, and $r = b - aq$ for some $q \geq 1$.

Now $r < a$ by the following argument: If $r = a$, then $a = b - aq$ so $b = a(q + 1)$, which implies that $a \mid b$, which contradicts the fact that a is not a divisor of b. Thus $r \neq a$.

If $r > a$, then $r - a \geq 1$. Also, $r - a = b - a(q+1)$ so $r - a$ is in S, contradicting the fact that r is the least member of S. So $r \not> a$. We conclude that $r < a$. This completes the existence part of the proof.

For uniqueness, suppose that q' and r' satisfy the equation $b = aq' + r'$. So we have $a(q - q') = r' - r$. This means that $r = r'$ by the following argument. Suppose that $r \neq r'$ and $r > r'$. Since both r and r' are positive numbers less than a, it follows that $r - r'$ is less than a. The only way for this to happen is for $q = q'$. Thus $r = r'$. If $r < r'$ we get the same conclusion. So q and r are unique. $\square$

Next we prove a most useful theorem. It will be used in proving the fundamental theorem of arithmetic in the next section and will be referred to extensively in Chapter 2. It says that the greatest common divisor of two arbitrary numbers a and b can be expressed as a linear combination of a and b. This means that $\gcd(a, b)$ can be expressed by a number of the form $by - ax$ or $ax - by$, where x and y are natural numbers. We begin with a lemma.

Lemma 1.1.10 *Let a and b be natural numbers. Then the set of natural numbers of the form $by - ax$ is the same as the set of natural numbers of the form $ax - by$, where x and y are natural numbers.*

Proof Suppose $r = by - ax$ for some natural numbers x and y. Then $r = aw - bv$, where $w = kb - x$ and $v = ka - y$ and where k is chosen large enough so that $ka > y$ and $kb > x$. Similarly, if r is of the form $ax - by$, it is also of the form $by - ax$. $\qquad\square$

Theorem 1.1.11 *Given natural numbers a and b, there exist natural numbers x and y such that $\gcd(a, b) = by - ax$.*

Proof Let a and b be natural numbers, and let

$$S = \{by - ax : x, y, \text{ and } by - ax \text{ are natural numbers}\}.$$

By the well-ordering theorem S has a smallest element, call it d. Suppose that $d = by_0 - ax_0$. We will prove that $d = \gcd(a, b)$. First notice that $d \le a$. This is because $a = by - ax$, where $y = 2a$ and $x = 2b - 1$. Next notice that $d \mid a$. This is true by the following reasoning. The division theorem assures us that $a = dq + r$, where $0 \le r < d$ and hence $a = (by_0 - ax_0)q + r$. Thus $r = ax - by$, where $x = x_0 q + 1$ and $y = y_0 q$. By the lemma we know that this number, if positive, is of the form of the numbers in S. But d is the smallest such number, so r must be 0 and therefore $d \mid a$. By a similar argument $d \mid b$ so d is a common divisor of a and b. Now every common divisor c of a and b divides every number in S because if $c \mid a$ and $c \mid b$ then $c \mid by - ax$. We conclude that every common divisor divides d so d is the greatest of the common divisors. $\qquad\square$

This theorem tells us nothing about how to find $\gcd(a, b)$ or how to find the numbers x and y in the preceding theorem. This is where we call upon the Euclidean algorithm. It will give us a method for finding gcd's and a method for finding the numbers x and y guaranteed by Theorem 1.1.11. The Euclidean algorithm is a procedure that we will use over and over again in subsequent chapters. It is fair to say it is the single most useful procedure employed in this book. As an aside, it may seem strange that Euclid, the most famous of all geometers, has an algebraic process named after him. But geometry and algebra were not separate subjects to Euclid. In his treatise, *Elements* (Book 7, Proposition 2), you will find his version of this algorithm.

The idea behind the Euclidean algorithm is to take two natural numbers, divide the smaller into the larger, obtain a remainder that is guaranteed by the division theorem to be smaller than the divisor, repeat the division operation, and obtain a new remainder that is even smaller. This procedure is continued until a remainder of 0 is obtained.

The Euclidean Algorithm *Let natural numbers a and b be given and suppose that a < b. Let r_i denote the successive remainders that are guaranteed by the division theorem.*

$$
\begin{aligned}
b &= aq + r_0 \\
a &= r_0 q_0 + r_1 \\
r_0 &= r_1 q_1 + r_2 \\
r_1 &= r_2 q_2 + r_3 \\
&\vdots \\
r_{n-2} &= r_{n-1} q_{n-1} + r_n \\
r_{n-1} &= r_n q_n + 0
\end{aligned}
$$

Example 1.1.12

Let $b = 1876$ and $a = 365$.

$$
\begin{aligned}
1876 &= 365(5) + 51 \\
365 &= 51(7) + 8 \\
51 &= 8(6) + 3 \\
8 &= 3(2) + 2 \\
3 &= 2(1) + 1 \\
2 &= 1(2)
\end{aligned}
$$

In this example 1 is $\gcd(1876, 365)$. Saying it another way, 1876 and 365 are relatively prime. ∎

This example is a specific instance of a general theorem. It says that last nonzero remainder, r_n in the Euclidean algorithm, is the greatest common divisor of a and b.

Theorem 1.1.13 *If r_n is the last nonzero remainder in the Euclidean algorithm for a and b, then $r_n = \gcd(a, b)$.*

Proof We use the first principle of induction to prove this. It is not obvious how the statement $P(n)$ should read. We will let the variable n stand for the length of the process.

Let $P(n)$ be the following: If the Euclidean algorithm of two numbers a and b terminates after $n + 1$ steps, then $r_{n-1} = \gcd(a, b)$. Let d denote $\gcd(a, b)$.

Let us consider $P(1)$. The algorithm ends after two steps: $b = aq + r_0$, and $a = r_0 q_0 + 0$. So $r_0 \mid a$ and by Theorem 1.1.6, $r_0 \mid b$ as well. Since d is the greatest of the common divisors, it follows that $r_0 \leq d$. But d must divide both a and b since it is a common divisor so, by Theorem 1.1.6 again, $d \mid r_0$. Thus $d \leq r_0$. We conclude that $d = r_0$.

Suppose that $P(n)$ is true and consider $P(n + 1)$. Notice that the algorithm that extends to $n + 2$ steps in length for a and b has the length of $n + 1$ steps for r_0 and a. By the induction hypothesis, $r_n = \gcd(a, r_0)$. Since $b = aq + r_0$ we use Theorem 1.1.6 to see that $r_n \mid b$ as well as $r_n \mid a$, so r_n is a common divisor of a and b. Thus $r_n \leq d$. Using the theorem again, since $d \mid b$, and $d \mid a$, it follows that $d \mid r_0$. Since $r_n = \gcd(a, r_0)$ it follows that $d \leq r_n$. Thus $d = r_n$. It follows that the theorem is true for algorithms of any length n and hence it is true for all pairs a and b.   □

We know that $\gcd(a, b) = by - ax$ for some pairs of numbers x and y and we are now ready to show how to find one of these pairs. We use the numbers from Example 1.1.12.

Example 1.1.14

Let $b = 1876$ and $a = 365$.

The lefthand side is the algorithm; the righthand side solves for the remainders.

$$
\begin{array}{rclcrcl}
1876 & = & 365(5) + 51 & \qquad & 51 & = & 1876 - 365(5) \\
365 & = & 51(7) + 8 & & 8 & = & 365 - 51(7) \\
51 & = & 8(6) + 3 & & 3 & = & 51 - 8(6) \\
8 & = & 3(2) + 2 & & 2 & = & 8 - 3(2) \\
3 & = & 2(1) + 1 & & 1 & = & 3 - 2(1)
\end{array}
$$

Successively substituting the remainders into the equation $1 = 3 - 2(1)$ starting with 2, we get

$$
\begin{aligned}
1 &= 3 - 2(1) \\
&= 3 - (8 - 3(2)) \\
&= -8 + 3(3) \\
&= -8 + (51 - 8(6))(3) \\
&= 51(3) - 8(19) \\
&= 51(3) - (365 - 51(7))(19) \\
&= -365(19) + 51(136) \\
&= -365(19) + (1876 - 365(5))(136) \\
&= 1876(136) - 365(699)
\end{aligned}
$$

so $y = 136$ and $x = 699$.

Notice that the 1 is the gcd of all the pairs (s, t) of numbers where (s, t) is, successively, (r_{n-2}, r_{n-1}), $\ldots$, (r_1, r_0), (r_0, a), (a, b). The above display shows how 1 can be represented in the form of $sy - tx$ or $sx - ty$, namely

$$1 = 3 - 2(1) = 3(3) - 8(1) = 51(3) - 8(19) = 51(136) - 365(19)$$
$$= 1876(136) - 365(699). \qquad \blacksquare$$

Example 1.1.15

Let us find $\gcd(1417, 4459)$ using the Euclidean algorithm.

$$
\begin{aligned}
4459 &= 1417(3) + 208 \\
1417 &= 208(6) + 169 \\
208 &= 169(1) + 39 \\
169 &= 39(4) + 13 \\
39 &= 13(3) + 0
\end{aligned}
$$

This shows that 13 is the greatest common divisor.

Working backward from the bottom up we get

$$
\begin{aligned}
13 &= 169 - 39(4) \\
&= 169 - (208 - 169(1))(4) \\
&= -208(4) + 169(5) \\
&= -208(4) + ((1417 - 208)(6))(5) \\
&= 1417(5) - 208(34) \\
&= 1417(5) - (4459 - 1417(3))(34) \\
&= -4459(34) + 1417(107)
\end{aligned}
$$

so $13 = 1417(107) - 4459(34)$. $\qquad \blacksquare$

While the Euclidean algorithm may seem lengthy for finding gcd's, it is programmable and the number of steps needed to calculate the answer is not that many. We leave it is an exercise to show that the remainders in the algorithm grow smaller at a relatively quick pace; in fact, $r_n < \frac{1}{2}r_{n-2}$. The French mathematician Gabriel Lamé (1795−1870) proved that the number of steps required in the Euclidean algorithm is at most five times the number of digits in the smaller number. The process of reversing the algorithm to find $\gcd(a, b)$ in the form $by - ax$ is cumbersome but it will be improved upon in Section 2.1.

EXERCISES

1. Make up and prove theorems for when a number N is evenly divisible by

   (a) 6

   (b) 8

   (c) 25

   (d) 60

2. Prove that the following theorem on divisibility by 11 works; that is $11 \mid N$ if and only if

   $$11 \mid ((-1)^n a_n + (-1)^{n-1} a_{n-1} + \cdots + (-1)^2 a_2 + (-1)^1 a_1 + (-1)^0 a_0).$$

3. Explain how you would represent the number 89 in base n notation where n is

   (a) five

   (b) seven

   (c) three

   (d) two

   (e) twelve

4. Prove Theorem 1.1.6: Let a, b, d, and r be natural numbers:

   (a) If $d \mid a$, then $d \mid ra$

   (b) If $d \mid a$ and $d \mid b$, then $d \mid (a + b)$

(c) If $d \,|\, (a+b)$ and $d \,|\, a$, then $d \,|\, b$.

5. Show by induction that the sum of the first several consecutive powers of 2 is one less than another power of 2.

6. Try a few cases for n and then show by induction that
$$2^n \,|\, (n+1)(n+2)\cdots(2n).$$

7. Show by induction that $1^2+2^2+\cdots+n^2 = (n)(n+1)(2n+1)/6$.

8. Arrange the odd numbers in a triangle like this:

$$1$$
$$3 \quad 5$$
$$7 \quad 9 \quad 11$$
$$13 \quad 15 \quad 17 \quad 19$$
$$\cdots$$

What can you say about the sum of the numbers:

(a) of the first n rows?

(b) of the nth row itself?

(c) What do parts (a) and (b) tell you?

9. Use induction to establish the validity of your observations in 8(a).

10. Use induction to establish the validity of your observations in 8(b).

11. Try a few cases for n and then show by induction that for all n, $n^5/5 + n^3/3 + 7n/15$ is a natural number.

12. (a) Show by induction that $n^5 + 4n + 10$ is always divisible by 5.

   (b) By altering the cofficients in this exercise, create several more theorems.

13. (a) Show by induction that $7^n + (2 \times 4^n) + 3$ is always divisible by 6.

(b) By altering the coefficients in this exercise, create several more true theorems.

14. Show by induction that, for all n,

(a) $9^n - 8n - 1$ is divisible by 64.

(b) $3^{2n+1} + 2^{n+2}$ is divisible by 7.

(c) $(2^n \times 3^{2n}) - 1$ is divisible by 17.

(d) $11^{n+2} + 12^{2n+1}$ is divisible by 133.

15. Show by induction, beginning with the appropriate initial number, that

(a) $n^2 < n!$

(b) $n^2 \le 2^n$

16. **The Tower of Hanoi**. This famous puzzle involves moving rings from peg to peg. There are three pegs, call them A, B, and C, and there are n rings each of a different diameter. Suppose that all the rings are on peg A arranged from small at the top to large at the bottom; that is, no larger ring is placed on top of a smaller ring. This task requires moving the rings, one at a time, from peg to peg, without ever placing a larger ring on top of a smaller, so that the rings end up on peg B in the same order as they were originally on peg A. What is the minimum number of moves that this task requires? Prove your assertion by induction.

17. Use the Euclidean algorithm to find $\gcd(a, b)$, where a and b are:

(a) 3234 and 5187

(b) 49104 and 66492

18. Given a and b, find x, y, v and w so that $by - ax = \gcd(a, b)$ and $aw - bv = \gcd(a, b)$.

(a) $a = 488$, $b = 223$

(b) $a = 1819$, $b = 3587$

19. If $by_0 - ax_0 = d$, give a general formula for other values of x and y so that $by - ax = d$.

20. (a) Show that the remainders in the Euclidean algorithm grow small in the following way: $r_n < \frac{1}{2}r_{n-2}$.

   (b) Using the result of part (a), find the maximum number of steps it would take for the Euclidean algorithm to produce $\gcd(a, b)$, where the smaller of a and b is a six-digit number.

21. Find two numbers a and b, both < 100, that require a maximum number of steps for the Euclidean algorithm to produce $\gcd(a, b)$. How many steps does it take?

1.2   The Fundamental Theorem of Arithmetic

There is another way of expressing numbers besides those mentioned in Section 1.1. This way does not use bases and polynomial expressions but rather a simple product of primes. That every natural number can be expressed uniquely in this fashion is not insignificant. In fact, it is so significant that it is called the fundamental theorem of arithmetic.

The Fundamental Theorem of Arithmetic *Any natural number greater than 1 can be written as the product of primes, and this representation is unique except for ordering.*

   To say it again, the numbers we use for counting can be constructed through multiplication of prime numbers. Furthermore, this building process is unique to that number. We shall call this unique way of representing a natural number its prime form.

Definition 1.2.1 *A natural number N is in* **prime form** *if it is represented as follows:*

$$N = p_1^{a_1} p_2^{a_2} \cdots p_n^{a_n}$$

where p_i are the distinct prime factors and $a_i > 0$ are the exponents for $1 \leq i \leq n$.

Here are the first 20 natural numbers in prime form (we assume 1 is already in prime form):

$1, \ 2, \ 3, \ 2^2, \ 5, \ 2 \times 3, \ 7, \ 2^3, \ 3^2, \ 2 \times 5, \ 11, \ 2^2 \times 3, \ 13, \ 2 \times 7,$
$3 \times 5, \ 2^4, \ 17, \ 2 \times 3^2, \ 19, \ 2^2 \times 5.$

Notice how this way of representing the first 20 numbers contrasts with the base 10 notation we are used to. Certainly base 10 notation is superior when it comes to adding, subtracting, or comparing sizes of numbers. In fact, these are virtually impossible tasks to do with numbers in prime form. However, multiplying, dividing, taking powers, taking roots, and identifying factors are all easier to do with numbers in prime form. Before analyzing the strengths and weaknesses of prime form representation, let us see why the fundamental theorem of arithmetic is true.

Theorem 1.2.2 *If n is a number greater than 1, then either n is a prime or n factors into the product of prime numbers.*

Proof We proceed by the second induction principle on the statement $\mathcal{P}(n)$: Either n is a prime or n is the product of primes. Clearly $\mathcal{P}(2)$ is true. Suppose that $\mathcal{P}(m)$ is true for all m such that $2 < m \le n$. Now consider $\mathcal{P}(n+1)$. If $n+1$ is prime, then $\mathcal{P}(n+1)$ is true. If $n+1$ is not prime, then $n+1 = ab$, where a and b are both less than n and greater than 1. Since $\mathcal{P}(a)$ is true, a is either prime or a product of primes. Likewise, b has the same property. But then $n+1$ is the product of primes too, so $\mathcal{P}(n+1)$ is true. It follows by induction that $\mathcal{P}(n)$ is true for all $n > 1$. □

As mentioned in our discussion of base n representation, uniqueness of representation is an important issue. The proof of uniqueness for the fundamental theorem is lengthy and takes us into mathematics that may seem far afield. Nevertheless, we include it here for two reasons: for the sake of completeness and because the ideas are fundamental for our study of integers in Chapter 2. Here are three lemmas that complete the proof.

Lemma 1.2.3 *If p is prime and $p \mid ab$, then $p \mid a$ or $p \mid b$.*

Proof Suppose that $p \mid ab$. Also suppose that p does not divide a. Then $\gcd(p, a) = 1$ so, by Theorem 1.1.11, there exists numbers x and y such that $ay - px = 1$. Multiplying both sides by b, we get $aby - pbx = b$. Now $p \mid aby$ and $p \mid pbx$, so $p \mid b$. □

Lemma 1.2.4 *If* $p \mid a_1 a_2 \cdots a_n$, *then* $p \mid a_i$ *for some* $i \leq n$.

The proof of this follows from the previous lemma using the first induction principle.

Lemma 1.2.5 *If* $n = p_1 p_2 \cdots p_k$, *then this representation is unique except for ordering.*

Proof We proceed by the second induction principle: Let $\mathcal{P}(n)$ be the statement for $n > 1$, the prime factorization of n is unique except for ordering. Clearly $\mathcal{P}(2)$ is true; the prime factorization of 2 is unique. Suppose that $\mathcal{P}(m)$ is true for all m such that $2 < m \leq n$. Suppose that $n + 1$ equals both $p_1 p_2 \cdots p_k$ and $q_1 q_2 \cdots q_j$. Now $p_1 \mid q_1 q_2 \cdots q_j$, so it follows from Lemma 1.2.4 that $p_1 \mid q_i$ for some i. Since q_i is a prime, it follows that $p_1 = q_i$. Thus we have

$$\frac{n+1}{p_1} = p_2 \cdots p_k = p_1 q_1 \cdots q_{i-1} q_{i+1} \cdots q_j.$$

Since $(n + 1)/p_1 \leq n$, our induction hypothesis tells us that these two sets of primes must be the same. Since $p_1 = q_i$, the two sets of primes in our original supposition must be the same also. □

Notice that in the representation of n in this lemma the same prime may be repeated.

The fundamental theorem of arithmetic provides an alternate way of expressing numbers; it is based on the prime numbers. Just as powers of 10 are the building blocks of base 10 notation and addition is the cement that binds them, the primes are the building blocks of the prime form notation and multiplication binds them together. Let us examine the strengths and weaknesses of prime form notation. Generally speaking, multiplicative ideas are more easily dealt with when numbers are written in prime form.

Example 1.2.6

Let $a = 2^3 \times 3 \times 5 \times 7 \times 17^2$ and $b = 2^2 \times 3 \times 7^4 \times 13$.

(a) It is virtually impossible to compare relative sizes. For example, which one do you think is bigger, a or b? And the prime form of $a + 1, b + 1, a + b$ and $a - b$ can't be computed from a and b.

(b) We can certainly multiply a and b together:

$$a \times b = 2^5 \times 3^2 \times 5 \times 7^5 \times 13 \times 17^2.$$

Notice that when we multiply two numbers in prime form, we simply add the exponents of the like primes. And we can divide too, but in this example $a \div b$ does not equal a natural number. While there are more 2s and an equal number of 3s and more 5s and more 17s in a than in b, there are more 7s and 13s in b. Division yields the fraction

$$a \div b = \frac{2 \times 5 \times 17^2}{7^3 \times 13}.$$

The division process leads to a natural number if the exponents of the like primes of the dividend are greater than or equal to the exponents of the like primes of the divisor. The exponents of the primes of the quotient are the respective differences of exponents of the dividend and divisor. For example,

$$\frac{2^4 \times 3^2 \times 5^8 \times 7}{2^3 \times 3^2 \times 5} = 2 \times 5^7 \times 7.$$

(c) We can find powers $a^3 = 2^9 \times 3^3 \times 5^3 \times 7^3 \times 17^6$. We simply multiply the exponents of each prime in a by 3. We can find roots, too, when they exist as natural numbers. Unfortunately, $\sqrt{a}$ is not a natural number because the exponents of each prime are not all even numbers. We can find $\sqrt{2^2 \times 3^6 \times 5^4}$, however. It is $2 \times 3^3 \times 5^2$, which we found by dividing the exponent of each prime by 2.   ■

We can also easily identify factors of numbers and greatest common divisors, and least common multiples of several numbers, when they are written in prime form.

Example 1.2.7

(a) Factors are easily identified. For example, $a = 2^4 \times 3^2$ is a factor of $b = 2^7 \times 3^2 \times 5^2 \times 11$ because every prime in a (that is, 2 and 3) is represented in b to an equal or greater power. Similarly, multiples are easily identified for the same reasons. Recall that b is a multiple of a exactly when a is a factor of b.

(b) Listing all the factors of a number is relatively easy. For example, for the number $2^2 \times 3$, the factors are 2, 2^2, 3, 2×3, $2^2 \times 3$, and, of course, 1. In base 10 notation we are saying that the factors of 12 are 2, 4, 3, 6, 12, and 1. There are six of them.

(c) Finding gcd's is easy. Consider the numbers $a = 2 \times 3 \times 5^4$ and $b = 2^3 \times 5^4 \times 7$. The $\gcd(a, b) = 2 \times 5^4$. We can find this by taking the product of all the shared primes. In this example, each number has a 2 and a 5. But they actually share four 5s, so we must use the exponent 4 for the prime 5. In general, each shared prime will be raised to the exponent that is the smaller (smallest) of those represented. In this case, 1 is the smaller exponent shared by the prime 2 (as found in a) and 4 is the smaller exponent shared by the prime 5 (as found in both a and b).

(d) Similarly, finding lcm's is convenient too. Letting a and b be the same as in part (c), we find $\text{lcm}(a,\ b) = 2^3 \times 3 \times 5^4 \times 7$. This is clearly a multiple of both a and b. While we won't list all of the multiples because there are infinitely many of them we can, nevertheless, tell that this is the smallest shared multiple because it includes all the primes in either number and it uses the larger of the exponents of these primes. In particular, we have the exponent of 3 for the prime 2 as found in b, the exponent 1 for the prime 3 as found in a, the exponent 4 for the prime 5 as found in both a and b, and the exponent 1 for the prime 7 as found in b.

(e) The $\gcd(2^2 \times 3,\ 2 \times 3^3 \times 5,\ 3^2 \times 5^2 \times 7) = 3$ because 3 is the only prime represented in all three numbers and the 1 is the smallest exponent of 3 (as found in $2^2 \times 3$).

(f) The numbers $2^2 \times 5$ and 3×7 are relatively prime because they do not share a common prime. ∎

There are lots of nice problems that can be approached by considering prime factorization. Here are four such problems.

Example 1.2.8
(a) The number N has 6 and 8 as factors. What other factors must N have?

Solution: $6 = 2 \times 3$, $8 = 2^3$. So N must have 2^m and 3^n in its prime makeup, where $m \geq 3$ and $n \geq 1$. It follows that any number that has 2^j and 3^k in its prime makeup, where $j \leq 3$ and $k \leq 1$, will be a factor of N. So, besides 6 and 8 the following will be factors of N: 2, 2^2, $2^2 \times 3$, and $2^3 \times 3$; that is, 2, 4, 12, and 24.

(b) If ab is a perfect square and a and b are relatively prime, then what can you say about a and b?

Solution: If ab is a perfect square, then the exponents of the primes within ab must all be even numbers. Since a and b are relatively prime, they do not share a prime. Thus the exponents of primes within a must be even and the same is true for b. Therefore, a and b must, themselves, be perfect squares. Notice that if we had not put the added condition that a and b be relatively prime, this would not be true. For example, $3 \times 12 = 36$, a perfect square, yet neither 3 nor 12 is a perfect square.

(c) How many zeros are at the end of the number 100! $(100! = 1 \times 2 \times 3 \times \cdots \times 100)$?

Solution: Since a 0 is produced in base 10 notation when a 2 is multiplied by a 5, the question becomes: "How many 2s and how many 5s are in the prime form of 100!?" Clearly there are a lot more 2s than 5s. In fact, there is at least one 2 for every even number less than or equal to 100. There is a single 5 in 5, 10, 15, 20, 30, 35, 40, 45, 55, 60, 65, 70, 80, 85, 90, and 95. There are two 5s, in 25, 50, 75, and 100. All in all, that makes 24 5s in the prime form of 100!. Thus 100! ends in 24 zeros.

(d) Find a number that is a perfect square when divided by 2, and a perfect cube when divided by 3.

Solution: Let the number N be written in prime form. If it is to be a perfect square after division by 2, it must have an odd number of 2s in its makeup. Similarly, if it is to be a perfect cube after division by 3 it must have $3k + 1$ 3s in its prime form, for some k. Furthermore, being a perfect square after division by 2, all the other primes must be raised to even powers; being a perfect cube after division by 3, all other primes must be raised to powers that are multiples of 3. Putting all of this together, we find that $2^3 \times 3^4 = 648$ fills the bill. Of course, many other numbers do too. ∎

One concluding thought about prime form notation: We know elementary arithmetic is awkward with numbers in prime form, but there is a more fundamental difficulty, as we shall discover in the next section—there are infinitely many prime numbers and only a finite few have been identified. While this may not be surprising (after all, harnessing an infinite number of numbers is difficult), it turns out that the task of identifying which numbers are prime and which are composite appears to be insuperable for large numbers. As of now there is no known procedure for doing it. Because of this lack of knowledge, the concept of expressing numbers in prime form, which is so helpful for analyzing certain mathematics problems, is not a useful alternative to the base 10 system of representation.

EXERCISES

1. Write the product of the first 20 numbers in prime notation; that is, write $1 \times 2 \times 3 \times 2^2 \times \cdots \times (2^2 \times 5)$ in prime notation.

2. (a) In writing 100! in prime form, what would be the exponent for the prime 2? How about for the prime 3?

   (b) How many zeros does 1000! end with when written in base 10 notation?

   (c) What is the smallest number n such that $n!$ ends in at least 1000 zeros in base 10 notation?

3. Finish the following:

   (a) If 5 and 6 are factors of N then other factors are . . .

   (b) If 8 and 12 are factors of N then other factors are . . .

4. (a) Find a number that when divided by 2 is a perfect square, when divided by 3 is a perfect cube, and when divided by 4 is a perfect fourth power.

   (b) Find a number that when divided by p is a perfect pth power, where $p = 2, 3, 5,$ and 7.

5. A mathematically inclined jailer has decided to let a few prisoners go. In the middle of the night he walks along a row of n cells n times. The first time he turns every lock, opening them all up. On the second pass, he turns every other lock beginning with the second, thus relocking the even numbered cells. On the third pass he turns the locks on every third cell beginning with the third. Thus the cells numbered $3k$ are affected. And so it goes until he turns the lock on the nth cell. Which prisoners are set free? Explain your answer.

6. Find gcd and the lcm of the following pairs of numbers written in prime form.

   (a) $2^2 \times 7 \times 23,\ \ 2^3 \times 5 \times 29$

   (b) $2^3 \times 3 \times 5 \times 7,\ \ 2^2 \times 5^3 \times 11$

   (c) $2^4 \times 3^3 \times 7 \times 19^2 \times 29^3,\ \ 2 \times 3^6 \times 5 \times 7^2 \times 13 \times 19^4 \times 31^7$

7. Finish the following and explain your answers.

(a) If $\gcd(n, m) = n$, then $\text{lcm}(n, m) =$

(b) If $\text{lcm}(n, m) = n$, then $\gcd(n, m) =$

(c) If n and m are relatively prime, then $\text{lcm}(n, m) =$

8. Answer true or false and explain your reasoning.

   (a) If n is a prime, then the pair n, m are relatively prime for any number m.

   (b) If $\gcd(n, m) = d$, then $\gcd(n^2, m^2) = d^2$.

9. If $\gcd(n, m) = 12$, what are the possible values for $\gcd(n^2, m)$?

10. Suppose that $a, b,$ and c are natural numbers.

    (a) Prove: If $\gcd(a, b) = 1$ and $a \mid bc$, then $a \mid c$.

    (b) Prove: If $\gcd(a, b) = 1$ and $a \mid c$ and $b \mid c$, then $ab \mid c$.

    (c) Show that if the condition of $\gcd(a, b) = 1$ is dropped in (a), then the statement is not necessarily true.

    (d) Show that if the condition $\gcd(a, b) = 1$ is dropped in (b), then the statement is not necessarily true.

11. List all the factors of

    (a) 735

    (b) 264

    (c) 1170

12. What is the maximum number of factors that a number N can have if

    (a) $N \leq 100$. What number(s) has (have) this many?

    (b) $N \leq 1000$. What number(s) has (have) this many?

    (c) $N \leq 1000000$. What number(s) has (have) this many?

13. Find a formula for the number of factors of N where

$$N = p_1^{a_1} p_2^{a_2} \cdots p_n^{a_n}.$$

14. Here is a slightly different number system. Let $\mathbb{E}$ be the set of even numbers: 2, 4, 6, ... Define a prime to be any number n in $\mathbb{E}$ that cannot be expressed as the product of two numbers in $\mathbb{E}$. For example, 6 is a prime in $\mathbb{E}$.

   (a) Find the first 10 primes in $\mathbb{E}$.
   (b) Find a number that can be expressed as the product of primes in two different ways; in three different ways; in four different ways.
   (c) What is $\gcd(8, 58)$?
   (d) Give examples showing that the division theorem does not work in $\mathbb{E}$.

1.3   Searching for Primes

In this section we take a glimpse at the majesty of the universe of prime numbers. Our glimpse will look in upon the findings of some of the greatest mathematicians of all time. The form, the distribution, the frequency, the spacing, the very nature of prime numbers has held an enduring fascination for mathematicians. Finding primes has been a quest of mathematicians for all time, and in this age of computers enormous strides are being made.

   Here is a list of the primes less than 1000.

Table of Primes

2	3	5	7	11	13	17	19	23	29	31	37
41	43	47	53	59	61	67	71	73	79	83	89
97	101	103	107	109	113	127	131	137	139	149	151
157	163	167	173	179	181	191	193	197	199	211	223
227	229	233	239	241	251	257	263	269	271	277	281
283	293	307	311	313	317	331	337	347	349	353	359
367	373	379	383	389	397	401	409	419	421	431	433
439	443	449	457	461	463	467	479	487	491	499	503
509	521	523	541	547	557	563	569	571	577	587	593
599	601	607	613	617	619	631	641	643	647	653	659
661	673	677	683	691	701	709	719	727	733	739	743
751	757	761	769	773	787	797	809	811	821	823	827
829	839	853	857	859	863	877	881	883	887	907	911
919	929	937	941	947	953	967	971	977	983	991	997

With this table we can find all of the primes up to 1 million. This is because the factors of a number, N, come in pairs, one of which is always less than or equal to $\sqrt{N}$. Thus we need only look at the prime factors up $\sqrt{N}$ in order to check whether a number is prime.

Theorem 1.3.1 *In order to find the prime factors of a composite number N, you need only divide N by primes $\leq \sqrt{N}$. The other prime factors can be obtained from the quotients of these divisions.*

Proof Suppose that p is prime, $p \mid N$ and $p \geq \sqrt{N}$. Let $q = N/p$. So $N = pq$. Now $q \leq \sqrt{N}$, because $p \geq \sqrt{N}$. Also we know, from the fundamental theorem of arithmetic, that q is the product of prime powers. Of course, the primes that make up q are all $\leq \sqrt{N}$ and since $p = N/q$, p must indeed arise as a quotient from successive divisions of primes $\leq \sqrt{N}$. $\square$

Example 1.3.2

(a) Is 91 a prime? Let us check the primes $\leq \sqrt{91}$. Since $\sqrt{91} \approx 9.54$, we must check 2, 3, 5, and 7. We find that $7 \mid 91$; in fact, $7 \times 13 = 91$. So 13 is the other prime that divides 91.

(b) Is 443207 a prime? If not, what number or numbers divide 443207? Since $\sqrt{443207} \approx 665.7$ we have to check out primes all the way to 665 to find out if 443207 is prime or not. With our table we can actually do this, but it wouldn't be pleasant. Luckily some small primes divide 443207. In fact, we find that 17, 29, and 31 all divide 443207. In its prime form $443207 = 17 \times 29^2 \times 31$. ∎

Let us address some general questions that have been explored in the long and celebrated history of prime numbers:

1. *How many primes are there?*

2. *How many primes are there less than a given number n?*

3. *What is the nature of primes? Are they distributed in an understandable way? Is there a formula that yields the primes? Do they follow any sort of pattern?*

In the course of answering these questions we will cite names of several mathematicians, some of whom are among the greatest mathematicians who ever lived. Here are brief introductions; for an in-depth study, include a course in History of Mathematics in your curriculum. Euclid (circa 350 B.C.) you have heard about in high school. Eratosthenes (276−196 B.C.) was one of the ancient Greek mathematicians. Marin Mersenne (1588−1648) was a reclusive French monk. Rene Descartes (1596−1650) was a French mathematician-philosopher for whom the Cartesian coordinate plane is named. Pierre de Fermat (1601−1665), also a French mathematician, has been called the greatest mathematician of the seventeenth century. This claim could be disputed since Isaac Newton (1642−1727) did much of his work in the seventeenth century. Newton was an English mathematician-physicist and founder of calculus; he is considered by some historians as the greatest mathematician who ever lived. Leonhard Euler (1707−1783) was a Swiss mathematician who is known as the most prolific mathematician in history. Carl Friedrich Gauss (1777−1855), a German mathematician, is generally considered the father of modern mathematics. Edouard Lucas (1842−1891) was a French number theorist. The Lucas sequence, a companion to the Fibonacci sequence, is named for him. D. H. Lehmer (1905−1991), a contemporary American number theorist, was a pioneer in developing computer techniques for finding primes.

The answers to our three questions, such as they are, perhaps are not satisfactory to the beginner, but the subject of primes is intriguing partly because some of these questions are so difficult to answer. Here are the short answers; we will then expand on them.

1. There are infinitely many primes.

2. The number of primes less than n is approximately $n/\ln(n)$.

3. Primes are of no particular form. They are not distributed in an algorithmic fashion. There is no known formula that will give all primes, nor even a large number of primes. There is no discernible pattern that the primes seem to follow.

Let us begin with the easiest of the questions.

1. Euclid knew that there were infinitely many primes, and he had a proof.

Theorem 1.3.3 *There are infinitely many prime numbers.*

Proof Suppose there are only a finite number of primes $p_1, p_2, \ldots, p_n$. Let $N = (p_1 p_2 \cdots p_n) + 1$; since N is larger than all of the primes, it is not a prime. So one of the primes, call it p_i, divides N. Since $p_i \mid N$ and also $p_i \mid p_1 p_2 \dots p_n$, it follows from Theorem 1.1.6 that p_i divides the difference of the two numbers. But $N - p_1 p_2 \dots p_n = 1$ and $p_i \nmid 1$. This is a contradiction. Thus there are infinitely many primes.   □

It is fun to check out numbers of the form $N = (p_1 p_2 \cdots p_n) + 1$ for the first few primes just to see what kinds of numbers result. The first four such numbers are $2 + 1$, $(2 \times 3) + 1$, $(2 \times 3 \times 5) + 1$, and $(2 \times 3 \times 5 \times 7) + 1$. These numbers are 3, 7, 31, and 211. They are all primes. Does this surprise you? Should it have been this way? You can pursue this in the exercises. Interestingly, the largest known prime number N, where $N = (p_1 p_2 \cdots p_n) + 1$, is $(2 \times 3 \times \cdots \times 24029) + 1$. It has $10,387$ digits.

2.   Looking at the table of primes, we can carry out our own count of how many primes there are; at least up to 1000. In fact, the total number is 168. The primes start out with a flourish; four of the first six numbers after 1 are prime. Their frequency drops off considerably as we proceed through the first 100 numbers. There are 8 primes in the first 20 numbers, but only 17 in the next 80 numbers. This makes a total of 25 primes in the first 100. After that there are 21 numbers in the next block of 100. Then there are 16, 16, 17, 14, 16, 14, 15, and 14 in the next eight blocks of 100. The dropoff is not that apparent after the first 200 numbers, but there *is* a dropoff. And this dropoff can be expressed in terms of logarithms. For a general answer we must deal with approximations. We give two famous approximations here. The following table shows this. Here $\pi(n)$ represents the number of primes less than n, $\ln n$ represents the natural logarithm of n, and $\text{Li}(n) = \int_2^n 1/\ln(x)\, dx$. The $n/\ln n$ and the $\text{Li}(n)$ columns are rounded to the nearest integer.

n	$\pi(n)$	$\ln(n)$	$n/\ln(n)$	$\frac{\pi(n)}{n/\ln(n)}$	$\text{Li}(n)$
10^2	25	4.605	22	1.15	29
10^3	168	6.908	145	1.16	177
10^4	1,229	9.21	1,086	1.13	1,245
10^5	9,592	11.413	8,686	1.10	9,630
10^6	78,498	13.816	72,380	1.08	78,628
10^7	664,579	16.118	620,420	1.07	664,918
10^8	5,761,455	18.421	5,428,681	1.06	5,762,209
10^9	50,847,543	20.723	48,254,942	1.05	50,849,244

Notice that the number of primes less than n, $\pi(n)$, is approximately $n/\ln n$. Another way of saying this is that the probability a number $\leq n$, chosen at random, is prime is about $1/\ln n$. Putting it more precisely, $\pi(n)/(n/\ln n)$ gets close to 1 as n grows large. This result is deep and significant. It is known as the **prime number theorem** and was stated by Gauss originally in 1792. We will not prove this, but we can check it out in the table for "small" values of n.

The $\text{Li}(n)$ function is even a better approximation for $\pi(n)$. As the table shows, the differences between the approximation and the reality are a mere 4, 9, 16, 38, 130, 339, 754, and 1701, respectively. The fraction $\text{Li}(n)/\pi(n) = 1.00003$ for $n = 10^9$ and, as n gets larger, this ratio also approaches 1. For example, for $n = 10^{15}$, $\text{Li}(n)/\pi(n) = 1.000000003$.

An interesting sidelight to this concerns a theoretical result that has been proved. While one might guess that $\text{Li}(n)$ always exceeds $\pi(n)$, it was proved by the great English number theorist J. E. Littlewood, in 1914, that for some really large n this is not true. In 1933 a number N was produced. It is called **Skewes' number** after the mathematician who found it and it is *really* big: $N = 10^{10^{10^{34}}}$. Let's write it again.

$$N = 10^{10^{10,000,000,000,000,000,000,000,000,000,000,000}}$$

So it is known that there is an $n < N$, where $\pi(n) > \text{Li}(n)$. It is not known, however, what that number n is. Skewes number makes the googolplex seem miniscule. It gives one a glimpse into the enormity of the universe of numbers that can be dealt with by the human mind, at least the minds of number theorists. Recently the bound has been lowered to a more manageable size, but the result is still bizarre. It is now known that in the interval between 6.62×10^{370} and 6.69×10^{370} there are 10^{180} consecutive integers where $\text{Li}(n)$ is less than $\pi(n)$.

3. In examining the form of primes we can say, without fear of contradiction, that they are all odd numbers except for 2. Putting it another way, except for 2 they are all of the form $2k + 1$ for some k, where k is a natural number. And, with the exception of the prime 3, they are all of the type $3k+1$ or $3k+2$. Also, they are all of the form $4k+1$ or $4k+3$. And they are all of the form $5k+1$, $5k+2$, $5k+3$, or $5k + 4$. Furthermore, there are infinitely many primes of each type.

These observations are a special case of a general theorem established
by the German mathematician Peter Gustav Dirichlet (1805−1859).

Dirichlet's Theorem *If a and b are relatively prime, then the series
of numbers*

$$a, \, a + b, \, a + 2b, \, a + 3b, \, , \ldots, \, a + kb, \ldots$$

contains infinitely many primes.

The sequence of numbers a, $a + b$, $a + 2b$, $a + 3b$, $, \ldots$, $a + kb$, $\ldots$
is called an **arithmetic sequence**. Dirichlet's theorem tells us, for
example, that there are infinitely many prime numbers ending in 9;
they appear in the sequence $9 + 10k$, where $k = 1, 2, 3, \ldots$. The first
few are 19, 29, 59, 79, and 89. Similarly, there are infinitely many
primes ending in 99, in 999, or any length sequence of 9s. But while
arithmetic sequences may contain infinitely many primes, none is
made up strictly of primes.

Theorem 1.3.4 *An arithmetic sequence contains infinitely many
composite numbers.*

Proof Consider the arithmetic sequence $a + kb$ and suppose that
$a + k_0b$ is a prime number, p. Then the sequence $a + lb$ is divisible
by p, where $l = k_0 + kp$. This is because

$$a + lb = a + (k_0 + kp)b = (a + k_0b) + (kpb)$$

and p divides both terms of the sum.   □

It is an old unsolved question whether there are arithmetic sequences of primes of arbitrary length. The longest arithmetic sequence of primes currently stands at 22. It is

$$11410377850553 + 4609098694200k; \;\; k = 0, 1, \ldots, 21.$$

The prime factorization of the common difference between the numbers is

$$4609098694200 = 2^3 \cdot 3 \cdot 5^2 \cdot 7 \cdot 11 \cdot 13 \cdot 17 \cdot 19 \cdot 23 \cdot 1033$$

You may notice that the primes $2, 3, 5, 7, 11, 13, 17$, and 19 are all
represented in the difference. With a bit of experimentation we see
that it must be true that, for an arithmetic sequence of k numbers to
be all prime, the common difference must be divisible by all primes
less than k.

Example 1.3.5

(a) A sequence of three primes is $3, 5, 7$. The common difference must be divisible by 2; in fact, it is 2.

(b) A sequence of five primes is $5, 11, 17, 23, 29$. The common difference, 6, is divisible by all the primes less than 5.

(c) A sequence of six primes is 7, 37, 67, 97, 127, 157. The common difference, 30, is divisible by all primes less than 6.

(d) Here is an arithmetic sequence of length 13:

$$13 + 9918821194590k; \ k = 0, 1, \ldots, 12.$$

You don't need to check it for primes; a computer has produced it. Incidentally,

$$9918821194590 = 2 \cdot 3 \cdot 5 \cdot 7 \cdot 11 \cdot 4293861989$$

so you can see that the common difference is divisible by all primes less than 13.   ■

We shall state our observations as a theorem and leave the proof to the student. It is a good exercise in wrestling with concepts of divisibility and remainders, which will be studied in detail in Chapter 2.

Theorem 1.3.6 *Suppose that $k > 2$. If all the terms of the arithmetic sequence*

$$p, p + d, p + 2d, p + 3d, \ldots, p + (k - 1)d$$

are primes, then the common difference d is divisible by every prime less than k.

The search for primes within arithmetic sequences is only the beginning of the search for patterns within the primes. Arithmetic sequences are linear functions: $f(n) = a + bn$. The search for prime-producing quadratic polynomials is also interesting and quite different from the linear case. If we take the simplest of the quadratic polynomials, $f(n) = p_0 + n + n^2; n = 0, 1, 2, \ldots$, we notice that the skips are of increasing size as n moves along. The skips themselves jump from 2 to 4 to 6 and so on.

Example 1.3.7

Consider $f(n) = p_0 + n + n^2; n = 0, 1, 2, \ldots$ for primes p_0.

(a) Let $p_0 = 5$. We get four primes: 5, 7, 11, 17. The sequence ends with 25.

(b) Let $p_0 = 11$. We get 10 primes: 11, 13, 17, 23, 31, 41, 53, 67, 83, 101. The sequence ends with 121.   ∎

It looks like the quadratic polynomial $f(n) = p_0 + n + n^2; n = 0, 1, 2, \ldots$ can produce, at most, $p_0 - 1$ primes. This is true because the p_0th term in the sequence that begins with p_0 is $p_0 + (p_0 - 1) + (p_0-1)^2$ and this equals p_0^2. Incredibly, the longest sequence of primes generated is largely contained in our small table of primes. You are asked to find it in an exercise.

The next logical question to ask is whether any polynomial can produce a string of primes that does not end. Putting it another way, is there a polynomial f defined on n that will produce an infinite succession of prime numbers for all $n > N$ for some number N? Let's examine an example.

Example 1.3.8

Consider the cubic $f(x) = x^3 - 15x^2 + 27x + 3$. Right away we discover that $f(2) = 5$, a prime number. A graphing calculator shows that this function becomes negative for $n = 3, 4, \ldots 12$ and becomes positive at 13; $f(13) = 16$, a composite number. And there are lots more composites, too. Notice that for $n = 17, 22, \ldots 2 + 5k$, where $k \geq 3$ that $f(n)$ is divisible by 5. This is because

$$f(2 + 5k) = (2 + 5k)^3 - 15(2 + 5k)^2 + 27(2 + 5k) + 3$$
$$= 2^3 - 15(2^2) + 27(2) + 3 + \text{multiples of 5}$$
$$= 5 + \text{multiples of 5}.$$

So this cubic generates an infinite sequence of composite numbers. It can't, therefore, generate a sequence of all primes after a certain number N.   ∎

The argument used in this example can be made general.

Theorem 1.3.9 *If f is a polynomial with integer coefficients with a positive leading coefficient, then there are infinitely many natural numbers $n \geq 1$ such that $f(n)$ is a composite number.*

Proof Let $f(x) = a_k x^k + a_{k-1} x^{k-1} + \cdots + a_0$ with $a_k \geq 1$, $k \geq 1$. If $f(n)$ is composite for all n, we are finished. If not, suppose that $f(n_0) = p$, a prime number. Choose N large enough so that $f(n) > p$ for all $n > N$. Such an N exists because $f(x) \to \infty$ as $x \to \infty$ Consider $f(n_0 + hp)$, where h is large enough so that $n = n_0 + hp > N$. Now

$$f(n_0 + hp) = a_k(n_0 + hp)^k + a_{k-1}(n_0 + hp)^{k-1} + \cdots + a_0$$
$$= (a_k n_0^k + a_{k-1} n_0^{k-1} + \cdots + a_0) + \text{multiples of } p$$
$$= p + \text{multiples of } p.$$

So $f(n)$ is a multiple of p. Since there are infinitely many choices for h each of which gives an n for which $f(n)$ is composite, the proof is complete. □

Another natural question to ask is if primes tend to be of one form more than another. We have learned that there are infinitely many primes of the form $a + bk$, where a and b are relatively prime. So, for example, there are infinitely many primes of the form $3k + 1$ and $3k + 2$. Is one form favored over the other? Our chart may help us in this matter. Checking it out, we may count the number of primes of the form $3k + 1$ and $3k + 2$; in fact, there are more primes of the form $3k + 2$ than $3k + 1$. It turns out that this lead by primes of the form $3k + 2$ persists until $608, 981, 813, 029$, a prime of the form $3k + 1$, puts primes of that form into the lead—a lead they hold onto for nearly the next 2 trillion numbers. This information could not have been found without computers. But it could not have been accomplished without mathematicians either. What is known is that there are infinitely many primes of the form $3k + 1$ and $3k + 2$. Furthermore, they occur "equally often" in the sense that the ratio of the number of primes of one type to the number of the other type approaches 1 as we scan further and further out into the universe of numbers. It has also been proved that the lead of one type over the other changes hands infinitely many times. This fact about $3k + 1$ and $3k + 2$ is an instance of a general theorem.

Theorem 1.3.10 *Let n be fixed. If a and b are relatively prime to n then the primes of the form $nk + a$ and $nk + b$ are equally distributed.*

This theorem says, in particular, that primes of the form $10k + 1, 10k + 3, 10k + 7$ and $10k + 9$ (that is, primes ending in 1 or 3 or 7 or 9) occur equally often. The proof of this theorem can be found in advanced books on number theory.

Studying our chart, we can see for ourselves how the primes are distributed. They are sometimes close together (for example 41, 43; and 881, 883) and they are sometimes far apart (for example, 113, 127 and 293, 307). It is believed that there are infinitely many twin primes (that is, primes separated by a single composite), but this has not been proved. One of the largest known pairs was discovered in 1995. It is

$$(242206083 \times 2^{38880}) \pm 1.$$

These numbers contain $11,713$ digits.

As we have noted, successive primes may also be very far apart. The biggest separation between consecutive primes occurring for numbers less than 3 trillion is 652. Putting it another way, there are 651 consecutive composites that lie between 2614941710599 and 2614941711251. In fact, it can be shown that primes can be separated by arbitrarily long strings of composites. For example, if you want to point explicitly to a separation of at least magnitude n between consecutive primes, simply consider the sequence of $n-1$ consecutive composites: $n! + 2, n! + 3, \ldots, n! + n$. The first number is divisible by 2, the second by 3, and so on. Actually these tend to be huge numbers; for example, if $n = 100$, then the consecutive composites are made up of 158 digits. A separation of 100 between primes is no big deal for numbers of this size, as we shall see. The first occurrence of 99 consecutive composites is between the two primes 396733 and 396833.

We can find a rough approximation of how far apart primes are spaced, on average, using the prime number theorem. Here is an example.

Example 1.3.11

(a) The prime number theorem tells us that, on average, primes between 1 and n are spaced approximately $\ln(n)$ numbers apart. So for numbers with 158 digits the primes should be, on average, about $\ln(10^{158}) = 158 \times \ln(10) \approx 364$ numbers apart. This is considerably larger than the separation of 100 that our factorial example guarantees.

(b) On the other hand, the first occurrence of a separation of 100 or more is the separation of 112 that occurs at the prime 370261.

There are 111 consecutive composites between 370261 and 370373. The average separation between primes for numbers up to 370261 is about $\ln(370261) = 12.821 \approx 13$ numbers. So this is a big separation for numbers of this size.   ∎

The primes don't seem to follow a pattern. Ideally there would be a function f that would take a natural number, n, and convert it into a prime number. In fact, there is: The function that lists the primes (that is, $f(1) = 2$, $f(2) = 3$, $f(3) = 5$ and so on) would be such a function. But this is not helpful; a simple listing of the primes does not reveal a pattern. There are more interesting functions that provide primes, but they are not simple functions and, generally speaking, they are not useful. For example, there exists a value of r so that the function $f(n) = [r^{3^n}]$ produces primes for $n = 1, 2, 3, \ldots$. Here the brackets indicate the greatest integer function. Unfortunately, this does not produce all the primes and the value for r is not known. It has also been proven that there is a polynomial of degree 25 with 26 unknowns that has the property that when the unknowns are replaced by natural numbers and the result is positive, the number is prime. And that's not all; every prime number is obtainable this way. Thus the set of prime numbers *is* the set of positive values of this polynomial when the unknowns take on natural numbers values. Unfortunately, this function is only known to exist; it is not known explicitly. This is another example of an existence proof of a bizarre fact that leaves one in amazement.

As you can see, the answers to questions **1**, **2**, and **3** are, at best, incomplete. Even in this day of powerful computers, the prime numbers hold many secrets. Understanding the primes has held the fascination of mathematicians from time immemorial. Arguably the greatest of them all, Gauss, observed in his *Disquisitiones Arithmeticae*:

> The problem of distinguishing prime numbers from composite numbers and of resolving the latter into their prime factors is known to be one of the most important and useful in arithmetic. It has engaged the industry and wisdom of ancient and modern geometers to such an extent . . . (that) the dignity of the science itself seems to require that every possible means be explored for the solution of a problem so elegant and so celebrated.

Here is a brief history of the search for prime numbers.

Eratosthenes, who lived in the third century B.C., developed a way of generating prime numbers. It is known as the **sieve of Eratosthenes**; here is how it works. From a listing of the numbers, beginning with 2, the multiples of successive primes are filtered out. It looks like this for numbers up to 30; parentheses around the numbers indicate multiples of primes that have passed through the sieve and been discarded.

We begin by filtering out all multiples of 2 larger than 2.

2 3 (4) 5 (6) 7 (8) 9 (10) 11 (12) 13 (14) 15 (16) 17 (18) 19 (20) 21 (22) 23 (24) 25 (26) 27 (28) 29 (30)...

Once the even numbers are gone, the multiples of 3 that are larger than 3 are filtered out.

2 3 5 7 (9) 11 13 (15) 17 19 (21) 23 25 (27) 29...

The multiples of 5 larger than 5 are filtered out next.

2 3 5 7 11 13 17 19 23 (25) 29...

Removing 25 from the list leaves 2, 3, 5, 7, 11, 13, 17, 19, 23, and 29, all the prime numbers less than 30. Theorem 1.3.1 tells us that we need not cross out multiples of 7 until we expand our list of numbers to 49.

Eratosthenes' sieve is easily programmable, but it is not an efficient way of generating primes. As numbers get large, even a computer bogs down.

There is evidence that the search for large prime numbers predates Eratosthenes. Twenty-five centuries ago the Chinese apparently believed that n was prime if and only if $n \mid 2^n - 2$. In fact, this was thought to be true for 23 centuries. The following chart confirms the statement for small numbers.

n	$2^n - 2$
2	$2^2 - 2 = 2$
3	$2^3 - 2 = 6\ = 3 \times 2$
4	$2^4 - 2 = 14$
5	$2^5 - 2 = 30 = 5 \times 6$
6	$2^6 - 2 = 62$
7	$2^7 - 2 = 126 = 7 \times 18$
8	$2^8 - 2 = 254.$

But for somewhat larger n the belief gets more difficult to check because, as n grows, numbers of the form 2^n grow very quickly. It turns out that $p \mid 2^p - 2$ for all primes p; we shall prove that in Chapter 2. It is also true that $n \nmid 2^n - 2$ for all composites up to 341. But $341 \mid 2^{341} - 2$ and 341 is composite; $341 = 11 \times 31$. That $341 \mid 2^{341} - 2$ will be proved in Chapter 2 using elementary number theoretic results of Fermat, Euler, and Gauss. It is lucky we have mathematicians to work their magic because $2^{341} - 2$ has 103 digits.

Factoring numbers of the size of $2^{341} - 2$ can now actually be done with modern high-speed computers using the latest of mathematical techniques, but this is only a very recent development. Numbers on the order of a googol are near the limit of those that can be represented completely in their prime form. Beyond this, finding at least one factor of much larger numbers has become routine. Unbelievably, the number that is one more than the googolplex (that is, $N = 10^{10^{100}} + 1$) has been found to be composite; it has a 36-digit factor. That factor (complete with commas to help you read it) is

$$316,912,650,057,057,350,374,175,801,344,000,001.$$

Numbers that have received the most scrutiny in the search for large primes are numbers close to 2^n; in particular, numbers of the form $2^n \pm 1$. The first few numbers of the form $2^n + 1$ are 2, 3, 5, 9, 17, 33, 65, 129, and 257 for $n = 0, 1, 2, 3, 4, 5, 6, 7, 8$. Notice that the only primes in this list are 2, 3, 5, 17, and 257. These correspond, respectively, to $n = 0, 1, 2, 4$, and 8. These numbers are powers of 2. In fact, it is the case that the number $2^n + 1$ can be prime only if n is a power of 2. Here is why.

Theorem 1.3.12 *If n has an odd factor greater than 1, then $2^n + 1$ is composite.*

Proof Suppose that $k > 1$ is an odd factor of n and $kl = n$. We may factor the algebraic expresssion $x^n + 1$ as follows:

$$x^n + 1 = (x^l + 1)(x^{n-l} - x^{n-2l} + \cdots + (-1)^{r-1} x^{n-rl} + \cdots + (-1)^{k-1} x^{n-kl}).$$

This last term is $+1$ since k is odd and $kl = n$. Now substituting 2 in for x, we obtain

$$2^n + 1 = (2^l + 1)(2^{n-l} - 2^{n-2l} + \cdots + (-1)^{r-1} 2^{n-rl} + \cdots + (-1)^{k-1} 2^{n-kl}).$$

Since neither factor is 1, we see that $2^n + 1$ is composite. $\qquad \square$

So numbers of the form $2^{2^n} + 1$ are the ones that are eligible for consideration as primes. These numbers are important enough to have a name; they are called Fermat numbers.

Definition 1.3.13 *A number of the form $2^{2^n} + 1$ is called a* **Fermat number**. *We denote this number by F_n.*

Numbers of the form $2^n - 1$ look like this: 1, 3, 7, 15, 31, 63, 127, and 255 for $n = 1, 2, 3, 4, 5, 6, 7$, and 8. Notice that the only primes in this list are 3, 7, 31, and 127. Those numbers correspond to $n = 2, 3, 5$, and 7. These numbers, n, are themselves prime numbers. It appears as if the quest to find primes of the form $2^n - 1$ should be limited to prime numbers n. The following theorem bears this out.

Theorem 1.3.14 *If n is composite, then $2^n - 1$ is composite.*

Proof Suppose n is composite; thus $n = kl$ and neither k nor l is 1. Then, from algebra, we may write $x^n - 1$ as follows:

$$x^n - 1 = (x^l - 1)(x^{n-l} + x^{n-2l} + \cdots + x^{n-kl}).$$

This last term is 1. Now substituting in 2 for x we obtain:

$$2^n - 1 = (2^l - 1)(2^{n-l} + 2^{n-2l} + \cdots + 2^{n-kl}).$$

Neither factor of $2^n - 1$ is 1 because neither k nor l is 1; hence $2^n - 1$ is composite. $\square$

So the source of prime numbers must come from numbers of the form $2^p - 1$ where p is a prime. Such numbers are called Mersenne numbers after a 17th century French monk, Father Marin Mersenne.

Definition 1.3.15 *The number $2^p - 1$, where p denotes a prime number, is called a* **Mersenne number** . *We denote it by M_p.*

Theorems 1.3.12 and 1.3.14 give us some direction on how to factor numbers of the form $2^n \pm 1$.

Example 1.3.16

Let us factor the number $N = 1073741823$ into its prime parts.

Now $1073741823 = 2^{30} - 1 = (2^{15} + 1)(2^{15} - 1)$ and

$$2^{15} + 1 = (2^5 + 1)(2^{10} - 2^5 + 1) = 33 \times 993 = 3^2 \times 11 \times 331$$

$$2^{15} - 1 = (2^5 - 1)(2^{10} + 2^5 + 1) = 31 \times 1057$$

$$2^{15} - 1 = (2^3 - 1)(2^{12} + 2^9 + 2^6 + 2^3 + 1) = 7 \times 4681 = 7 \times 31 \times 151.$$

So we have found the primes that make up N; they are: 3, 7, 11, 31, 151, and 331. In fact; $1073741823 = 3^2 \times 7 \times 11 \times 31 \times 151 \times 331$. ∎

It turns out, in the search for primes, that the promise held by Fermat is quite different from the promise held by Mersenne. Here are the stories of the two types of numbers.

Fermat conjectured that numbers of the form $F_n = 2^{2^n} + 1$ were primes. In fact, they are for $n = 0, 1, 2, 3$, and 4. These numbers are 2, 3, 5, 17, 257, and 65537, respectively. $F_5 = 4294967297$ and it was too big for him to check. Unfortunately (or fortunately), Euler, in 1732, found that $641 \mid F_5$; $F_5 = 641 \times 6700417$. Euler also proved that if F_n is composite then one of the factors is of the form $(k \cdot 2^{n+1}) + 1$. Notice that $641 = (10 \times 2^6) + 1$. Since Euler's time, no Fermat number, F_n, for $n > 4$, has been found to be prime. Since these numbers get huge very quickly, it was virtually impossible for mathematicians to find prime factors for Fermat numbers until computers came along. For example, F_6 has 20 digits and F_7 has 39. In 1905 it was found that F_7 was composite, but it took 66 years before Brillhart, Morrison, and a computer in 1970 provided its prime factorization:

$$2^{128} + 1 = (59649589127497217) \times (5704689200685129054721).$$

For F_8, compositeness was determined in 1909, but the factors were not found until 1981. There are two prime factors for F_8. The smaller has 16 digits: 1238926361552897. The larger prime factor has 62 digits. You can find it by dividing F_8 by the 16 digit prime.

Now complete factorization of F_n has been found for $n = 5, 6, 7, 8, 9$, and 11. It is known that F_n is composite for all $n > 4$ up to $n = 22$. There are several Fermat numbers larger than F_{22} that are known to be composite. For example, in 1980, it was shown that F_{9449} has $(38 \times 2^{9449}) + 1$ as a factor. But the general state of knowledge concerning the primality of Fermat numbers is meager indeed. It is suspected that there are no Fermat primes after $n = 4$ but that has not been proven. It is not even known if there are a finite number of Fermat primes. Amazingly, it is not even known if there are infinitely many composite Fermat numbers.

In 1644, Mersenne made the claim that if p were 2, 3, 5, 7, 13, 17, 19, 31, 67, 127, or 257, then $2^p - 1$ is prime. Proof of this was not forthcoming for a long time because these numbers grow very large very quickly. As we saw previously, the claim for 2, 3, 5, and 7 is correct. In 1536, Hudalrichus Regius exhibited the factorization of $2^{11} - 1$: $2^{11} - 1 = 2047 = 23 \times 89$. That is no feat today, but 460 years ago he was using the Roman numeral system rather our base 10 system. In 1588 Pietro Cataldo correctly verified that $2^{17} - 1 = 131071$ and $2^{19} - 1 = 524287$ were both prime. It wasn't until 1772 that Euler confirmed that Mersenne was correct for $p = 31$; $2^{31} - 1 = 2147483647$. More than a century later in 1876 Lucas discovered, however, that Mersenne was wrong for $p = 67$; it turns out that $2^{67} - 1$ is composite. He could not, however, produce the factors. In 1903, at a meeting of the American Mathematical Society, an American mathematician, Frank Cole, delivered a paper entitled "On the Factorization of Large Numbers". His lecture consisted of walking to the board, raising 2 to the sixty-seventh power, and subtracting off 1 by hand; then writing down $(193707721) \times (761838257287)$ and multiplying by hand. The results were the same. Upon completion of the two calculations, he returned to his seat without saying a word. He supposedly received the first standing ovation ever given at a meeting of the AMS. It wasn't until 1947, more than 300 years after Mersenne made his conjecture, that the complete story was known about the 55 primes less than 257. He missed on only five of them. The primes that yield Mersenne primes are those on his original list without 67 and 257 and with 61, 89 and 107 added. The twelfth Mersenne number, M_{127}, the largest prime on Mersenne's original list, held the official record as the largest known prime for 75 years. The number $2^{127} - 1$ has 39 digits. It is

$$170, 141, 183, 460, 469, 231, 731, 687, 303, 715, 884, 105, 727.$$

Amazingly, Lucas confirmed its primality in 1876 and he did it all with pencil and paper!

Three quarters of a century after Lucas, computers came into the picture and new records for the largest prime began being set almost every year. In 1952, a non-Mersenne prime, $(2^{148} + 1)/17$, found using a mechanical desk calculator, had 44 digits. It held the record for only a few months. In the same year Raphael Robinson discovered five new Mersenne primes using the SWAC (Standards Western Automatic Computer). The largest, M_{2281} had 687 digits.

By 1978 seven more had been found when, surprisingly, two 18-year-old students in Hayward, California, Laura Nickel and Curt Noll, used 440 hours of computer time and found the twenty-fifth, and next, Mersenne prime, M_{21701}. A year later Noll found the twenty-sixth. Since then 12 more Mersenne primes have been discovered, the most recent being $M_{6972593}$ found June 1, 1999. It is the first known prime with more than 1 million digits. In fact, it has more than 2 million digits. It will likely be the largest prime number found by the end of the second millennium.

The state of knowledge concerning the primality of Mersenne numbers is as meager as it is for Fermat numbers. Unlike Fermat primes, it is suspected that there are infinitely many Mersenne primes but this has not been proven. In fact, it is suspected that, for any integer n, there are at least two Mersenne primes between M_n and M_{2n}. Like Fermat numbers it is strongly suspected that there are infinitely many composite Mersenne numbers but that too is unproved.

For the record here are the 38 generating primes that have been found. The first 31 are consecutive Mersenne primes. We list the prime p for the Mersenne prime $2^p - 1$:

$p = 2, 3, 5, 7, 13, 17, 19, 31, 61, 89, 107, 127, 521, 607, 1279, 2203,$
$2281, 3217, 4253, 4423, 9689, 9941, 11213, 19937, 21701, 23209,$
$44497, 86243, 110503, 132049, 216091, 756839, 859433, 1257787,$
$1398269, 2976221, 3021377$ and 6972593.

No doubt larger primes will be found; in fact, at the rate new Mersenne primes are being discovered, it is predicted that a Mersenne prime with at least 1 billion digits will be found in the next 6 years. All of the record holders for the largest prime number have been Mersenne primes since 1952, with the exception of the prime $391581 \times 2^{216193} - 1$. It was found in 1989 and surpassed the current record holder of the time, M_{216091}, by a mere 37 digits. It held the record from 1989 to 1992. It is not the largest known non-Mersenne prime, however; that record belongs to $302627325 \times 2^{530101} + 1$, a prime found in 1999. All of the Mersenne prime record holders since 1952 have been found by computers using the Lucas-Lehmer test. This test is named for the nineteenth century French number theorist, Edouard Lucas and the twentieth century American number theorist and computer specialist, D. H. Lehmer.

Perhaps computers will eventually find a prime with a trillion digits, perhaps the gaps between the record holder primes will eventually be filled in with consecutive Mersenne primes, but there is a limit to what technology do. Luckily the human mind is not bound by limits, and predicting the findings of creative mathematicians of the future is impossible.

We have thrown around the number of digits that large numbers have. We can figure this out with logarithms base 10. The number of digits that N has is the least integer greater than the integral part of $\log N$. Recall that $N = 10^{\log_{10} N}$, that $\log(ab) = \log a + \log b$, and that $\log a^n = n \log a$. Here is how we use this information to find the number of digits in the number $(38 \times 2^{9449}) + 1$, one of the factors of the Fermat number, F_{9449}.

Example 1.3.17

Since $(38 \times 2^{9449}) + 1$ is almost the same as 38×2^{9449}, we will examine the size of the latter.

$$\log(38 \times 2^{9449}) = \log 38 + \log 2^{9449} \approx 1.4798 + 9449(\log 2) \approx 2846.012$$

So $N \approx 10^{2846.012}$. This means that N has 2847 digits.   ∎

We conclude this section with some large prime numbers whose digits that you can remember. Most large primes do not have a pattern within their digits, but here are three examples that you can commit to memory and impress your friends:

(1) $828180\ldots321$. These are just the numbers counting backward from 82 to 1.

(2) $111\ldots1$. This is a sequence of 1031 ones.

(3) $1999\ldots9$. This sequence begins with a single one and ends with 3020 nines.

EXERCISES

1. Numbers of the form $(p_1 p_2 \cdots p_n) + 1$, are not prime for all n. Examine the following numbers N; if they are prime; say so, if they are composite, factor them: $N = (2 \times 3 \times \cdots \times p) + 1$, where p is

   (a) 11

   (b) 13

   (c) 17

2. Consult the table of primes. What is the biggest separation between

   (a) consecutive primes?
   (b) consecutive pairs of twin primes?

3. There are $8,169$ pair of twin primes less than 1 million and 3424506 pair of twin primes less than 1 billion. How many twin primes are less than 1000?

4. Most numbers do not have many distinct primes in their makeup. For the numbers $2, 3, \ldots 10$ the average number of distinct primes contained in these nine numbers is $11/9 \approx 1.2$. The average number of distinct primes that make up numbers less than 10^8 is only 2.9. Amazingly, the average number of distinct primes that make up numbers less than a googolplex is just 5.4. What is the average number of distinct primes in the makeup of numbers n where

   (a) $2 \le n \le 100$
   (b) $2 \le n \le 1000$

5. What is the longest arithmetic sequence of primes where the common difference is 30?

6. Find an arithmetic sequence of eight primes. What is the common difference?

7. Prove Theorem 1.3.6. If all the $n > 2$ terms of the arithmetic sequence

$$p, p + d, p + 2d, p + 3d, \ldots, p + (n - 1)d$$

are primes, then the common difference d is divisible by every prime less than n.

8. Consult the table of primes. What is the longest arithmetic sequence of primes you can find in the table?

9. What is the longest sequence of primes (not necessarily consecutive) you can generate using the quadratic function of the form $f(n) = n^2 + n + p$, where p is a prime ($n = 0, 1, 2, \ldots$)?

10. Consider the quadratic $f(n) = n^2 + n + p$ for various primes p, where $n = 0, 1, 2, \ldots$. What can you say about the nature of numbers, $f(n)$, for $n > p$?

11. What is the longest sequence of primes (not necessarily consecutive) you can generate using the quadratic function of the form $f(n)$, where $f(n)$ is

    (a) $n^2 + 21n + 1$
    (b) $3n^2 + 3n + 23$

12. If you have a computer, determine how many primes the following functions produce before yielding a composite number.

    (a) $f(n) = n^2 + n + 27941$
    (b) $f(n) = 103n^2 - 3945n + 34381$

13. Write these numbers in their prime form.

    (a) 341380
    (b) 454542
    (c) 704529
    (d) 869498

14. Using ideas from Euclid's proof of the infinitude of primes, prove that there are infinitely many primes of the form

    (a) $4k + 3$
    (b) $6k + 5$
    (c) You can also try proving that there are infinitely many primes of the form $4k + 1$, but this is much harder.

15. Show the following:

    (a) Any prime of the form $3n + 1$ is of the form $6n + 1$.
    (b) Any composite number of the form $3n + 2$ has a factor of this form.

16. What can you say about primes of the form $n^2 + 2$?

17. Show the following:

   (a) The product of twin primes is one less than a perfect square.

   (b) The sum of twin primes other than $3, 5$ is always divisible by 12.

18. The product of twin primes, other than 3 and 5, is of the form $9n - 1$.

19. Explain why the sum of consecutive odd primes (not necessarily twin primes) always has at least three, not necessarily different, prime factors. For example, 31 and 37 are consecutive odd primes. $31 + 37 = 68$ and $68 = 2^2 \times 17$. This makes three primes: two 2s and a 17.

20. Find as many prime factors as you can for the following numbers:

   (a) $2^{21} - 1$

   (b) $2^{64} - 1$

   (c) $2^{45} - 1$

   (d) $2^{44} - 1$

   (e) $2^{60} - 1$

21. Filling in the following hypotheses, state a theorem and then prove it.

   (a) $3 \mid 2^n - 1$ if n is . . .

   (b) $7 \mid 2^n - 1$ if n is . . .

   (c) $3 \mid 2^n + 1$ if n is . . .

   (d) $5 \mid 2^n + 1$ if n is . . .

22. Prove by induction that p_n, the nth prime number, is $\leq 2^{2^{n-1}}$

23. We learned that for numbers up to 3 trillion the largest separation between consecutive primes is 652 while the largest separation between consecutive primes for numbers up to 5 trillion is 778.

(a) What is the average separation between consecutive primes for numbers up to 3 trillion?

(b) What is the average separation between consecutive primes for numbers up to 5 trillion?

24. **Stirling's formula** is a good approximation of $n!$ for large values of n: $n! \approx \sqrt{2\pi n}\ n^n e^{-n}$.

   (a) 100! has 158 digits. How well does Stirling's formula approximate this?

   (b) Approximately how many digits does 1000! have?

   (c) There are at least 999 consecutive composites between $1000!+2$ and $1000!+1000$. What is the average separation between primes for numbers that are of the size of 1000!?

   (d) What are the sizes of the numbers where the average separation between primes is 1000?

25. Exactly how many digits are in the non-Mersenne primes:

   (a) $391581 \times 2^{216193} - 1$.

   (b) $302627325 \times 2^{530101} + 1$.

26. Exactly how many digits are in:

   (a) the thirty-eighth Mersenne prime?

   (b) the Fermat number F_{9449}?

27. What percent of the numbers are prime in the range between 1 and

   (a) the thirty-eighth Mersenne prime?

   (b) F_{9449}?

28. Approximately how many primes are there between the thirty-seventh and the thirty-eighth Mersenne primes?

1.4 Number Fascinations

Besides prime numbers, mathematicians have been fascinated with certain other types of numbers. In fact, questions about the additive and multiplicative makeup of numbers have been asked and answered to an astonishing degree. There are triangular numbers, Fibonacci numbers, perfect numbers, multiply perfect numbers, amicable numbers, sociable numbers, deficient numbers, abundant numbers, normal numbers, and weird numbers. We shall indulge ourselves a bit in the hope that some reader will become enthusiastic. We begin with two types of numbers that are curious because of their additive makeup: triangular numbers and Fibonacci numbers.

Triangular numbers are the numbers that represent the number of dots that can form an equilateral triangle. These numbers have been an object of fascination dating back at least 2500 years to the Pythagoreans in Greece. As you can see, the numbers 1, 3, 6, and 10 count the number of dots in these triangles. It is easy to see that these numbers are made up of the sum of the first few consecutive numbers; 1, $1 + 2$, $1 + 2 + 3$, and $1 + 2 + 3 + 4$.

Definition 1.4.1 *A* **triangular number** *is a number of the form*

$$1 + 2 + 3 + \cdots + n.$$

Here are some facts that have been known for a long time about triangular numbers. We leave further discoveries for the exercises.

Theorem 1.4.2

1. *A triangular number is of the form $n(n + 1)/2$. (Pythagoras, about 550 B.C.*

2. *The sum of two consecutive triangular numbers is a perfect square. (Nicomachus, about 100 A.D.*

3. *The difference of squares of consecutive triangular numbers is a perfect cube.*

Proof (1) Let $N = 1 + 2 + \ldots + n$. So

$$2N = (1 + 2 + \cdots + n - 1 + n) + (1 + 2 + \cdots + n - 1 + n)$$
$$= (1 + n) + (2 + n - 1) + \cdots + (n - 1 + 2) + (n + 1)$$
$$= n(n + 1).$$

So $N = n(n + 1)/2$.

(2) Let $(1 + 2 + \cdots + n - 1 + n)$ and $(1 + 2 + \cdots + n - 1 + n + n + 1)$ represent consecutive triangular numbers. Separating out the term $n + 1$, we may add as in (1) and get $n(n + 1)$. Then putting the term $n + 1$ back into the sum, we get $n(n + 1) + (n + 1) = (n + 1)^2$.

(3) We may write consecutive triangular numbers as $(n - 1)n/2$ and $n(n + 1)/2$. Squaring these numbers and subtracting, we get

$$[n^2(n + 1)^2/4] - [n^2(n - 1)^2/4]$$
$$= (n^2/4)[(n^2 + 2n + 1) - (n^2 - 2n + 1)]$$
$$= (n^2/4)(4n) = n^3. \qquad \square$$

Another type of number known for its additive makeup is the Fibonacci number. The Fibonacci numbers hold a special charm for both professional and amateur mathematicians. There is even a journal, *The Fibonacci Quarterly*, that delves into the magic of these numbers. Fibonacci was the pen name of the mathematician Leonardo of Pisa (1180−1250), perhaps the greatest mathematician of the Middle Ages. The sequence was named after him in the nineteenth century after it was discovered in his famous work *Liber Abaci* by the number theorist Edouard Lucas. Ironically the sequence was based on a simple problem concerning a population of rabbits:

> A man puts one pair of rabbits in a certain place entirely surrounded by a wall. How many pair of rabbits can be produced from that pair in a year if the nature of these rabbits is such that every month each pair bears a new pair which, from the second month on, becomes productive?

The sequence of pairs can be figured to be 1, 1, 2, 3, 5, 8, 13, ... continuing indefinitely. After the first two numbers, 1, 1, the succeeding numbers are simply the sum of the previous two.

Definition 1.4.3 *The sequence a_n of numbers $a_1, a_2, \ldots, a_n$ is the* **Fibonacci sequence** *if $a_1 = a_2 = 1$ and $a_{n+2} = a_n + a_{n+1}$.*

There are hundreds of results known about Fibonacci numbers; several are offered in the exercises. Here is one. Notice that if we add up the first few Fibonacci numbers we get another one—almost: $1 + 1 + 2 + 3 = 7$. That is $8 - 1$. And $1 + 1 + 2 + 3 + 5 + 8 = 20$. That is $21 - 1$. This is true in general.

Theorem 1.4.4 *In the Fibonacci sequence a_n the sum $a_1 + a_2 + \cdots + a_n = a_{n+2} - 1$.*

Proof Notice that since $a_i = a_{i+2} - a_{i+1}$, we may write the sum $a_1 + a_2 + \ldots + a_n$ as $(a_3 - a_2) + (a_4 - a_3) + \cdots + (a_{n+2} - a_{n+1})$, which equals $a_{n+2} - a_2$. Since $a_2 = 1$, we have $a_{n+2} - 1$. □

An alternate proof might be more appealing for those who are attracted to proof by induction. This equality is tailor made for this.

Alternate Proof Let $\mathcal{P}(n)$ be $a_1 + a_2 + \cdots + a_n = a_{n+2} - 1$.
$\mathcal{P}(1)$ is true because $a_1 = a_3 - 1 = 1$.
Suppose $\mathcal{P}(n)$ is true and consider $\mathcal{P}(n+1) : a_1 + a_2 + \cdots + a_n + a_{n+1} = a_{n+3} - 1$.
The following equalities show $\mathcal{P}(n+1)$ is true: $a_1 + \cdots + a_n + a_{n+1} = (a_1 + \cdots + a_n) + a_{n+1} = (a_{n+2} - 1) + a_{n+1} = (a_{n+1} + a_{n+2}) - 1 = a_{n+3} - 1$.
So the equality follows by induction. □

Since we have spoken of primes in Section 1.3, you might be interested in knowing that the 571st Fibonacci number is the largest known Fibonacci prime. It has 119 digits. We shall continue to encounter the Fibonacci sequence as we progress through the book (in particular in Sections 3.2 and 4.3 on continued fractions). There is also a project in Chapter 5 on the subject.

Another rich source of inquiry into number fascinations is based on summing up the divisors of numbers. For example, the divisors of 8 are 1, 2, 4, and 8. The sum of these numbers is 15. The divisors of 6 are 1, 2, 3, and 6. The sum of these numbers is 12. The divisors of 12 are 1, 2, 3, 4, 6, and 12. These divisors add up to 28. We may list our findings like this:

Number, N	Divisors	Sum of divisors, $\sigma(N)$
8	$1, 2, 4, 8$	15
6	$1, 2, 3, 6$	12
12	$1, 2, 3, 4, 6, 12$	28

Notice that we have demonstrated three different situations here: $\sigma(N) < 2N$ for $N = 8$, $\sigma(N) = 2N$ for $N = 6$, and $\sigma(N) > 2N$ for $N = 12$. There are names for these three types of numbers. We shall adopt the standard definitions that disregard the number, N, itself as a divisor. Such divisors, called the aliquot parts, include all the proper divisors of a number along with the number 1, as well.

Definition 1.4.5

(a) The divisors of a number, N, that are less than N are called the **aliquot parts** *of N.*

(b) A **perfect number** *is a number that equals the sum of its aliquot parts.*

(c) An **abundant number** *is a number whose aliquot parts add to more than the number itself.*

(d) A **deficient number** *is a number whose aliquot parts add to less than the number itself.*

So we have discovered a number of each type: 8 is deficient, 6 is perfect, and 12 is abundant.

By far the most fascinating numbers of these three types are the perfect numbers. The use of the word "perfect" shows the veneration that such numbers were given. Saint Augustine linked the perfection of God and the number 6 when he explained that God chose to create the earth in 6 days rather than in a single moment because 6 represented perfection.

> Six is a number perfect in itself, and not because God created all things in six days; rather the inverse is true; God created all things in six days because this number is perfect. And it would remain perfect even if the work of the six days did not exist.

Evidence of the perfection of the universe was noted by early Biblical scholars, who cite that 28 days is the time it takes the moon to orbit the earth; 28 is also a perfect number. The name "perfect" is attributed to the Pythagoreans. Four perfect numbers were known by the Greeks: 6, 28, 496, and 8128. In Euclid's *Elements* there is a proof that any number of the form $2^{p-1}(2^p - 1)$ is a perfect number, so long as $2^p - 1$ is a Mersenne prime. These four perfect numbers correspond to $p = 2, 3, 5,$ and 7. Here is a proof of Euclid's theorem.

Theorem 1.4.6 *A number of the form* $2^{p-1}(2^p - 1)$ *is perfect if* $2^p - 1$ *is a Mersenne prime.*

Proof One key is to recall geometric series from algebra.

$$(a - 1)(1 + a + a^2 + a^3 + \cdots + a^n) = (a^{n+1} - 1)$$

So, for $a \neq 1$ we have the formula

$$1 + a + a^2 + a^3 + \cdots + a^n = (a^{n+1} - 1)/(a - 1).$$

Here, with $a = 2$, we have the series

$$1 + 2 + 2^2 + 2^3 + \cdots + 2^n = (2^{n+1} - 1).$$

Let us look at the number $2^{p-1}(2^p - 1)$. Since both p and $2^p - 1$ are prime, the aliquot parts of $2^{p-1}(2^p - 1)$ are $1, 2, 2^2, 2^3, \ldots, 2^{p-1}$ along with $2^p - 1, 2(2^p - 1), 2^2(2^p - 1), 2^3(2^p - 1), \ldots, 2^{p-2}(2^p - 1)$. Using geometric series formulas, we find

$$1 + 2 + 2^2 + 2^3 + \cdots + 2^{p-1} = 2^p - 1, \text{ and also}$$

$$1(2^p - 1) + 2(2^p - 1) + 2^2(2^p - 1) + \cdots + 2^{p-2}(2^p - 1) = (2^{p-1} - 1)(2^p - 1).$$

Adding these two numbers, we get

$$2^p - 1 + (2^{p-1} - 1)(2^p - 1) = 2^{p-1}(2^p - 1). \qquad \square$$

Leonhard Euler proved a partial converse to this theorem; he proved that all even perfect numbers must be of this form. In fact, all known perfect numbers are even and it is suspected, but not yet proven, that there are no odd perfect numbers. So since there are 38 known Mersenne primes there are 38 known perfect numbers. As of June 1999, the largest known Mersenne prime was generated by the prime $6,972,593$, hence the largest known perfect number is

$$2^{6972592}(2^{6972593} - 1).$$

Since it has more than 4 million digits, we will not bother to write it down. In fact, only 12 perfect numbers have fewer than 100 digits, so writing them down explicitly is not a convenient exercise. Here are the first seven perfect numbers:

$$6, 28, 496, 8128, 33550336, 8589869056, 137438691328.$$

The eighth perfect number $2^{30}(2^{31}-1)$ was published in 1811 by
Peter Barlow in his book *Theory of Numbers*. He says that this
number "is the greatest that ever will be discovered; for, as they are
curious, without being useful it is not likely that any person will ever
attempt to find one beyond it." In fact, human curiosity seems to
know no bounds; 30 more perfect numbers have been discovered and
the search is very much in progress. And there is a lot known about
these "not very useful" numbers. For example, while no odd perfect
number has ever been found, it is known that if n is an odd perfect
number with k distinct prime factors, then $n < 4^{4^k}$. So, if $k = 7$,
for example, n would have to have fewer than 9865 digits. It is also
known that such an odd perfect number would have to have more
than 200 digits. So it has been narrowed down.

Marin Mersenne discovered that the factors of 120, other than
120 (that is, $1, 2, 3, 4, 5, 6, 8, 10, 12, 15, 20, 24, 30, 40$, and 60) sum up
to 240 which is twice 120. He wrote this discovery to Descartes
who promptly found nine other numbers that shared this multiple
property. Such numbers have become known as multiply perfect
numbers.

Definition 1.4.7 *A* **multiply perfect number** *is a number whose
aliquot parts add to a multiple of the number. That multiple is called
the* **multiplicity** *of the number.*

While it has not become a major goal of mathematicians to discover multiply perfect numbers, a fair amount of time has been
spent over the past few centuries searching for them. With computers, many new multiply perfect numbers are being found every
year. Here is a brief account of the state of affairs. More than 3300
multiply perfect numbers are now known. Curiously, only six numbers of multiplicity 2 have been found and they have been known
for a long time. The smallest is $120 = 2^3 \times 3 \times 5$, the largest has
11 digits. Only 36 numbers of multiplicity 3 are known; the smallest
is $30240 = 2^5 \times 3^3 \times 5 \times 7$, the largest has 46 digits. There are 65
known numbers of multiplicity 4; the smallest is

$$14182439040 = 2^7 \times 3^4 \times 5 \times 7 \times 11^2 \times 17 \times 19$$

and the largest has 101 digits. There are 245 known numbers of
multiplicity 5; the smallest is 154345556085770649600. Its prime

form is

$$2^{15} \times 3^5 \times 5^2 \times 7^2 \times 11 \times 13 \times 17 \times 19 \times 31 \times 43 \times 257.$$

The largest has 193 digits. There are 516 known numbers of multiplicity 6, 1118 of multiplicity 7, 1259 of multiplicity 8, and 55 of multiplicity 9. These counts are felt to be definitive for multiplicities 2, 3, 4, and 5. The count for multiplicity 6 may be complete as well, but for higher multiplicities more numbers are expected. In fact, new numbers with large multiplicities are being discovered monthly. It is hypothesized that there is only a finite number of multiply perfect numbers of any given multiplicity and that there is no limit on the size of the multiple. So far the largest multiplicity is 9 and the smallest of these numbers has 685 digits. Stay tuned.

One curious fact about multiply perfect numbers relates to perfect numbers. As stated, there is no known odd perfect number. If one were to exist it would have to have more than 200 digits. It is also known that there are six multiply perfect numbers of multiplicity 2. It is virtually guaranteed that there are no more. Yet if there were an odd perfect number, then there would be another number of multiplicity 2. This is because of the following relationship.

Theorem 1.4.8 *If N is an odd perfect number, then $2N$ is a multiply perfect number of multiplicity 2.*

The proof is left as an exercise.

There are numbers whose divisors, other than themselves, add to a second number and, in return, the aliquot parts of the second add to the first. For example, the sum of 220's aliquot parts is 284, the sum of 284's aliquot parts is 220. When this happens the numbers are called amicable.

Definition 1.4.9 *Two numbers are **amicable** if the sum of the aliquot parts of one equals the other number.*

The Greeks believed that amicable numbers had a special influence on establishing friendships between individuals. The problem was that they knew of only one amicable pair. It wasn't until 1636 that a second pair was found; Fermat found the pair of amicable numbers: 17296 and 18416. In 1638, Descartes found a third pair, 9363584 and 9437056. A hundred years later, Euler discovered 59 new pairs, several of which are smaller than Fermat's and Descartes's. Euler did such a thorough job that for the next 150 years only four more examples were found. Amazingly, in 1867, a 16-year-old Italian boy discovered an amicable pair that was smaller than any previously known pair except 220 and 284. This pair, 1184 and 1210, is now known to be the second smallest pair. Since computers have taken over, it is known that there are 42 pairs smaller than 1 million, 108 pairs smaller than 10 million, and 236 pairs smaller than 100 million. Thus amicable numbers are not nearly as rare as perfect numbers.

A "natural" extension of amicable numbers is sociable numbers.

Definition 1.4.10 *The numbers* $n_1, n_2, \ldots, n_k$ *are* **sociable numbers** *if* $k \geq 3$ *and the sum of the respective aliquot parts equals another number in the group.*

Notice that summing aliquot parts can be a game. Just start with a number and proceed through its various sums of aliquot parts and notice what happens. If the beginning number is 6, then the sum is simply 6 again. If the beginning number is 220, then we get 284 and then 220 back again. It turns into a cycle. And presumably with sociable numbers, if we begin with such a number we will eventually cycle back to that number again. Let us set this up more formally.

Definition 1.4.11 *An* **aliquot sequence** *of numbers,* $s_k(n)$, $n = 0, 1, \ldots$ *is a sequence formed as follows:* $s_0(n) = n$, $s_1(n) =$*the sum of the aliquot parts of* $s_0(n)$*; generally* $s_{k+1}(n)$ *is the sum of the aliquot parts of* $s_k(n)$*.*

Example 1.4.12

(a) Let $n = 20$. The aliquot sequence for n is 20, 22, 14, 10, 8, 7, 1. It ends here. The sequence is of length 7.

(b) Let $n = 1184$. The aliquot sequence for n is 1184, 1210, 1184. It repeats in a cycle of length 2. We know that 1184 and 1210 are amicable numbers.

(c) Let $n = 1264460$. Its aliquot sequence is 2115324, 2784580, 4938136, 1264460. It repeats in a cycle of length 4. These four numbers form a sociable group. You do not need to check the authenticity of this group unless you like. The numbers are large. ■

This example indicates that either the aliquot sequence ends with a 1, or it cycles back. It is not known if these are the only possibilities, but it is suspected that there are sequences that are unbounded. With the advent of computers all sorts of data can be gathered about these sequences. For example, what is the longest cycle that has been found? What is the longest sequence that returns to 1? What is the largest member of a sequence, relative to the original number? Here are a few answers. The smallest number that has an extraordinarily long aliquot sequence is $n = 138$. It returns to 1 on the 177th term. The 128th term of the aliquot sequence is 179931895322. Another small number, 840, ends on the 488th term and reaches a 49-digit number on its 287th term. The number 276 is the smallest number whose destination is unknown. After 469 steps the numbers had reached 45 digits in length and there is no immediate end in sight. The longest known cycle is 28. Putting it another way, the largest group of sociable numbers is 28. Only two sociable groups were known before computers; since computers have set to work many more have been found. Strangely enough, only recently, after extensive computer searches, has a sociable group of size 3 been found. It is

$$2^5 \times 3 \times 13 \times 293 \times 337, \ 2^5 \times 3 \times 5 \times 13 \times 16561, \ 2^5 \times 3 \times 13 \times 99371.$$

Finally, we finish up with "normal" and "weird" numbers. While these numbers may be whimsical, they probably aren't any more whimsical than the mathematicians who study them.

Definition 1.4.13 *A number is called* **normal** *if it is abundant and contains a subset of aliquot parts that adds to itself. A number is called* **weird** *if it is abundant and not normal. A weird number is* **primitive** *if none of its factors are weird.*

Normal numbers are the norm; for example, $12 = 1+2+3+6$ and $30 = 5 + 10 + 15$. In fact, it is not easy to find an abundant number that is not normal. While weird numbers are rare, they are more plentiful than perfect numbers. There is one primitive weird number less than 100, one more less than 1000, three more less than 10,000. It has been proved that there are infinitely many weird numbers.

EXERCISES

1. Prove the following:

   (a) (Plutarch, around 100 A.D.) The number n is triangular
       if and only if $8n + 1$ is a perfect square.

   (b) (Euler, 1775) If n is a triangular number, then so are the
       numbers $9n + 1, 25n + 3$, and $49n + 6$.

   (c) (Arabhatta, 500 A.D.) The sum of the first n triangular
       numbers is $n(n + 1)(n + 2)/6$.

2. In the sequence of triangular numbers; 1, 3, 6, 10, ... find

   (a) two triangular numbers whose sum and difference are both
       triangular numbers

   (b) three successive triangular numbers whose product is a
       perfect square

   (c) three successive triangular numbers whose sum is a perfect
       square

3. Show that all known perfect numbers are triangular numbers.

4. Define a **product perfect number** to be a number that equals
   the product of its aliquot parts. For example 6, is a product
   perfect number because $6 = 1 \times 2 \times 3$. Completely characterize
   all product perfect numbers.

5. If a and b are Fibonacci numbers what can you say about
   $\gcd(a, b)$?

6. Experiment with terms of the Fibonacci sequence and see what
   relationships you can find. Let a, b, c, and d represent consecutive Fibonacci terms for this problem. Look for arguments to
   substantiate the relationships you discover.

   (a) Compare ac with b^2.

   (b) Compare ad with bc.

   (c) Compare $b^2 - a^2$ and ab.

   (d) What can you say about $(a - c)b$?

   (e) What can you say about $c^2 - a^2$?

(f) What can you say about $a^2 + b^2$?

(g) What can you say about $(ad)^2 + (2bc)^2$?

7. Discover formulas for the following series. We denote the terms of the Fibonacci sequence by $a_1, a_2, \ldots, a_n$. Prove your findings using induction.

   (a) $a_1 + a_3 + \cdots + a_{2n-1}$

   (b) $a_2 + a_4 + \cdots + a_{2n}$

   (c) $a_1 - a_2 + a_3 - a_4 + \cdots + (-1)^{n+1}a_n$

   (d) $a_1 a_2 + a_2 a_3 + \cdots + a_{2n-1} a_{2n}$

   (e) $a_1 a_2 + a_2 a_3 + \cdots + a_{2n} a_{2n+1}$

   (f) $a_1^2 + a_2^2 + \cdots + a_n^2$

8. Find a formula relating a_{n+m} to a_n, a_m and the predecessors and successors of a_n and a_m. Prove your formula by induction.

9. Form the sequence of numbers a_2/a_1, a_4/a_2, $\ldots$, a_{2n}/a_n, $\ldots$. Find some characteristics of this sequence; in particular, say how it may be analogous to the Fibonacci sequence.

10. Prove or disprove:

    (a) If n is an abundant number, then kn is also abundant, where k is a natural number.

    (b) If n is deficient, then $2n$ is also deficient.

11. Show the following:

    (a) 2^n is always deficient.

    (b) $2^n \times 3$ is abundant for $n > 1$.

    (c) $3^n \times 5$ is always deficient.

12. Find an odd number less than 1000 that is abundant.

13. The smallest multiply perfect number of multiplicity 2 is 120. Find the next larger multiply perfect number of multiplicity 2.

14. The third multiply perfect number of multiplicity 2, in order of size, is smaller than 1 million. Find it.

15. Write out the following numbers in prime form and find their multiplicity.

   (a) 459818240
   (b) 1476304896
   (c) 51001180160
   (d) 518666803200

16. Form the aliquot sequences for the following numbers. Find the length of the cycle or the length of the sequence, if it ends.

   (a) $n = 30$
   (b) $n = 60$
   (c) $n = 2620$
   (d) $n = 12496$
   (e) $n = 14316$

17. Find a number whose aliquot sequence has length longer than 20.

18. Find two weird numbers. One has two digits and one has three digits.

19. Prove Theorem 1.4.8 If N is an odd perfect number, then $2N$ is a multiply perfect number of multiplicity 2.

20. Prove or disprove: If N is a multiply perfect number of multiplicity $p - 1$, where p is a prime that does not divide N, then pN is a multiply perfect number of multiplicity p.

21. Prove the following:

   (a) If $N \neq 1$, then $6N$ is normal.
   (b) If p is a prime and $p > 144$, then $70p$ is weird.

22. Show that the sum of consecutive odd cubes, beginning with 1, is a triangular number. Do this by first finding a formula and then establishing your formula by induction.

23. Suppose that N is a perfect number larger than 6.

    (a) Give a formula for N as the sum of consecutive odd cubes.

    (b) How many odd cubes are needed to add up to the thirty-eighth perfect number?

24. Find a formula for the sum of the factors of a number N where

$$N = p_1^{a_1} p_2^{a_2} \cdots p_n^{a_n}.$$

25. Prove or disprove the following "converses" of Theorem 1.4.6.

    (a) If N is a perfect number and $N = 2^m p$, where p is an odd prime, then $p = 2^{m+1} - 1$.

    (b) If N is an even perfect number and $N = 2^m q$, where q is odd, then q is prime and of the form $2^{m+1} - 1$.

Chapter 2

The Integers

The integers are made up of the natural numbers, 0, and the negatives of the natural numbers. Many of the numerical problems from antiquity were problems that required integer solutions. This makes sense; originally mathematics was used to solve real-world problems and such problems generally look for whole number answers. It turns out that requiring answers in whole numbers can be a very difficult challenge. Here are two classical examples. The first apparently originated in ancient China.

> A band of 17 pirates stole a sack of gold coins. When they tried to divide the fortune into equal portions, 3 coins were left over. In the fight that followed one pirate was killed. The coins were redistributed into equal portions and this time 10 coins were left over. Another fight broke out and another pirate was killed. Now the coins could evenly be distributed. What is the number of coins that were stolen?

The second example is a puzzle that the French mathematician Bernhard Frénicle de Bessy (1605−1675) sent to the great English mathematician John Wallis (1616−1703) in the mid-1600s regarding the famous Battle of Hastings.

> The men of Harold stood well, as was their wont, and formed sixty and one squares, with a like number of men in every square thereof, and woe to the hardy Norman who ventured to enter their redoubts; for a single blow of a Saxon war-hatchet would break his

lance and cut through his coat of mail ... . When
Harold threw himself into the fray the Saxons were
one mighty square of men, ... How many men were
there?

These examples come under the general heading of Diophantine
equations. Mathematicians have been intrigued with solving Diophantine equations for more than three millennia. Every civilization,
both ancient and modern, has posed such problems; in this chapter
we offer examples from China, India, Greece, Egypt, England, and
France.

The most famous of all Diophantine equations, $x^n + y^n = z^n$,
appeared to be impossible to solve with nonzero integers for $n > 2$.
In 1621 Fermat claimed he proved that it is indeed impossible, a
theorem that has become known as Fermat's last theorem. But the
"proof" was too long to fit in the margin of the book where he
made the claim. In 1995 it was finally proved to be impossible in a
landmark paper by Englishman Andrew Wiles. This result caught
the attention of the world, both mathematical and nonmathematical
alike, although it must be wondered how strongly nonmathematicians
felt about it. For the case $n = 2$, we have the equation $x^2 + y^2 = z^2$,
which has the form of the famous Pythagorean theorem. Finding
integers x, y, z that make up the sides of a right triangle was studied
as long ago as 1500 B.C.by the Babylonians and completely solved
around 500 B.C. by the Pythagoreans.

In this chapter we study both the theoretical and the practical
problems that the integers present. We shall begin the chapter with
a look at Diophantine equations and then move to the concept of
congruence, a powerful tool developed by Gauss, that will not only
help in approaching Diophantine equations but will lead to powerful
theoretical results that answer questions that arose in our study of
the natural numbers.

2.1   Diophantine Equations

Diophantus of Alexandria probably lived in the third century A.D.but
almost nothing is known about his life. The following epigram, dating
from the fourth century, gives the only indication of how long he
lived. Typically it is in the form of a puzzle.

The gods granted him childhood for a sixth of his life,

and a twelfth for his adolescence. A barren marriage took up a seventh of his life. Five years passed and then a child was born to him. The child lived for half of his father's life. After his child's death, Diophantus lived four more years, drowning his pain in the study of numbers. Then he gave up his life.

Letting n stand for his age at death we may write the preceding information in the equation

$$(1/6)n + (1/12)n + (1/7)n + 5 + (1/2)n + 4 = n.$$

The answer is 84.

Diophantus is credited with writing the first treatise on algebra, *Arithmetica*. Only six of the original 13 books remain, but in these books Diophantus invented a system of notation complete with a recognition of unknowns that allowed him to address algebra problems. A **Diophantine equation** is an indeterminate equation that admits only integer solutions. It is roughly of the form $f(x) + g(y) = h(z)$, where f, g, and h are polynomials but more polynomials and more variables may be present. Let us list a few examples. We begin with Ramanujan's quote at the beginning of the book: "1729 is a very interesting number; it is the smallest number expressible as the sum of two cubes in two different ways." In Diophantine terms Ramanujan's observation may be posed as a puzzle: What is the smallest positive integer c such that $x^3 + y^3 = c$ has two solutions?

This leads to lots of problems of this form. For example, consider the following equations:

1. $x^2 + y^2 = c$. Find the smallest positive integer c such that this equation has two solutions, three solutions, four solutions, ...

2. $x^3 + y^3 = c$. Find the smallest positive integer c such that this equation has three solutions, four solutions, five solutions, ...

3. $x^3 - y^3 = c$. Find the smallest positive integer c such that this equation has two solutions.

4. $x^n + y^n = c$. Find the smallest positive integer c such that this equation has two solutions where $n = 4, 5, \ldots$.

For (2) the answers get large fast. Without computers many of these Diophantine questions would not be answered. The smallest c such that the equation $x^3 + y^3 = c$ has three solutions is 87539319. The three sums are $167^3 + 436^3$, $228^3 + 423^3$, and $255^3 + 414^3$. For four solutions the answer is 6963472309248. Its sums are

$$2421^3 + 19083^3, \ 5436^3 + 18948^3, \ 10200^3 + 18072^3, \ 13322^3 + 16630^3.$$

For five solutions the answer is 48988659276962496. Its sums are

$$38787^3 + 365757^3, \ 107839^3 + 362753^3, \ 205292^3 + 342952^3,$$
$$221424^3 + 336588^3, \ 231518^3 + 331954^3.$$

In addressing (4) we can easily say that $c = 50$ is the smallest number such that $x^2 + y^2 = c$ in two ways: $50 = 7^2 + 1^2 = 5^2 + 5^2$. If the further condition is added that the numbers in the solution must be different, then 65 is the smallest such number: $65 = 7^2 + 4^2 = 8^2 + 1^2$. Ramanujan gives the answer for $x^3 + y^3 = c$. For quartics, it is said that Euler knew the answer: 635318657 is the smallest number that can be expressed as the sum of two quartics in two different ways: $635318657 = 59^4 + 158^4 = 133^4 + 134^4$. There is no known answer for $n = 5$ or $n = 6$.

Then there is the storied Diophantine equation: Find all the nontrivial integer solutions to the equation $x^n + y^n = z^n$. That there are no nontrivial integer solutions for $n > 2$ is called Fermat's last theorem. We can generalize this in lots of ways: Here are three.

Find all nonzero integral solutions to

5. $x^n + y^n = kz^n$

6. $x^n - y^n = z^n$

7. $x_1^n + x_2^n + \cdots + x_k^n = z^n$

Regarding (7), generally speaking there are many, usually infinitely many, solutions to this type of equation when $k \geq n$. For example $x^3 + y^3 + z^3 = a^3$ has infinitely many solutions; the smallest is $3^3 + 4^3 + 5^3 = 6^3$. In 1769 Euler conjectured that no nth power of an integer is the sum of fewer than n nth powers of integers. It was shown false in 1965: $27^5 + 84^5 + 110^5 + 133^5 = 144^5$. It has now been proved that there are infinitely many such counterexamples for fourth powers; the smallest one is $95800^4 + 217519^4 + 414560^4 = 422481^4$.

Fermat was instrumental in sparking interest in several types of Diophantine equations. In 1657, he issued a challenge to other mathematicians of the day to find a square number that is one more than 61 times another square number. This is the Diophantine equation: $x^2 - dy^2 = 1$, where $d = 61$. You recognize it as the equation describing the battle of Hastings puzzle sent by Frénicle to Wallis. In fact, Frénicle supplied many answers for $d \leq 150$ and challenged Wallis to extend the table to 200. While Frénicle did not offer a solution to Fermat's problem, he issued the challenge for $d = 151$ and 313, apparently solutions he already knew. Wallis found a solution right away (or so he claimed): $(126862368)^2 - 313(7170685)^2 = 1$. The other equation took longer: $(1728148040)^2 - 151(140634693)^2 = 1$. The calculation necessary to find these numbers is awesome. The answer for $d = 61$ is slightly larger than the numbers for $d = 151$. We can add two more types of Diophantine equations to the list.

8. $x^2 - dy^2 = \pm 1$

9. $x^n - dy^n = c$, where $n \geq 2$ and c is any integer

Equations of the form $x^2 - dy^2 = \pm 1$ are called **Pell equations**, and we will look at them further in Section 2.3 and again in Chapter 4.

We have barely scratched the surface. The possibilities are literally endless. Here are some exotic results.

Euler showed in 1738 that $x^2 - y^3 = 1$ has only one solution $x = 3$, $y = 2$. It is also true that $x^3 - y^2 = 2$ has just one positive solution: $x = 3$, $y = 5$; and $x^3 - y^2 = 4$ has just one positive solution: $x = 5$, $y = 11$. In 1932, Atle Selberg, a Norwegian mathematician who achieved fame 17 years later for his purely arithmetic proof of the prime number theorem, showed that $x^4 - y^3 = 1$ has no solution.

It has been shown that there are only four solutions to this puzzle: For what n and k does $1 + 2 + ... + n = 1^2 + 2^2 + ... + k^2$? The largest of these is $n = 645, k = 85$. We leave finding the first three as an exercise.

It is known that there are solutions to the equation $x^3 - 117y^3 = 5$, but it is not known how many there are. It is known, however, that there are at most 18 of them.

And then there is this. The polynomial $x^2 + 2y^2 + 7z^2 + 13w^2$ using integers x, y, z, w can represent every positive integer with the exception of 5. Furthermore, there are exactly 87 other polynomial forms such as this that can represent every positive integer but one.

As has been mentioned, the most famous of all Diophantine equations is $x^n + y^n = z^n$. Searching for nontrivial integral solutions for the Diophantine equation $x^n + y^n = z^n$, where n is a natural number greater than 1, has created as much interest and new mathematics throughout the centuries as the study of any mathematical idea. Here is a brief accounting of its story. Methods for systematically solving the equation $x^2 + y^2 = z^2$ were known to the Babylonians around 1500 B.C. and solutions to this equation were known to the Pythagoreans around 500 B.C. To the Pythagoreans, solutions represented examples of right triangles with integer sides, important indeed for the discoverers of the famous Pythagorean theorem. In the third century A.D., Diophantus knew of the complete set of solutions for the case $n = 2$. But for cases $n > 2$ the situation was not simple. After 13 centuries of failing to find even a single integral solution to $x^n + y^n = z^n$ for a single case of $n > 2$, Fermat, in the early seventeenth century, wrote that there were no nontrivial solutions at all. That there were no solutions was suspected. Six centuries before, Arab mathematicians were convinced that there were no integral solutions for the case $x^3 + y^3 = z^3$. But this was the first time a proof was claimed for the general case. Unfortunately, Fermat did not have room enough to show his "proof." This is what he wrote in the margin of his copy of Bachet's 1621 edition Diophantus work, *Arithmetica*, next to the proposition "To divide a given square number into 2 squares":

> On the other hand it is impossible to separate a cube into 2 cubes or a biquadrate into 2 biquadrates or generally any power except a square into 2 powers with the same exponent. I have discovered a truly marvelous proof of this, which, however the margin is not large enough to contain.

This note was written more than 20 years before his death and was not discovered until several years after his death. It is now known as Fermat's last theorem.

Fermat's Last Theorem *The equation $x^n + y^n = z^n$ has no nontrivial integral solutions for whole numbers $n > 2$.*

While it seems to be true that Fermat had proved his assertion for biquadrates, or quartics as we call them, there is no evidence that he had proved his assertion for general exponents n. Nevertheless, his statement focused even more interest on the problem and it intensified the search for a proof or a counterexample to his theorem. A century and a half later Euler produced a proof of Fermat's last theorem for cubics. By the mid-1800s there was a great deal of activity surrounding the theorem, especially in France and Germany. Proofs were regularly submitted at conferences and flaws were uncovered. In search for a solution a whole new approach to algebra was created by the German mathematician Ernst Kummer (1810–1893). In 1908 a prize of 100,000 marks was left in the will of a German mathematician named Wolfskehl. It would go to the person who could prove Fermat's theorem. More than one thousand false proofs were submitted during the next three years. Throughout the twentieth century work toward a proof was steady, and in the 1980s and early 1990s there was a flurry of activity. By that time, powerful computers and some of the finest number theorists had combined to show that Fermat's last theorem was true for all exponents up to about 2 million. Since the truth of the theorem for a prime p implies its truth for all multiples of p, they could say that there were no integer solutions to $x^n + y^n = z^n$, for all multiples of prime exponents p less than 1 million.

Then on June 23, 1993, Andrew Wiles, a British-born Professor of Mathematics at Princeton University, finished a series of lectures at Cambridge's Isaac Newton Institute with a theorem that included Fermat's last theorem as a corollary. There was huge celebration in the mathematical community and notice was even taken of this achievement on the front page of the New York Times. But initially all was not well with Wiles's proof and another year and a half would go by before he and a fellow British mathematician Richard Taylor could fix it up properly and deliver it in manuscript form. The proof has now been accepted even though it is based on results and methods in several distinct highly technical areas of mathematics and few mathematicians have thoroughly read it. These areas had not been created in Fermat's time and, while there may be elementary proofs found in the future, there is little doubt that Fermat did not have a proof of the theorem that bears his name. But Fermat deserves his due. He is the originator of the modern study of Diophantine equations.

Let us now examine some elementary Diophantine equations. The simplest type of Diophantine equation is the linear equation

$ax + by = c$ for integers a, b, and c. Apparently these were too simple for Diophantus himself because he did not address these kinds of problems. We will address them here, however.

Definition 2.1.1 *A linear* **Diophantine equation** *is an equation of the form* $ax + by = c$ *where* a, b, *and* c *are integers and* $a, b \neq 0$. *If* x, y *is a* **solution** *to a linear Diophantine equation, then* x *and* y *must be integers.*

Here is a simple example of a problem that leads to a linear Diophantine equation.

Example 2.1.2

In a barnyard of chickens and pigs I counted both the heads and the legs of the animals and came up with 70. How many of each animal is there?

We shall let c stand for the number of chickens and p for the number of pigs. Then $c + p$ counts the number heads and $2c + 4p$ counts the number of legs. We get the equation $c + p + 2c + 4p = 3c + 5p = 70$. By trial and error we can come up with five solutions. They are the following pairs (c, p):

$(0, 14)$, $(5, 11)$, $(10, 8)$, $(15, 5)$, and $(20, 2)$.

Of course, if we allowed for negative numbers of chickens and pigs we could find infinitely many solutions to the mathematical equation $3c + 5p = 70$. The answers would be of the form $c = 5k$, $p = 14 - 3k$, where k stands for any integer.   ∎

The following example shows what sorts of solutions can be expected for linear Diophantine equations.

Example 2.1.3

(a) Solve: $6x + 7y = 4$. We can proceed by trial and error and find that $x = 3$ and $y = -2$ works. Furthermore, $x = 10$, $y = -8$ and $x = -4$, $y = 4$ work. And there are other solutions, too. In fact, $x = 3 + 7k$, $y = -2 - 6k$ are solutions, where k stands for any integer.

(b) Solve: $6x + 8y = 4$. Proceeding again by trial and error, we find $x = 2$, $y = -1$ works. So also does $x = 6$, $y = -4$. In fact, all pairs of the form $x = 2 + 4k$, $y = -1 - 3k$ are solutions.

(c) Solve: $6x + 9y = 4$. Proceeding by trial and error, we have trouble finding any pairs, x, y, that work. And none will because the term $6x + 9y$ is divisible by 3 while the number 4 is not. ∎

So, as you can see, different equations allow for different sets of solutions. Here is a general theorem that tells the story.

Theorem 2.1.4 *The linear equation $ax + by = c$ has integral solutions if and only if d divides c where $d = \gcd(a, b)$. Furthermore, if x_0, y_0 is a solution, then the other solutions are of the form $x = x_0 + (b/d)k$, $y = y_0 - (a/d)k$, where k stands for any integer.*

Proof Suppose that $ax + by = c$ has the solution x_0, y_0. Since $d \mid a$ and $d \mid b$, it follows from Theorem 1.1.6 that $d \mid c$.

Suppose that $d \mid c$ and suppose that $dl = c$. By Theorem 1.1.11 there exist integers x_0, y_0 such that $ax_0 + by_0 = d$. Thus

$$a(lx_0) + b(ly_0) = dl = c.$$

Now let x_0, y_0 be a solution to $ax + by = c$. Then

$$a(x_0 + (b/d)k) + b(y_0 - (a/d)k) = ax_0 + by_0 + abk/d - bak/d = c$$

so $x_0 + (b/d)k$, $y_0 - (a/d)k$ represent solutions as well.

All that is left to show is that there are no other solutions. Suppose x', y' is a solution. Let z and w be such that $x' = x_0 + z$, $y' = y_0 - w$. So $a(x_0 + z) + b(y_0 - w) = c$. Since x_0, y_0 is a solution, it follows that $az = bw$. Thus $az/d = bw/d$. This means that since a/d and b/d are relatively prime $b/d \mid z$. Thus $z = (b/d)k$ for some k. But this implies that $w = (a/d)k$. Thus the solutions above are the only solutions. □

Now that we know what the solutions of a linear Diophantine equation look like the question turns to finding them. In particular, how do we find one solution? We can build the others from that one. Trial and error works only for equations with small coefficients. The method we introduce here is based on the Euclidean algorithm. It was introduced in Section 1.1 and we review it here. Notice that we have slightly altered the coefficients for this version.

Let a and b be natural numbers such that $a < b$. Now apply the division theorem over and over again until we reach exact divisibility.

$$b = aq_1 + r_1$$
$$a = r_1 q_2 + r_2$$
$$r_1 = r_2 q_3 + r_3$$
$$\vdots$$
$$r_{n-2} = r_{n-1} q_n + r_n$$
$$r_{n-1} = r_n q_{n+1} + 0$$

Theorem 1.1.13 tells us that r_n is $\gcd(a, b)$.

Theorem 1.1.11 says that the $\gcd(a, b)$ can be represented by the linear expression $by - ax$ for some natural numbers x and y. Let us recall Example 1.1.12 in the context of Diophantine equations and solve $365x + 1876y = 1$.

Performing the Euclidean algorithm for $b = 1876$ and $a = 365$, we get

$$1876 = 365(5) + 51$$
$$365 = 51(7) + 8$$
$$51 = 8(6) + 3$$
$$8 = 3(2) + 2$$
$$3 = 2(1) + 1$$
$$2 = 1(2).$$

Recall that we reversed the process and found the answer $by - ax = 1876y - 365x = 1$, where $y = 136$ and $x = 699$. Since the Diophantine equation here is written as $365x + 1876y = 1$, we have $x = -699$, $y = 136$ as the solution.

The process of finding solutions in Section 1.1 is tedious but, thankfully, there is an easier way. Here is how it works. Study the following table; we shall refer to it as the **table of quotients** or the **quotient table**. It is associated with the Euclidean algorithm.

		5	7	6	2	1	2	
1876	1	0	1	7	43	93	136	365
365	0	1	5	36	221	478	699	1876

Let us see how this table of quotients was constructed. Notice that we have listed the quotients from the Euclidean algorithm in the row at the top. We may label them as q_1, q_2, q_3, q_4, q_5 and q_6. Beginning with 1, 0 the numbers that follow in the row labeled "1876" are, respectively, 1, 7, 43, 93, 136, and 365. Labeling these numbers by b_1, b_2, b_3, b_4, b_5 and b_6 and letting the initial "1" be b_{-1}, and the following "0" be b_0, the sequence follows the recursive relation: $b_{n+1} = q_{n+1}b_n + b_{n-1}$. Similarly, labeling the numbers next to 365 by a_1, a_2, a_3, a_4, a_5 and a_6 and letting the initial "0" be a_{-1}, and the following "1" be a_0, we have $a_{n+1} = q_{n+1}a_n + a_{n-1}$.

Notice that the answers x and y appear in the next to the last column: $y = 136$ and $x = 699$ form the solution to the equation $1876y - 365x = 1$. In the final column the numbers 1876 and 365 are flipped over.

Here is the general formulation.

Theorem 2.1.5 *Let a and b be given and suppose that $b > a$ and the Euclidean algorithm has been performed. Suppose the resulting table of quotients looks like this:*

	q_1	q_2	$\cdots$	q_i	$\cdots$	q_n	q_{n+1}
b	1 0	b_1 b_2		b_i		b_n	b_{n+1}
a	0 1	a_1 a_2		a_i		a_n	a_{n+1}

The q_i represent the quotients resulting from the Euclidean algorithm, and $a_{i+2} = q_{i+2}a_{i+1} + a_i$ and $b_{i+2} = q_{i+2}b_{i+1} + b_i$. Notice that we begin this with $a_{-1} = 0$, $a_0 = 1$, $b_{-1} = 1$, $b_0 = 0$.

Then, letting $d = \gcd(a, b)$ and $c = \text{lcm}(a, b)$, we have

$$bb_n - aa_n = (-1)^{n-1}d \text{ and } bb_{n+1} = aa_{n+1} = c.$$

Proof The proof is by strong induction on k for $k \leq n + 1$. Let $\mathcal{P}(k)$ be the proposition $bb_k - aa_k = (-1)^{k-1}r_k$. Now $\mathcal{P}(1)$ reads $bb_1 - aa_1 = (-1)^0 r_1$. Since $a_1 = q_1$ and $b_1 = 1$, this reads $b - aq_1 = r_1$. This is true.

Now suppose that the statement $\mathcal{P}(l)$ is true for all $1 \leq l < k$ and consider $\mathcal{P}(k)$. We know that

$bb_k - aa_k = bq_kb_{k-1} + bb_{k-2} - aq_ka_{k-1} - aa_{k-2} =$

$q_k(bb_{k-1} - aa_{k-1}) + (bb_{k-2} - aa_{k-2}) =$ (by the induction hypothesis)

$q_k(-1)^k r_{k-1} + (-1)^{k-1}r_{k-2} = (-1)^k(q_k r_{k-1} - r_{k-2}) =$

$(-1)^k(-1)r_k = (-1)^{k-1}r_k.$

Since we know that $r_n = d$, it follows that $bb_n - aa_n = (-1)^{n-1}d$. Since we know that $r_{n+1} = 0$, we know that $bb_{n+1} = aa_{n+1}$. The proof that this is the $\text{lcm}(a,b)$ is left as an exercise. $\qquad\square$

Example 2.1.6

(a) Solve the linear Diophantine equation $289x + 123y = 1$. We use the Euclidean algorithm on the numbers 289 and 123.

$$
\begin{aligned}
289 &= 123(2) + 43 \\
123 &= 43(2) + 37 \\
43 &= 37(1) + 6 \\
37 &= 6(6) + 1 \\
6 &= 1(6)
\end{aligned}
$$

Next we form the quotient table:

		2	2	1	6	6	
$b = 289$	1	0	1	2	3	20	123
$a = 123$	0	1	2	5	7	47	289

We know that $bb_n - aa_n = (-1)^{n-1}$, and since $n = 4$ we have $n-1 = 3$ so one solution is $289(20) - 123(47) = (-1)^3 = -1$. Thus $x_0 = -20$ and $y_0 = 47$ is a solution to $289x + 123y = 1$. The other solutions are of the form $x = -20 + 123k$, and $y = 47 - 289k$.

(b) Solve the equation $208x + 159y = 117$. We use the Euclidean algorithm on the numbers 208 and 159.

$$
\begin{aligned}
208 &= 159(1) + 49 \\
159 &= 49(3) + 12 \\
49 &= 12(4) + 1 \\
12 &= 1(12)
\end{aligned}
$$

Next we form the quotient table:

		1	3	4	12	
208	1	0	1	3	13	159
159	0	1	1	4	17	208

We have $n = 3$ so $n - 1 = 2$ and thus $208(13) - 159(17) = (-1)^2 = 1$. Since we are solving the equation $208x + 159y = 117$, we must multiply through by 117. So our particular solution will be

$$x_0 = 13(117) = 1521, \ y_0 = 17(117) = 1989.$$

The general solution, therefore, is

$$x = 1521 + 159k, \ y = -1989 - 208k.$$

We can find smaller particular solutions by finding appropriate values for k. For example, if we let $k = -9$, then we have the solution $x_0 = 90$, $y_0 = -117$.

(c) Let us return to our example 2.1.2 with the chickens and the pigs. We want to solve $3x + 5y = 70$ with our newly acquired method. As you can see, using this powerful method for this elementary problem adds humor to the book. Here's the Euclidean algorithm and the resulting quotient table.

$$
\begin{aligned}
5 &= 3(1) + 2 \\
3 &= 2(1) + 1 \\
2 &= 1(2)
\end{aligned}
$$

		1	1	2	
5	1	0	1	1	3
3	0	1	1	2	5

Now $n = 2$ so $n - 1 = 1$ and thus $5(1) - 3(2) = -1$. Thus $2, -1$ is a solution to the equation $3x + 5y = 1$. Multiplying through by 70, we have the equation $3(140) + 5(-70) = 70$. The general solution, therefore, is

$$x = 140 + 5k, \ y = -70 - 3k.$$

This doesn't look much like our answer in Example 2.1.2, but if we let $k = -28, -27, \ldots -24$ we get the solutions: 0, 14; 5, 11; 10, 8; 15, 5; and 20, 2. ∎

Examples (b) and (c) of Example 2.1.6 show that our method for solving linear Diophantine equations, $ax + by = c$, is most effective when we have large coefficients for x and y and when c is either

gcd(a, b) or a small multiple of it. In the next section we introduce a
different method of solving these equations that is more appropriate
for large values of c. We shall now include here some real-life problems that concern simultaneity and "near" simultaneity both in time
and in space. The examples deal with distance, weight, time, and
volume. These problems generally have gcd(a, b) for c.

Example 2.1.7

Suppose that my unit of measurement is my stride, which is 105
centimeters, and your unit of measure is your stride, which is 87
centimeters, and we take a walk in the snow.

Suppose we begin at the same place; I'll go first and make my
foot prints in the snow and then you follow.

(a) Will your footprints ever land precisely in mine? If so, where?

(b) Between instances of overlapping footprints, where are the
occurrences of near misses?

Suppose I begin my walk 2 centimeters ahead of you.

(c) Will the footprints ever overlap precisely? If not, where will
be the first instance of the nearest of near misses?

As we shall see, part (a) is an lcm problem, and part (b) is a gcd
problem. We use the Euclidean algorithm to find the gcd$(105, 87) =$
3.

$$
\begin{aligned}
105 &= 87(1) + 18 \\
87 &= 18(4) + 15 \\
18 &= 15(1) + 3 \\
15 &= 3(5)
\end{aligned}
$$

Here is the associated quotient table.

		1	4	1	5
105	1	0	1	4	5	29
87	0	1	1	5	6	35

(a) Obviously my prints occur at $105x$, yours at $87y$, and your
105th step will land exactly on my 87th step. But this happens
earlier, as our quotient table shows; it tells us that $29 \times 105 = 87 \times
35 = 3045$ which is the lcm$(105, 87)$. So your 35th step lands directly

on my 29th step. So also does your 70th step on my 58th step and, generally, your $35k$th step on my $29k$th step.

(b) The table also tells us that $105(5) - 87(6) = 3 = \gcd(105, 87)$. So your 6th step is just 3 centimeters behind my 5th step. We know, from (a), that your $(6 + 35k)$th step lags behind my $(5 + 29k)$th step by 3 centimeters. We can ask if one of your steps is ever exactly 3 centimeters ahead of one of mine; that is, does $105x - 87y = -3$ ever occur? The answer is yes. If we back up from the identical steps at 3045 centimeters by 5 of my steps and 6 of yours, we find your 29th step is 3 centimeters ahead of my 24th step. This would happen generally at my $(24 + 29k)$th step and your $(29 + 35k)$th step.

(c) If I begin my walk 2 centimeters ahead, there will never be a precise overlap because of the results of (a) and (b). At the points of overlap in (a) I will be 2 centimeters ahead. At the near misses in (b) I will be either 5 centimeters ahead or 1 centimeter behind. The closest of these would be 1 centimeter behind, which occurs first at your 29th step and my 24th step. ∎

Example 2.1.8

Here is an example with time. Two adjacent stars in the sky brighten up briefly at fixed time intervals. Star A brightens every 6 days, 2 hours, and 18 minutes. Star B brightens every 2 days, 6 hours, and 36 minutes.

Just now star A and star B are flashing simultaneously.

(a) Will this ever happen again? If so, when?

(b) Between simultaneous flashings, how close together can the stars flash. When is the first "near" simultaneous flashing and when is the second?

Changing this to minutes, we see that star A brightens every 8778 minutes and star B brightens every 3276 minutes. As with the previous example, we shall find our answers from the Euclidean algorithm and its associated table of quotients.

$$
\begin{aligned}
8778 &= 3276(2) + 2226 \\
3276 &= 2226(1) + 1050 \\
2226 &= 1050(2) + 126 \\
1050 &= 126(8) + 42 \\
126 &= 42(3) + 0
\end{aligned}
$$

		2	1	2	8	3
8778	1	0	1	1	3	25	78
3276	0	1	2	3	8	67	209

(a) The table shows that stars A and B will flash simultaneously when $8778(78) = 3276(209) = 684{,}684$ minutes; that is, 1 year, 110 days, 11 hours, 24 minutes.

(b) The table also shows that $8778(25) - 3276(67) = (-1)^3 42 = -42$. So after $8778(25) = 219{,}450$ minutes star A flashes and 42 minutes later, at the $3276(67) = 219{,}492$ mark, star B flashes. Measuring back from the first simultaneous flash, we find that after $8778(78-25) = 465{,}234$ minutes A flashes and after $3267(209-67) = 465{,}192$ minutes star B flashes. So here B flashes 42 minutes before A. ∎

Example 2.1.9

Here is an example with weights on a balance scale. Using units of 28 pounds 7 ounces on one pan and 19 pounds 8 ounces on the other,

(a) How can we arrange the weights so that they exactly balance?

(b) What is the smallest weight we can accurately measure on a balance scale? How can we measure it? What is the fewest number of weights needed to make this measurement?

Let us change the units to ounces. So we have weights of order 455 ounces and 312 ounces. Let us use the 455-ounce weights on pan A and 312-ounce weights on pan B. Once again we use the Euclidean algorithm and its associated table of quotients.

$$
\begin{aligned}
455 &= 312(1) + 143 \\
312 &= 143(2) + 26 \\
143 &= 26(5) + 13 \\
26 &= 13(2) + 0
\end{aligned}
$$

		1	2	5	2
455	1	0	1	2	11	24
312	0	1	1	3	16	35

(a) The table tells us that the scale will exactly balance if we put 24 455-ounce weights on pan A and 35 312-ounce weights on pan B.

(b) Let us see how close we can get to a balance without actually getting a perfect balance. This will indicate the smallest amount we can accurately weigh. The table shows that $455(11) - 312(16) = 13$. So, if we put 11 455-ounce weights on pan A and 16 312-ounce weights on pan B along with a 13-ounce weight, the pans will balance. This answer uses $11 + 16 = 27$ total weights. The general solution is $x = 11 + 24k$, $y = -16 - 35k$. Now, if we play with the integer k we can come up with different answers. For example, if $k = 1$, then we have a balance with 35 455-ounce weights on pan A and 51 312-ounce weights on pan B along with a 13-ounce weight. This takes 86 weights. If $k = -1$ we have $x = -13$, $y = 19$. So $312(19) - 455(13) = 13$. This means that if 13 455-ounce weights are put on pan A and 19 312-ounce weights on pan B then the extra 13-ounce weight would be placed on pan A to balance the scales. This takes 32 weights. It appears as if 27 weights is the fewest. ■

EXERCISES

1. It has been proved that $1^2 + 2^2 + \cdots + k^2$ is a perfect square for only two values of k. One of those values is 1, what is the other?

2. Find integral values of x and y such that $1 + x + x^2 + x^3 = y^2$, where $x \neq 0, \pm 1$.

3. Find three positive integral values of n and k such that $1 + 2 + \cdots + n = 1^2 + 2^2 + \cdots + k^2$.

4. Find the smallest positive c so that $x^2 + y^2 + z^2 = c$ has

   (a) one solution
   (b) two solutions
   (c) three solutions

   (We assume, in this problem, that none of x, y, or z is 0 and the multiple solutions must be different threesomes.)

5. Find the smallest positive number that is the sum of three cubes in two different ways.

6. The Diophantine equation $x^4 + y^4 + z^4 = 2673$ can be solved with positive integral values in two different ways; find them.

7. Find the number of different ways that the following numbers can be written as the sum of two cubes.

   (a) 4104

   (b) 46683

8. From *Arithmetica* by Diophantos. Express these questions as Diophantine systems and try to find solutions.

   (a) Find two square numbers such that when one forms their product and adds either of the numbers to it, the result is a square.

   (b) Find three numbers in an arithmetic series such that the sum of any two of them is a perfect square.

   (c) Find three numbers such that their sum is a square and the sum of any two of them is a square.

9. Find four solutions to the Diophantine equation $x^2 + y^2 = z^3$. Do you see a pattern you can extend?

10. Find five primitive solutions to each of these equations. A solution x, y, z is primitive if $\gcd(x, y, z) = 1$.

    (a) $x^2 + 2y^2 = z^2$
    (b) $x^2 + y^2 = 2z^2$

11. Examine consecutive squares in the following vein: $10^2 + 11^2 + 12^3 = 13^2 + 14^2$ is an example of five consecutive squares where the sum of the first three equals the sum of the last two.

    (a) Find an example of three consecutive squares where the sum of the first two equals the third.

    (b) Find an example of seven consecutive squares where the sum of the first four equals the sum of the final three.

    (c) Using the examples, look for a pattern and try to extend it.

12. Find the $\gcd(a, b)$ and then solve the linear Diophantine equation $ax + by = \gcd(a, b)$ where

    (a) $a = 28644$, $b = 13566$
    (b) $a = 57171$, $b = 116109$
    (c) $a = 101556$, $b = 605682$

13. In each of the following equations, find the smallest x and y such that

    (a) $889x + 511y = 63$
    (b) $5720x + 6171y = 77$
    (c) $9702x + 2873y = 103$

14. A friend of mine owns several 10-ounce silver pieces, each worth \$57. I own several 1-ounce gold pieces each worth \$375.

    (a) If I owed her money, what is the smallest amount I could pay off with coin swapping? Describe how I would pay this amount.
    (b) If she owed me money, what is the smallest amount she could pay off? Describe how she would do it.

15. Two different types of weights are used to ascertain the weight of a 6-ounce rock on a balance scale. One type of weight is a 20-pound 5-ounce weight, the other is a 33-pound 11-ounce weight. Find the smallest number of these two types of weights that are necessary to do the weighing.

16. Two stars are unaccountably flaring up periodically. Star A flares at intervals of 44, hours 39 minutes, and 57 seconds. Star B flares at intervals of 72 hours, 27 minutes, and 45 seconds. Star A just flared and star B followed just a minute later. Is this the closest interval of successive flares? If not, find a closer interval and tell when it will occur.

17. Finish the following statements and then prove them: Assume that a, b, and c are integers.

    (a) The line $ax + by = c$ in the xy plane passes through the point (x_0, y_0), where x_0 and y_0 are integers, if and only if

(b) The line $ax + by = c$ passes through at least one point (x_0, y_0) in the first quadrant, where x_0 and y_0 are integers, if the following conditions on a, b, and c are satisfied:

(c) The line $ax + by = c$ passes through exactly one integer point in the first quadrant if the following conditions on a, b, and c are satisfied:

(d) The line $ax + by = c$ passes through exactly k integer points in the first quadrant if the following conditions on a, b, and c are satisfied:

18. Finish the proof of Theorem 2.1.5; that is, $bb_{n+1} = aa_{n+1} = c$, where $c = \text{lcm}(a, b)$.

2.2   Congruence Arithmetic

The theory of congruences is one of the most powerful, yet accessible, tools for studying the natural numbers and the integers. We will use it here to study Diophantine equations. It was first introduced by Gauss in his *Disquisitiones Arithmeticae*, a monumental work that appeared in 1801. Gauss was 24 years old at the time of its publication. Kronecker said of this work that "it is really astonishing to think that a single man of such young years was able to bring to light such a wealth of results, and, above all, to present such a profound and well-organized treatment of an entirely new discipline." That new discipline was modern number theory. In Chapter 1 of *Disquisitiones* Gauss introduces the concept of congruence. "If a number n measures the difference of two numbers a and b, then a and b are said to be congruent with respect to n; if not, they are not congruent."

First we review basic definitions for integers.

Definition 2.2.1

(a) An **integer** *is a number of the form n, where n is a natural number, n is 0, or $n = -m$, where m is a natural number and $-m + m = 0$.*

(b) The **absolute value** *of n, written $|n|$, has the following relationship to the integer n: $|n| = n$ if $n \geq 0$, $|n| = -n$ if $n < 0$.*

*(c) Let n be a nonzero integer. Then n is a **factor** of m and m is a **multiple** of n if there exists an integer r such that nr = m.*

Definition 2.2.2 *The integer a is **congruent** to integer b **modulo** n if n divides a − b.*

Notation: If a is congruent to b modulo n, we write $a \equiv b(\mathrm{mod}\ n)$ and say "a is congruent to b mod n."

Example 2.2.3

(a) $73 \equiv 46 \pmod 9$, because $73 - 46 = 27$ and $9 \mid 27$. Similarly: $73 \equiv 94(\mathrm{mod}\ 7)$ because $73 - 94 = -21$ and $7 \mid -21$.

(b) $5362 \equiv 2 \pmod{10}$, $57 \equiv -3 \pmod{12}$, $457 \equiv 147 \pmod{60}$, $77 \equiv -14 \pmod{13}$ ∎

Congruence is a viable tool for mathematicians because of the division theorem. It states that if a and b are natural numbers and $a < b$ and a is not a divisor of b, then there exist unique natural numbers q and r such that $b = aq + r$ and $1 \le r < a$. We saw in Section 1.1 that this theorem implies the fundamental theorem of arithmetic for natural numbers. While the fundamental theorem gives us uniqueness of representation under multiplication, the division theorem gives us uniqueness of positive remainders after division. Here is a version of the division theorem for integers that does this.

Theorem 2.2.4 (A Division Theorem for Integers) *If a is a natural number and b is an integer, then there exist a unique integer q and nonnegative integer r such that $b = aq + r$ and $0 \le r < a$.*

Proof If b is a natural number, then the conclusion follows from the division theorem for natural numbers. If b is a negative integer, then $-b$ is positive and there exist unique natural numbers q and nonnegative r such that $-b = aq + r$ and $0 \le r < a$. So $b = a(-q-1)+(a-r)$. Once again the conclusion of the theorem is true, with the quotient being $(-q-1)$ and the remainder being $(a-r)$. □

Notation: The unique remainder, r, that Theorem 2.2.4 assures us will be denoted here by $[b\ (\mathrm{mod}\ a)]$. It is read "b mod a" or "b modulo a." It is a nonnegative number less than a.

As a side note we should say there is a generalization of the division theorem for integers that is stated this way: If a is a non-zero integer and b is an integer, then there exist integers q and r such that $b = aq + r$ and $0 \le |r| < |a|$. For this version, if $b < 0$ there can be two possible remainders. For example, let $b = -43$ and $a = 6$; then we may let $q = -7$, in which case $r = -1$, or we may let $q = -8$, in which case $r = 5$. Since our interest is in uniqueness, we shall cite Theorem 2.2.4.

Using congruence terminology, Theorem 2.2.4 says that given a nonzero integer b and a natural number n, there exists a unique integer r such that $0 \le r < n$ and $b \equiv r \pmod{n}$. It follows that every integer is congruent, modulo n, to exactly one of the numbers $0, 1, 2, \ldots, n - 1$.

The following theorem follows immediately from the definitions and ties together the two notions of congruence and modulus.

Theorem 2.2.5 *For two integers a and b, $a \equiv b \pmod{n}$ if and only if $[a \pmod{n}] = [b \pmod{n}]$.*

Example 2.2.6

   (a) $73 \equiv 1 \pmod 9$ so $[73 \pmod 9] = 1$. Also, $[87 \pmod 9] = 6$, and $[-16 \pmod 9] = 2$.

   b) $[5362 \pmod{10}] = [202 \pmod{10}] = [2 \pmod{10}] = 2$. Also $[77 \pmod{10}] = [-13 \pmod{10}] = 7$.   ∎

Congruence is very much like equality. The following theorems state that all the relationships that you would like to be true are, in fact, true for the congruence relation.

Theorem 2.2.7 *Congruence modulo n satisfies the following three properties:*

   1. $a \equiv a \pmod{n}$. (reflexive property)

   2. if $a \equiv b \pmod{n}$, then $b \equiv a \pmod{n}$. (symmetric property)

   3. if $a \equiv b \pmod{n}$ and $b \equiv c \pmod{n}$, then $a \equiv c \pmod{n}$. (transitive property)

Proof The proof follows directly from the definitions. Here is the proof for the transitive property. Since $a \equiv b \pmod{n}$ there is an integer k such that $a - b = kn$. Similarly since $b \equiv c \pmod{n}$ there is an integer l such that $b - c = ln$. So

$$a - c = (a - b) + (b - c) = kn + ln = (k + l)n.$$

Thus $a \equiv c \pmod{n}$. □

A binary relation that is reflexive, symmetric, and transitive is called an **equivalence relation**. An equivalence relation is a generalization of the relation of equality and the theorem says that congruence is such a relation.

Theorem 2.2.8 *Congruence preserves addition, multiplication, and exponentiation. Let n be a natural number and let a, b, c, and d be integers.*

1. *If $a \equiv b \pmod{n}$ and $c \equiv d \pmod{n}$, then $(a + c) \equiv (b + d) \pmod{n}$.*

2. *If $a \equiv b \pmod{n}$ and $c \equiv d \pmod{n}$, then $ac \equiv bd \pmod{n}$.*

3. *If $a \equiv b \pmod{n}$, then $a^m \equiv b^m \pmod{n}$ for any natural number, m.*

Proof

(1) Since $a \equiv b \pmod{n}$, then $a - b = kn$ for some integer k. Similarly, $c - d = ln$ for some integer l. So

$$(a + c) - (b + d) = (a - b) + (c - d) = kn - ln = (k - l)n.$$

Thus $a + c \equiv b + d \pmod{n}$.

(2) We have the following equalities:

$$ac - bd = (a - b)(c - d) + ad + bc - 2bd$$
$$= (a - b)(c - d) + d(a - b) + b(c - d)$$
$$= (kn)(ln) + d(kn) + b(ln) = (kln + dk + bl)n.$$

So $ac \equiv bd \pmod{n}$.

(3) Since exponentiation is repetitive multiplication, we may prove this from (2) using the first induction principle. □

It would be nice to have cancellation for congruence as well; that is, if $ca \equiv cb \pmod{n}$, then $a \equiv b \pmod{n}$. However, this is not true.

Example 2.2.9

Consider the congruence $2 \times 7 \equiv 2 \times 1 \pmod{4}$. This is true because $14 \equiv 2 \pmod{4}$. But we cannot cancel the 2 from both sides of the congruence because $7 \not\equiv 1 \pmod{4}$. We can, however, say that $7 \equiv 1 \pmod{2}$. So if we write our original congruence as

$$2 \times 7 \equiv 2 \times 1 \pmod{2 \times 2}$$

then we can cancel the 2's all across the congruence and get $7 \equiv 1 \pmod{2}$. ∎

Theorem 2.2.10 *If $ca \equiv cb \pmod{n}$, then $a \equiv b \pmod{n/d}$, where d is the greatest common factor of c and n.*

Proof Since $ca \equiv cb \pmod{n}$, we may write $c(a - b) = kn$ for some integer k. Since $d = \gcd(c, n)$, we know that there are relatively prime numbers r, s such that $c = dr$, $n = ds$. It follows that $dr(a - b) = kds$, and canceling the d yields $r(a - b) = ks$. But then s divides $r(a - b)$ and since $\gcd(r, s) = 1$, it follows that $s \mid (a - b)$. So $a \equiv b \pmod{s}$; that is, $a \equiv b \pmod{n/d}$. □

Corollary 2.2.11

1. *If $ak \equiv bk \pmod{nk}$, where $k > 0$, then $a \equiv b \pmod{n}$.*

2. *If $ca \equiv cb \pmod{n}$ and $\gcd(c, n) = 1$, then $a \equiv b \pmod{n}$.*

3. *If $ca \equiv cb \pmod{p}$, p is a prime, and $p \nmid c$, then $a \equiv b \pmod{p}$.*

Example 2.2.12

(a) Find the final digit of $1^2 + 2^2 + 3^2 + \cdots + 100^2$. This translates into the question, what is $[(1^2 + \ldots + 100^2) \pmod{10}]$? It follows from Theorem 2.2.9 (1) that the answer to this can be written as

$[[1^2 \pmod{10}] + [2^2 \pmod{10}] + [3^2 \pmod{10}] + \cdots + [100^2 \pmod{10}] \pmod{10}]$.

Now the final digits of the successive squares are 1, 4, 9, 6, 5, 6, 9, 1, 0, and so on. So we have 10 series of

$$[(1 + 4 + 9 + 6 + 5 + 6 + 9 + 1 + 0) \pmod{10}].$$

Adding this up, we get $10(41)$, which is $[0 \pmod{10}]$. So the answer is 0.

(b) Find the final digit of 2^{100}. This translates into the question, what is $[2^{100} \pmod{10}]$? From Theorem 2.2.9 (3) and the fact that $2^5 \equiv 2 \pmod{10}$, we obtain

$$2^{100} \equiv (2^5)^{20} \equiv 2^{20} \equiv (2^5)^4 \equiv 2^4 \equiv 6 \pmod{10}.$$

So the answer is 6.

Here is another way of approaching this. Notice that the successive powers of 2 (mod 10) are, respectively, 2, 4, 8, 6, 2, 4, 8, 6,.... Since the pattern repeats every four terms, simply divide 100 by 4 and you get 25 full cycles of this pattern. Thus $2^{100} \pmod{10}$ is the fourth term of the 25th cycle. This is 6. ∎

In Chapter 1 we were interested in finding factors of a number. In Section 1.1, we discussed rules for divisibility by 2, 3, 4, 5, 6, 8, 9, 10, and 11. In this chapter we are interested in remainders after numbers are divided by a given number. Of course, when a number is a factor of another it leaves a remainder of 0 upon division. If a number is not a factor, there are also rules to tell what the remainder would be after division by 2, 3, 4, 5, 6, 8, 9, 10, and 11. For example, any odd number leaves a remainder of 1, after division by 2. Any number ending in 1 or 6 leaves a remainder of 1 after division by 5. Similarly, numbers ending in a 2 or 7 leave a remainder of 2, numbers ending in 3 or 8 leave a remainder of 3, and numbers ending in 4 or 9 leave a remainder of 4 after division by 5. For division by 3 the rule is based on the one for exact divisibility. Simply add up the digits and look at the sum. The remainder of the sum, after division by 3, is the remainder of the original number after division by 3. Similarly, the rule for division by 11 is also based on the one for exact divisibility. The remainder of the sum of the alternating series of digits, after division by 11, is the same as the remainder of the original number after division by 11.

Example 2.2.13

Let us consider the number $N = 2846367$. After division by 2 it leaves a remainder of 1. After division by 5 it leaves a remainder of 2. For division by 3, we find the sum $2 + 8 + 4 + 6 + 3 + 6 + 7 = 36$. Since 36 leaves a remainder of 0, $3 \mid N$. For division by 11 we evaluate

$2 - 8 + 4 - 6 + 3 - 6 + 7 = -4$. Since $-4 \equiv 7 \pmod{11}$, N leaves a remainder of 7 when divided by 11. Try it and you will see.   ∎

One of the most elementary and familiar uses of remainders in mathematics is the procedure of **casting out nines**. It is based on a rule for finding the remainder of a number upon division by 9. This rule is like the one for division by 3; simply add the digits of the number and the remainder of this sum, after division by 9, is the remainder of the original number after division by 9. For example, 328 leaves a remainder of 4 upon division by 9 because $328 = 36 \times 9 + 4$. We can find 4 by adding $3 + 2 + 8 = 13$ and, adding again, $1 + 3 = 4$. The term "casting out nines" refers to the fact that you can, while in the process of adding digits, throw aside sums that total 9. For example 45261934 leaves a remainder of 7 upon division by 9. As you add the digits from left to right you get $4 + 5 = 9$ —cast it out—, $2 + 6 + 1 = 9$ —cast it out—, 9 —cast it out—, and finally, $3 + 4 = 7$.

This rule works because

$$[a_n a_{n-1} \ldots a_2 a_1 a_0 \pmod{9}] = $$
$$[(a_n + a_{n-1} + \cdots + a_2 + a_1 + a_0) \pmod{9}].$$

Let us look into this a bit more carefully. N can be represented as

$$a_n a_{n-1} \ldots a_2 a_1 a_0 = 10^n a_n + 10^{n-1} a_{n-1} + \cdots + 10^2 a_2 + 10 a_1 + a_0.$$

This can be written as

$$(10^n - 1)a_n + a_n + (10^{n-1} - 1)a_{n-1} + a_{n-1} + \cdots + (10^2 - 1)a_2 + a_2 +$$
$$(10 - 1)a_1 + a_1 + a_0.$$

Now $9 \mid (10^i - 1)$ for each i in the expression and so when the expression is divided by 9 the remainder is the same as the remainder as when $a_n + a_{n-1} + \cdots + a_2 + a_1 + a_0$ is divided by 9.

Casting out nines is a well-known procedure that is used to check arithmetic problems. For example, if we find the product $45261934 \times 328 = 14945914352$ we can check it out like this. Simply cast out nines in the multiplier, the multiplicand, and in the product of your new multiplier and multiplicand. Compare this answer to the original product after nines have been cast from it. If they are the same, your answer has an $(8/9)$th chance of being correct. If they are different, your answer is wrong. In this particular example we have these results. Casting out nines in the multiplier and the multiplicand yields

7 and 4. Casting out nines in the product $7 \times 4 = 28$ leaves 1. Now casting out nines in the original product 14945914352 leaves a 2. So our answer must be wrong. This analysis works also for addition, subtraction, division, and exponentiation. This procedure works because arithmetic with remainders is consistent with arithmetic with integers, as we proved in Theorem 2.2.8.

Now let us apply the theory of congruence arithmetic to the study of linear Diophantine equations.

Definition 2.2.14 *A* **linear congruence (mod n)** *is a congruence of the form $ax \equiv b \pmod{n}$, where a and b are integers and n is a natural number greater than 1.*

Notice that integral solutions to the linear equation $ax + by = c$ can be found by solving the linear congruence $ax \equiv c \pmod{b}$. In fact, x_0, y_0 is a solution to $ax + by = c$ exactly when $ax_0 - c = (-y_0)b$, which occurs exactly when $ax_0 \equiv c \pmod{b}$. Thus the following theorem follows directly from Theorem 2.1.4.

Theorem 2.2.15 *The linear congruence $ax \equiv b \pmod{n}$ has a solution if and only if d divides b, where $d = \gcd(a, n)$. If $d \mid b$, then the congruence has d distinct solutions modulo n. If x_0 is a particular solution, the general form of the solutions is $[x_0 + (n/d)k \pmod{n}]$ for $k = 0, 1, 2, \ldots d - 1$.*

Corollary 2.2.16

1. *If $\gcd(a, n) = 1$, then the linear congruence $ax \equiv b \pmod{n}$ has a single solution mod n.*

2. *If $\gcd(a, n) = d$, then the linear congruence $ax \equiv b \pmod{n}$ has a single solution of the form $x_0 (\bmod\, n/d)$.*

Example 2.2.17

Recall Example 2.1.3. Let us examine the following congruences:

(a) $6x \equiv 4 \pmod{7}$

(b) $6x \equiv 4 \pmod{8}$

(c) $6x \equiv 4 \pmod{9}$

Without reference to Theorem 2.2.15, we can proceed by searching for solutions. The first several multiples of 6 are 6, 12, 18, 24, 30,

36, 42, and 48. These numbers modulo 7 are, respectively, 6, 5, 4, 3, 2, 1, 0, and 6. These numbers modulo 8 are, respectively, 6, 4, 2, 0, 6, 4, 2, and 0. These numbers modulo 9 are, respectively, 6, 3, 0, 6, 3, 0, 6, and 3.

Notice that in the case of the remainders modulo 7, all possible numbers less than 7 are represented. In particular, 4 is there. The solution to $6x \equiv 4 \pmod 7$ for $x < 7$ is $x = 3$.

The remainders modulo 8 are limited to four numbers and 4 is among them. The solution to $6x \equiv 4 \pmod 8$ occurs twice for $x < 8$; $x = 2$ and $x = 6$.

The remainders modulo 9 consist of three numbers and 4 is not among them. So there is no solution to $6x \equiv 4 \pmod 9$.   ∎

Now, reviewing Theorem 2.2.15, we see that this is the scenario we could have expected.

For (a), since $\gcd(6, 7) = 1$, we would expect there to be a single solution modulo 7. There is; it is $x = 3$. All other solutions are congruent to 3 (mod 7); that is, $7k + 3$.

For (b), $\gcd(6, 8) = 2$ and $2 \mid 4$, so we would expect two distinct solutions modulo 8. There are two: $x = 2$ and $x = 6$. All other solutions are equivalent to 2 (mod 8) and 6 (mod 8); that is, $8k + 2$ and $8k + 6$. Using Corollary 2.2.11, we can reduce $6x \equiv 4 \pmod 8$ to $3x \equiv 2 \pmod 4$ and arrive at the same solution but written in the form $4k + 2$.

For (c) $\gcd(6, 9) = 3$ and $3 \nmid 4$, so there are no solutions. And this is what we observed.

Example 2.2.18

(a) Recall the chicken, pig example; Example 2.1.2 and 2.1.6 (c). Here is a more sophisticated way of solving it (though trial and error would probably be preferable). The equation is $3x + 5y = 70$. This can be written in congruence language as $5y \equiv 70 \pmod 3$, which reduces to $2y \equiv 1 \pmod 3$. Thus $y \equiv 2 \pmod 3$; that is, $y = 2 + 3k$. So $3x + 5(2 + 3k) = 3x + 10 + 15k = 70$. Solving for x yields $x = 20 - 5k$. We get our five positive solutions by letting $k = 0, 1, 2, 3, 4$.

(b) Recall Example 2.1.6 (b): $208x + 159y = 117$. We shall approach this using congruence language. It will obviously be more complicated than (a). We can either solve the congruence $159y \equiv 117 \pmod{208}$ or the congruence $208x \equiv 117 \pmod{159}$. Either choice is fine; we shall choose the former. The method is called **reduction of moduli**.

The congruence $159y \equiv 117 \pmod{208}$ means

$$\text{(i)} \quad 159y = 117 + 208k \quad \text{for some } k.$$

Rewriting, we get $208k \equiv -117 \pmod{159}$; reducing mod 159 yields $49k \equiv 42 \pmod{159}$, which means

$$\text{(ii)} \quad 49k = 42 + 159l \quad \text{for some } l.$$

Rewriting this, we get $159l \equiv -42 \pmod{49}$; reducing mod 49 yields $12l \equiv 7 \pmod{49}$, which means

$$\text{(iii)} \quad 12l = 7 + 49m \quad \text{for some } m.$$

Rewriting, we get $49m \equiv -7 \pmod{12}$; reducing mod 12 yields $m \equiv 5 \pmod{12}$, which means

$$\text{(iv)} \quad m = 5 + 12n \quad \text{for some } n.$$

Now we unwind this. Using (iii) and (iv), we get

$$12l = 7 + 49m = 7 + 49(5 + 12n) = 252 + 588n, \text{ so } l = 21 + 49n.$$

Applying (ii), we get

$$49k = 42 + 159l = 42 + 159(21 + 49n) = 3381 + 7791n, \text{ so } k = 69 + 159n.$$

From (i) we finally obtain

$$159y = 117 + 208k = 117 + 208(69 + 159n) = 14469 + 33072n, \text{ so}$$
$y = 91 + 208n$ and $x = -69 - 159n$.

This agrees with our finding in Example 2.1.6 (b) that $y = -1989 - 208k$ and $x = 1521 + 159k$, where $n = -k - 10$. ∎

Reduction of moduli affords an alternate way of solving linear Diophantine equations. It is an easier approach than the Euclidean algorithm for certain problems, specifically those that have a large constant on the righthand side. This method has wider application as well. It is the approach used for attacking systems of linear Diophantine equations and systems of linear congruences.

Definition 2.2.19 *A system of linear congruences* (mod n_k)
is a set of congruences of the form

$$a_1 x \equiv b_1 \ (\text{mod } n_1)$$
$$a_2 x \equiv b_2 \ (\text{mod } n_2)$$
$$\vdots$$
$$a_r x \equiv b_r \ (\text{mod } n_r)$$

where a_k, b_k *and* n_k *are integers and* $n_k > 1$ *for* $1 \leq k \leq r$.

Recall Example 2.2.13. There we found that 2846367 left a remainder of 1 when divided by 2, 0 when divided by 3, 2 when divided by 5, and 7 when divided by 11. In congruence language we found that 2846367 is a solution to every one of four different linear congruences:

$$x \equiv 1 \ (\text{mod } 2)$$
$$x \equiv 0 \ (\text{mod } 3)$$
$$x \equiv 2 \ (\text{mod } 5)$$
$$x \equiv 7 \ (\text{mod } 11).$$

Example 2.2.20

Let us search for a smaller solution for the four linear congruences: $x \equiv 1 \ (\text{mod } 2)$, $x \equiv 0 \ (\text{mod } 3)$, $x \equiv 2 \ (\text{mod } 5)$, and $x \equiv 7 \ (\text{mod } 11)$.

Here is the straightforward, no-frills approach. Numbers that satisfy the congruence $x \equiv 1 \ (\text{mod } 2)$ are obviously odd numbers. If x satisfies the congruence $x \equiv 0 \ (\text{mod } 3)$, then x is divisible by 3. A list of the numbers that are both odd and divisible by 3 begins like this: 3, 9, 15, 21, 27, 33, 39, ... Also, x must satisfy the congruence $x \equiv 2 \ (\text{mod } 5)$ so the number must end in a 2 or 7. Since it is odd it must end in a 7. The first number that satisfies all three congruences is 27. Finally, we want x to satisfy the congruence $x \equiv 7$ (mod 11). The list of these numbers begins like this: 7, 18, 29, 40, 51, 62, 73, 84, 95, 106, 117. Stop. We have found an answer; it is 117. This satisfies all four linear congruences and it is the smallest positive number that does. Now we know two solutions to this series of congruences: 117 and 2846367.

Here is a our approach. From the system of congruences we obtain

$$(\text{i}) \quad x = 1 + 2k$$

(ii) $x = 3l$

(iii) $x = 2 + 5m$

(iv) $x = 7 + 11n$

Solving (i) and (ii), we get $1 + 2k = 3l$. Thus $3l \equiv 1 \pmod 2$. It follows that $l \equiv 1 \pmod 2$ and so $l = 1 + 2r$ for some r. Thus

(v) $x = 3(1 + 2r) = 3 + 6r.$

Solving (iii) and (v), we get $2 + 5m = 3 + 6r$. In congruence language we may write $6r \equiv -1 \pmod 5$ so $r \equiv 4 \pmod 5$. Thus $r = 4 + 5s$ for some s and therefore

(vi) $x = 3 + 6(4 + 5s) = 27 + 30s.$

Finally, solving (iv) and (vi), we get $7 + 11n = 27 + 30s$. So it follows that $30s \equiv -20 \pmod{11}$ and therefore $8s \equiv 2 \pmod{11}$, or $4s \equiv 1 \pmod{11}$. Thus $s \equiv 3 \pmod{11}$. We now have $s = 3 + 11t$ and hence

(vii) $x = 30(3 + 11t) + 27 = 117 + 330t$

where $t = 0, 1, 2, \ldots$.

Note that $2846367 = 117 + 330 \times 8625$. ∎

Just as there are congruences without solutions [see 2.2.17 (c)] so, too, are there systems of congruences without solutions even though each individual congruence within the system has a solution.

Example 2.2.21

(a) Suppose we have the system

(i) $x \equiv 2 \pmod 4$

(ii) $x \equiv 3 \pmod 6$.

Even without working this out, we see that solutions to (i) must be even numbers while solutions to (ii) must be odd. This is impossible.

(b) Here is a less obvious example.

(i) $x \equiv 1 \pmod 2$

(ii) $x \equiv 2 \pmod 3$

(iii) $x \equiv 4 \pmod 6$

So we have (i) $x = 1 + 2k$ and (ii) $x = 2 + 3l$. This tells us that $1 + 2k = 2 + 3l$, so $2k = 1 + 3l$. In congruence language we have $2k \equiv 1 \pmod 3$, so $k \equiv 2 \pmod 3$; that is, $k = 2 + 3m$. Thus, using (i) we get $x = 1 + 2(2 + 3m) = 5 + 6m$. But (iii) tells us that

$x = 4 + 6n$ for some n. This is a contradiction. Thus the system has no solution.   ∎

Some of the most famous mathematical puzzles have been problems that translate into solving linear congruences or systems of linear congruences. In particular, the ancient Indian, Chinese, and Greek mathematicians have explored such problems. Here are two examples. Several more are included in the exercises.

Example 2.2.22

This puzzle is attributed to Mahaviracarya, an Indian mathematician (about 850 A.D.).

There are 63 equal piles of plantain fruit along with 7 single fruits. They are evenly divided among 23 travelers. What is the number of fruits in each pile? Let x represent the number of fruits each traveler gets. Let y be the number of fruits in each pile. The following equation represents the situation.

$$\text{(i)} \quad 23x = 63y + 7$$

We may write this as a linear congruence $23x \equiv 7 \pmod{63}$. Let us reverse this and write $63y \equiv -7 \pmod{23}$. This simplifies to $17y \equiv 16 \pmod{23}$. We may substitute values in for $y, y = 0, 1, 2, \ldots, 22$ until we solve this or we may use the method of reduction of moduli. Rewriting this yields

$$\text{(ii)} \quad 17y = 16 + 23k \text{ for some integer } k.$$

Again using congruence terminology we get $23k \equiv -16 \pmod{17}$, which simplifies to $6k \equiv 1 \pmod{17}$. At this point we may solve the congruence empirically; $k = 3$. Substituting back in equation (ii), we have $17y = 16 + 23(3)$, so $y = 5$. Substituting this back into equation (i), we have $23x = 63(5) + 7$, so $x = 14$. This means that each traveler gets 14 fruits and there are 5 fruits in each pile.

The general solution can be found this way: $k = 3 + 17l$. Substituting this back in (ii), we have $17y = 16 + 23(3 + 17l) = 85 + 23(17)l$. Dividing by 17, we get $y = 5 + 23l$. And substituting in (i) $23x = 63(5 + 23l) + 7$ yields $x = 14 + 63l$. If we let $l = 1$, we have 77 fruits for each traveler, and 28 fruits in each pile. This is also a reasonable possibility for this problem. If $l = 2$ the answers of 130 fruits for each traveler and 51 fruits in each pile is a bit excessive.   ∎

Here is another problem that has been around, in some form, for centuries. It supposedly originated in ancient China. It involves a system of linear congruences. We repeat it from the introduction.

Example 2.2.23

A band of 17 pirates stole a sack of gold coins. When they tried to divide the fortune into equal portions, 3 coins were left over. In the fight that followed one pirate was killed. The coins were redistributed into equal portions and this time 10 coins were left over. Another fight broke out and another pirate was killed. Now the coins could evenly be distributed. What is the number of coins that were stolen?

Letting x represent the number of coins, the congruences that describe this situation are $x \equiv 3 \pmod{17}$ $x \equiv 10 \pmod{16}$ $x \equiv 0 \pmod{15}$. This means

$$\begin{aligned} &\text{(i)} & x &= 3 + 17k \\ &\text{(ii)} & x &= 10 + 16j \\ &\text{(iii)} & x &= 15h. \end{aligned}$$

Solving (i) and (ii) simultaneously, we get $17k = 7 + 16j$. Reducing modulo 16, we get $k \equiv 7 \pmod{16}$; or $k = 7 + 16l$. Now plugging this into (i), we get $x = 3 + 17k = 3 + 17(7 + 16l)$. So

$$\text{(iv)} \qquad x = 122 + 272l.$$

Solving (iii) and (iv) simultaneously, we get $272l = -122 + 15h$. Reducing, modulo 15, we get $2l \equiv 13 \pmod{15}$; so $l \equiv 14 \pmod{15}$; that is, $l = 14 + 15m$. Thus $x = 122 + 272(14 + 15m)$. So

$$\text{(v)} \qquad x = 3930 + 4080m.$$

Letting $m = 0$, we have the smallest number of coins: $x = 3930$. ∎

Notice that in both of the preceding examples, 2.2.20 and 2.2.23, the difference between successive solutions to the linear congruences follows a formula. We have, in Example 2.2.20, the term $330t$ and $330 = 2 \times 3 \times 5 \times 11$, the product of the respective moduli. Also in Example 2.2.23 we have the term $4080m$ and $4080 = 15 \times 16 \times 17$, the product of the respective moduli. These example suggest a general theorem about the solution to linear congruences. It is called, appropriately, the Chinese remainder theorem after the many mathematicians who posed such problems long ago.

Theorem 2.2.24 (Chinese Remainder Theorem) *If* n_1, n_2, n_3, *..., n_r, are relatively prime numbers in pairs [that is, $\gcd(n_i, n_j) = 1$ for all i and j, $i \neq j$] and b_1, b_2, b_3, ..., b_r, are any r integers, the system of congruences*

$$x \equiv b_1 \ (\text{mod } n_1)$$
$$x \equiv b_2 \ (\text{mod } n_2)$$
$$x \equiv b_3 \ (\text{mod } n_3)$$
$$\vdots$$
$$x \equiv b_r \ (\text{mod } n_r)$$

has a unique solution modulo $n_1 n_2 n_3 \cdots n_r$.

Proof This is done by induction on r, the number of different moduli. For two moduli, call them n and m, notice that if $x \equiv a \ (\text{mod } n)$, then $x = a + nt$ for integers t. Since $\gcd(n, m) = 1$, it follows from 2.2.16 (1) that the congruence $a + nt \equiv b \ (\text{mod } m)$ has a unique solution modulo m. Letting that solution be t_0 and letting $x_0 = a + nt_0$, we have $x_0 \equiv a \ (\text{mod } n) \equiv b \ (\text{mod } m)$. Now if y_0 also satisfies both congruences, then $n \mid (x_0 - y_0)$ and $m \mid (x_0 - y_0)$. Since $\gcd(n, m) = 1$, $nm \mid (x_0 - y_0)$. Thus $x_0 \equiv y_0 \ (\text{mod } nm)$. We leave the remainder of the proof to the interested student. □

There is an alternate method for solving linear congruences. It is preferable to the preceding method if the moduli are especially large because we can employ some labor saving tricks involving the Euclidean algorithm. Here is the theorem that applies. We leave it to the reader to provide a proof.

Theorem 2.2.25 *Suppose that $x \equiv r \ (\text{mod } m)$ and $x \equiv s \ (\text{mod } n)$. Suppose also that m and n are relatively prime. Then $x \equiv rny - smx$ (mod mn), where $ny - mx = 1$.*

Example 2.2.26

Let us solve the system of congruences:
$$x \equiv 118 \ (\text{mod } 365)$$
$$x \equiv 1345 \ (\text{mod } 1876).$$

We know from Example 1.1.14 that $1876(136) - 365(699) = 1$. Now solving for $x \equiv rny - smx$ (mod mn), we get

$$118 \times 136 \times 1876 - 1345 \times 365 \times 699 = -313050527; \ 365 \times 1876 = 684740$$

and $[-313050527 \pmod{684740}] = 560393$. So our answer is $x \equiv 560393 \pmod{684740}$. ∎

Finally, we should point out that even if the moduli in a system of linear congruences are not pairwise relatively prime, there may still be solutions. Example 2.2.21 shows that there need not be solutions; the following example shows that there can be solutions and they follow a certain form.

Example 2.2.27

(a) Consider the following system, where the moduli are 4 and 6.

$$\text{(i)} \quad x \equiv 3 \pmod{4}$$
$$\text{(ii)} \quad x \equiv 5 \pmod{6}$$

We see that (i) implies that $x = 3 + 4k$ and (ii) implies $x = 5 + 6j$. Thus $6j + 5 = 3 + 4k$. In congruence terminology we have $6j \equiv 3 - 5 \pmod{4}$. Reducing, we get $2j \equiv 2 \pmod{4}$. This reduces further to $j \equiv 1 \pmod{2}$. Thus $j = 1 + 2l$. Plugging this into (ii), we get $x = 5 + 6(1 + 2l) = 11 + 12l$. So $x \equiv 11 \pmod{12}$.

(b) Here is a more complicated example.

$$\text{(i)} \quad x \equiv 5 \pmod{6}$$
$$\text{(ii)} \quad x \equiv 5 \pmod{8}$$
$$\text{(iii)} \quad x \equiv 7 \pmod{10}$$

We see that (i) implies that $x = 5 + 6k$ and (ii) implies $x = 5 + 8j$. Thus $8j + 5 = 5 + 6k$. In congruence terminology, $8j \equiv 0 \pmod{6}$. Reducing, we get $2j \equiv 0 \pmod{6}$, so $j \equiv 0 \pmod{3}$. Thus $j = 3l$. Plugging this into (ii), we get

$$\text{(iv)} \quad x = 5 + 8j = 5 + 8(3l) = 5 + 24l.$$

Plugging this into (iii) yields $5 + 24l \equiv 7 \pmod{10}$. So we have $4l \equiv 2 \pmod{10}$. Reducing, this yields $2l \equiv 1 \pmod{5}$, so $l \equiv 3 \pmod{5}$. Thus $l = 3 + 5m$ and so

$$\text{(v)} \quad x = 5 + 24(3 + 5m) = 77 + 120m.$$

So our answer is $x \equiv 77 \pmod{120}$. ∎

This suggests the following theorem, which is similar to the Chinese remainder theorem.

Theorem 2.2.28 *If b_1, b_2, b_3, ..., b_r, are integers and the system of congruences*

$$x \equiv b_1 \pmod{n_1}$$
$$x \equiv b_2 \pmod{n_2}$$
$$x \equiv b_3 \pmod{n_3}$$
$$\vdots$$
$$x \equiv b_r \pmod{n_r}$$

has a solution, then the solution is unique modulo $[\mathrm{lcm}(n_1, n_2, n_3, \ldots, n_r)]$.

The proof is left as an exercise.

EXERCISES

1. Find the remainder of $12, 345, 678, 910, 111, 213, 141, 516, 171, 819$ upon division by

   (a) 2
   (b) 3
   (c) 5
   (d) 11

2. Make up rules for finding remainders of numbers when divided by

   (a) 4
   (b) 6
   (c) 12

3. What is the remainder of the sum of the squares: $1^2 + 2^2 + \cdots + 100^2$ when divided by

   (a) 7
   (b) 3
   (c) 11

4. Find the final digit of

(a) 2^{400}

(b) 3^{400}

(c) 7^{400}

(d) 8^{400}

5. For each natural number n, analyze the final digit of the following numbers:

(a) n^2

(b) n^3

(c) n^4

(d) n^5

(e) n^6

6. Find all the solutions to the following six congruences. The congruences are to be solved separately, not simultaneously.

(a) $5x \equiv 1 \pmod 7$

(b) $4x \equiv 6 \pmod{15}$

(c) $8x \equiv 14 \pmod{21}$

(d) $6x + 6 \equiv 12 \pmod{18}$

(e) $9x + 3 \equiv 17 \pmod{24}$

(f) $15x \equiv 8 \pmod{17}$

7. Find the general solution to the following three congruences using reduction of moduli. The congruences are to be solved separately, not simultaneously.

(a) $53x \equiv 42 \pmod{79}$

(b) $457x \equiv 189 \pmod{722}$

(c) $3906x \equiv 833 \pmod{10465}$

8. Find two numbers that add up to 1000. One number is a multiple of 23, the other a multiple of 17. How many such pairs of numbers are both positive? What are they?

9. A museum charges $2.90 for adults and $1.75 for children. It $505.05 in one day.

(a) How many children and how many adults might have visited that day?

(b) Answer (a) if there were at least 200 children.

(c) Answer (b) if there were at least 100 adults.

10. Given the line $11x + 8y = c$, find all the possible values of the integer c such that

(a) there is exactly one integer lattice point [an integer lattice point is a point (x, y) in the plane where both x and y are integers] on the line in the first quadrant

(b) there are exactly ten integer lattice points on the line in the first quadrant

11. Consider the Diophantine equation $ax + by = c$. Find conditions on integers a, b, and c such that

(a) there are solutions but none are positive. A solution x, y is positive if both x and y are positive.

(b) there are a finite number of positive solutions.

(c) there are infinitely many positive solutions.

(d) there are exactly n positive solutions where $n > 0$.

12. Given weights A, B, and C, where A is 57 lb., B is 74 lb., and C is 16 lb.,

(a) Using weights A and B, how many are needed to make a ton (2000 lb)?

(b) Using weights A, B, and C, how many are needed to make a ton?

13. Solve the Diophantine equation $9x + 13y + 21z = 1000$.

(a) Find all positive solutions.

(b) Find a general formula.

14. Find two distinct solutions modulo 210 for this system of congruences:

$$2x \equiv 3 \pmod 5$$
$$4x \equiv 2 \pmod 6$$
$$3x \equiv 2 \pmod 7$$

15. Find all the solutions for the system of congruences.

    (a)
    $$x \equiv 3 \pmod 4$$
    $$x \equiv 6 \pmod 7$$
    $$x \equiv 8 \pmod 9$$

    (b)
    $$2x \equiv 2 \pmod 4$$
    $$3x \equiv 4 \pmod 5$$
    $$2x \equiv 4 \pmod 6$$

16. Find all the solutions for the system of congruences.

    (a)
    $$x \equiv 5 \pmod{24}$$
    $$x \equiv 14 \pmod{15}$$
    $$x \equiv 13 \pmod{28}$$

    (b)
    $$x \equiv 20 \pmod{33}$$
    $$x \equiv 32 \pmod{45}$$
    $$x \equiv 5 \pmod{27}$$

17. Sun-Tsu (around the first century A.D.. Find a number which leaves a remainder of 2, 3, and 2 when divided by 3, 5, and 7, respectively.

18. Brahmagupta (around the seventh century A.D.. When eggs are removed 2, 3, 4, 5, and 6 at a time there remain, respectively, 1, 2, 3, 4, and 5, eggs. When they are taken out 7 at a time, none are left over. Find the smallest number of eggs that could be in the basket.

19. Old Chinese puzzle. Three rice farmers raised their rice collectively and divided it equally at harvest time. One year each of them went to a different market to sell his share of the rice. Each of the three markets only bought rice in multiples of a certain base weight, which differed at each of the three markets. The first farmer sold his rice at a market where the base weight was 87 lb. He sold all he could and returned with 18 lb. of rice.

The second farmer sold all the rice he could at a market where the base weight was 170 lb. and he returned with 58 lb. The third farmer sold all the rice he could at a market where the base weight was 143 lb. and returned with 40 lb. How much rice did they raise together?

20. My friend and I are counting a pride of lions. I find that counting by 3s there are 2 left over and counting by 4s there are 3 left over. She tells me that counting by 5s there are 2 left over.

   (a) What is the smallest number of lions there might be?

   (b) She also tells me that there are 3 left over when counting by 7s. Now what is the smallest number of lions there might be?

   (c) What if another person tells us that there are 5 left over when counting by 8s? Is this possible? If not, which count makes this impossible?

21. Suppose the number $207,266,909$ has been divided, respectively, by 2, by 3, by 5, by 7, and by 11.

   (a) What are the respective remainders after these divisions?

   (b) What is the smallest positive number that has the very same remainders that you found in the answers from (a)?

22. Captain Stevenson is in command of a squadron of soldiers.

   (a) He does not know exactly how many, but he believes there are fewer than 2000. He wants to line them up so that they can march in columns, but he is not having any luck. When lined up in columns of 10 soldiers there is 1 soldier left over, when lined up in columns of 11 there are 7 left over, when lined up in columns of 9 there are 4 left over, and when lined up in columns of 8 there is 1 left over. Is there some number of soldiers per column that will work out perfectly? If so, what is it?

   (b) Does the situation change if the number of soldiers is fewer than 5000? If so, how?

   (c) What are the possibilities if the number is less than 10000?

23. Three beacons are flashing at different intervals. The red one flashes every 25 seconds and it flashed just this moment. The blue one flashes in intervals of 21 seconds and it will flash 10 seconds from now. The green beacon flashes at intervals of 35 seconds and it will flash 5 seconds after the blue one. Beginning at this moment, how long will it take before the three beacons all flash simultaneously? Or will they every flash together?

24. What can you say about the statement "if $x \equiv 1 \pmod{n}$ and $x \equiv 1 \pmod{m}$ then $x \equiv 1 \pmod{nm}$"? Give examples for when it is false. Find conditions for under which it is true. Give reasons for your answers.

25. Prove Theorem 2.2.25. Suppose that $x \equiv r \pmod{m}$ and $x \equiv s \pmod{n}$. Suppose also that m and n are relatively prime. Then $x = rny - smx + kmn$ represents solutions where k is any integer and $ny - mx = 1$.

26. Find solutions for the following systems of congruences.

(a)
$$x \equiv 13266 \pmod{29749}$$
$$x \equiv 21025 \pmod{63011}$$

(b)
$$x \equiv 251 \pmod{977}$$
$$x \equiv 261 \pmod{455}$$
$$x \equiv 799 \pmod{821}$$

27. Prove Theorem 2.2.28. If b_1, b_2, b_3, $\ldots$, b_r are integers and the system of congruences

$$x \equiv b_1 \pmod{n_1}$$

$$x \equiv b_2 \pmod{n_2}$$

$$x \equiv b_3 \pmod{n_3}$$

$$\vdots$$

$$x \equiv b_r \pmod{n_r}$$

has a solution, then the solution is unique mod $(\operatorname{lcm}(n_1, n_2, \ldots, n_r))$.

28. Suppose you have an unlimited number of 29 cent and 33 cent stamps.

(a) What is the largest denomination of postage you can *not* form with these stamps? *Hint:* Solve this problem first for smaller numbers.

(b) If 29 and 33 were changed to a and b, what is the answer to (a)?

2.3   Pell and Pythagorus

Now we move to the study of quadratic Diophantine equations. Its general form can be $ax^2 + by^2 = cz^2$ or $ax^2 + by^2 = c$. We devote this section to two classic equations: the Pell equation $x^2 - ky^2 = \pm 1$, and the Pythagorean equation $x^2 + y^2 = z^2$. We consider more general forms in Section 3.4.

Definition 2.3.1 *An equation* $x^2 - ky^2 = \pm 1$, *where k is a natural number and where it is understood that the solutions are to be integers, is known as a* **Pell equation** *or* **Pell's equation**.

Pell equations, named for John Pell (1611-1685), predate Pell by, perhaps, 2000 years. One example of a Pell equation dates back to about 250BC. It was contained in a letter sent by Archimedes to Eratosthenes as a challenge to Alexandrian scholars. It involved cows and bulls of four different colors and involved eight unknowns connected by nine equations. The problem boils down to solving the Pell equation: $x^2 - 4729494y^2 = 1$. The solution to the cattle problem contains enormous numbers; one of the unknowns is 206545 digits long. Naturally it wasn't solved. Around 250 AD, Diophantus worked on such equations. About 650 AD, Brahmagupta, a Hindu mathematician who is known for his work on general linear Diophantine equations, stated that "A person who can, within a year, solve the equation $x^2 - 92y^2 = 1$ is a mathematician". As with all Diophantine equations, Pell equations were worked on extensively during Fermat's time. We have mentioned that Frénicle had challenged Wallis to find solutions for $x^2 - ky^2 = 1$ for k up to 200. One of the equations that had not been solved was for the case when $k = 61$. The story associated with this case is given in the introduction. Here it is again. The year was 1066.

> The men of Harold stood well, as was their wont, and formed sixty and one squares, with a like number of men in every square thereof, and woe to the hardy Norman who ventured to enter their redoubts; for a single blow of a Saxon war-hatchet would break his lance and cut through his coat of mail .... When Harold threw himself into the fray the Saxons were one mighty square of men, ...

Apparently the Saxons had 61 phalanxes each containing a square number of men. The addition of Harold enabled them to rearrange themselves in a single square. So $61y^2 + 1$ would be a perfect square, x^2. It turns out that the smallest solution is $x = 1766319049$ and $y = 226153980$. This would require more than a billion soldiers. More than a third of the present-day population of the earth would have had to be conscripted for this.

Now if the men of Harold still insisted on forming square phalanxes with a like number in each phalanx, it would have been possible if the number of phalanxes were 65 rather than 61. This is because $65(16^2) + 1 = 129^2$. That is, it would have taken $16,640$ men arranged in 65 squares of 256 men apiece. Each square would have 16 men on the side. With Harold added, the new huge square would have 129 men on a side. This makes sense for the population of Harold's day. Actually, if Harold wanted lots of phalanxes with few men in each he might have chosen the number 1620. We may solve the equation $1620y^2 + 1 = x^2$ with $y = 4$ and $x = 161$; that is, $1620(4^2)+1 = 161^2$. Thus 1620 regiments each arranged in squares of 16 soldiers can be, with the addition of one general, rearranged into a the huge square with $25,921$ soldiers. However, if Harold slipped up and chose 1621 squares, instead of 1620 squares, he would have a serious problem. It would take nearly all the atoms in the universe to effect such an arrangement since the smallest solution, x, has 76 digits.

As you can see from the above examples, finding solutions to Pell equations is not simple. Indeed solutions do exist and in Section 4.3 we will see how they may be found. For small values of k we may find them by inspection.

Example 2.3.2

(a) With some work we can find solutions to $x^2 - 2y^2 = 1$ and $x^2 - 2y^2 = -1$. By trying a few numbers we see that the following values of x and y work for $x^2 - 2y^2 = 1$: $x = 3$, $y = 2$; and $x = 17$, $y = 12$. For $x^2 - 2y^2 = -1$ we find $x = 7$, $y = 5$ and $x = 41$, $y = 29$.

(b) For $x^2 - 3y^2 = 1$ we find that $x = 2$, $y = 1$; $x = 7$, $y = 4$; and $x = 26$, $y = 15$ work. For $x^2 - 3y^2 = -1$ nothing seems to work. And we can prove nothing will work. If we write the equation $x^2 - 3y^2 = -1$ as a congruence mod 3, we see that $x^2 - 3y^2 \equiv -1$ (mod 3) becomes $x^2 \equiv 2$ (mod 3). This cannot happen since $x^2 \equiv 0$ or 1 (mod 3). ∎

Two things stand out from this example: first, there may be lots of solutions to a Pell equation; and second, there may be no solutions at all and congruence arithmetic is a useful tool for showing this. Let us look into the first observation. As with linear Diophantine equations it is possible, from a single solution, to generate infinitely many other solutions to a Pell equation. Here is how it works.

Theorem 2.3.3

1. *Consider the Pell equation $x^2 - ky^2 = 1$. If $x = r$, $y = s$ is a solution then x, y is also a solution, where*

$$x = (\tfrac{1}{2})[(r + s\sqrt{k})^n + (r - s\sqrt{k})^n] \text{ and}$$

$$y = (\tfrac{1}{2\sqrt{k}})[(r + s\sqrt{k})^n - (r - s\sqrt{k})^n].$$

2. *Consider the Pell equation $x^2 - ky^2 = -1$. If r, s is a solution, then x, y is also a solution, where*

$$x = (\tfrac{1}{2})[(r + s\sqrt{k})^{2n-1} + (r - s\sqrt{k})^{2n-1}] \text{ and}$$

$$y = (\tfrac{1}{2\sqrt{k}})[(r + s\sqrt{k})^{2n-1} - (r - s\sqrt{k})^{2n-1}].$$

Proof Here is a proof of (1); the proof of (2) is similar. Let $x = r$ and $y = s$ be values satisfying the equation $x^2 - ky^2 = 1$. Then $(r^2 - ks^2)^n = (1)^n = 1$. Then, letting $x^2 - ky^2 = (r^2 - ks^2)^n$, we may factor both sides and get

$$(x + y\sqrt{k})(x - y\sqrt{k}) = (r + s\sqrt{k})^n(r - s\sqrt{k})^n.$$

Setting the positive $\sqrt{k}$ factors equal and the negative $\sqrt{k}$ factors equal, we obtain $x + y\sqrt{k} = (r + s\sqrt{k})^n$ and $x - y\sqrt{k} = (r - s\sqrt{k})^n$. Solving for x and y gives our result:

$$x = (\tfrac{1}{2})[(r + s\sqrt{k})^n + (r - s\sqrt{k})^n],$$
$$y = (\tfrac{1}{2\sqrt{k}})[(r + s\sqrt{k})^n - (r - s\sqrt{k})^n]. \qquad \square$$

Example 2.3.4

(a) We found that $r = 3$ and $s = 2$ satisfies the Pell equation $x^2 - 2y^2 = 1$. Another solution is $r = 17$, $s = 12$. Let us use our formula above and find some more. Notice that if $n = 1$, then $x = r$ and $y = s$. Letting $n = 2$, we get

$$x = (\tfrac{1}{2})[(3 + 2\sqrt{2})^2 + (3 - 2\sqrt{2})^2] =$$
$$(\tfrac{1}{2})(9 + 12\sqrt{2} + 8 + 9 - 12\sqrt{2} + 8) = 17.$$

$$y = (\tfrac{1}{2\sqrt{2}})[(3 + 2\sqrt{2})^2 - (3 - 2\sqrt{2})^2] =$$
$$(\tfrac{1}{2\sqrt{2}})(9 + 12\sqrt{2} + 8 - 9 + 12\sqrt{2} - 8) = 12.$$

Letting $n = 3$, we get

$$x = (\tfrac{1}{2})[(3 + 2\sqrt{2})^3 + (3 - 2\sqrt{2})^3] =$$
$$(\tfrac{1}{2})(27 + 54\sqrt{2} + 72 + 16\sqrt{2} + 27 - 54\sqrt{2} + 72 - 16\sqrt{2}) = 99$$

$$y = (\tfrac{1}{2\sqrt{2}})[(3 + 2\sqrt{2})^3 - (3 - 2\sqrt{2})^3] =$$
$$(\tfrac{1}{2\sqrt{2}})(27 + 54\sqrt{2} + 72 + 16\sqrt{2} - 27 + 54\sqrt{2} - 72 + 16\sqrt{2}) = 70.$$

So $99^2 - 2(70^2) = 1$.

(b) $r = 1$, $s = 1$ is the least solution to $x^2 - 2y^2 = -1$. Notice that this is what you get by letting $n = 1$ in the formula. Let $n = 2$. Then we get

$$x = (\tfrac{1}{2})[(1 + \sqrt{2})^3 + (1 - \sqrt{2})^3] =$$
$$(\tfrac{1}{2})(1 + 3\sqrt{2} + 6 + 2\sqrt{2} + 1 - 3\sqrt{2} + 6 - 2\sqrt{2}) = 7.$$

$$y = (\tfrac{1}{2\sqrt{2}})[(1 + \sqrt{2})^3 - (1 - 1\sqrt{2})^3] =$$
$$(\tfrac{1}{2\sqrt{2}})(1 + 3\sqrt{2} + 6 + 2\sqrt{2} - 1 + 3\sqrt{2} - 6 + 2\sqrt{2}) = 5.$$

So $7^2 - 2(5^2) = -1$. $\blacksquare$

Luckily there are simpler ways of generating these solutions. Coincidentally, one of those ways involves the table of quotients we used to find solutions to linear Diophantine equations in Section 2.1. The following example and theorem give a taste; Sections 3.2 and 3.3 offer a first course, and Section 4.3 provides a full meal.

Example 2.3.5

Let us consider the solutions to $x^2 - 2y^2 = \pm 1$. They are, in order of size, $1, 1$; $3, 2$; $7, 5$; $17, 12$; $41, 29$. Notice that there is a pattern here. Beginning with $x = 1$, $y = 1$ the x value of the following term in the series is the sum of twice the previous x value plus the one previous to that. That is, $x_{n+2} = 2x_{n+1} + x_n$. This also holds true of the y values: $y_{n+2} = 2y_{n+1} + y_n$. This can be captured with the following table, which is so much like a quotient table from Section 2.1 that we will still use that name.

	1	2	2	2	2		
y_n	1	0	1	2	5	12	29
x_n	0	1	1	3	7	17	41

Notice that there is another relationship that holds between the numbers x_n and y_n: $x_{n+1} = x_n + 2y_n$ and $y_{n+1} = x_n + y_n$. The proof that these two relationships are equivalent is done by induction, and we leave it as an exercise. We use this relationship to prove the following theorem about the solutions to the Pell equation $x^2 - 2y^2 = \pm 1$.

Theorem 2.3.6 *Beginning with $x_1 = 1$, $y_1 = 1$, define the following sequence x_n, y_n by the recurrence relation: $x_{n+1} = x_n + 2y_n$, $y_{n+1} = x_n + y_n$. Then $x_n^2 - 2y_n^2 = (-1)^n$.*

Proof We proceed by induction on n. Let $\mathcal{P}(n)$ be $x_n^2 - 2y_n^2 = (-1)^n$. Clearly $\mathcal{P}(1)$ is true. Let us suppose that $\mathcal{P}(n)$ is true and consider $\mathcal{P}(n+1)$. By the recurrence relation,

$$x_{n+1}^2 - 2y_{n+1}^2 = (x_n + 2y_n)^2 - 2(x_n + y_n)^2 =$$
$$x_n^2 + 4x_n y_n + 4y_n^2 - 2x_n^2 - 4x_n y_n - 2y_n^2 = 2y_n^2 - x_n^2 = -(-1^n) = (-1)^{n+1}.$$
$\square$

Thus the table of quotients can be used to generate an infinite number of solutions to the Pell equation $x^2 - 2y^2 = \pm 1$.

We saw that there was no solution to the Pell equation $x^2 - 3y^2 = -1$. This holds true for many Pell equations. The methods used for showing this can be very effective for Diophantine equations in general.

Example 2.3.7

(a) The Pell equation $x^2 - 7y^2 = -1$ has no solution. We can show this as we did for its counterpart $x^2 - 3y^2 = -1$ in Example 2.3.2 (b).

If $x^2 - 7y^2 = -1$, then $x^2 \equiv -1 \pmod 7$. This cannot happen since $x^2 \equiv 0, 1, 2,$ or $4 \pmod 7$. And the Pell equation $x^2 - 11y^2 = -1$ has no solution either. If $x^2 - 11y^2 = -1$, then $x^2 \equiv -1 \pmod{11}$. This cannot happen since $x^2 \equiv 0, 1, 3, 4, 5,$ or $9 \pmod{11}$.

(b) There is no solution to the Diophantine equation $x^2 - 3y^2 = 2$. This is because if $x^2 - 3y^2 = 2$, then $x^2 \equiv 2 \pmod 3$. This is impossible because $x^2 \equiv 0$ or $1 \pmod 3$. Similarly, there are no solutions to the Diophantine equations $x^2 - 5y^2 = 2$ or $x^2 - 5y^2 = 3$ because $x^2 \equiv 0, 1,$ or $4 \pmod 5$. ∎

Example 2.3.7 (a) leads us to conjecture that Pell equations of the form $x^2 - ky^2 = -1$ have no solution if $k \equiv 3 \pmod 4$. This is the case.

Theorem 2.3.8 *Pell equations of the form $x^2 - ky^2 = -1$ have no solution if $k \equiv 3 \pmod 4$.*

Proof If $x^2 - ky^2 = -1$ has solutions, then $x^2 - ky^2 \equiv -1 \pmod 4$ has solutions. Since $k \equiv 3 \pmod 4$, we may rewrite our congruence as $x^2 + y^2 \equiv 3 \pmod 4$. But $x^2 \equiv 0$ or $1 \pmod 4$, and the same goes for y^2 so $x^2 + y^2 \equiv 0, 1,$ or $2 \pmod 4$ only. □

So we know that some Pell equations have no solutions, and others have infinitely many solutions, but to find them you must know one of them. Here is short list of the solutions of smallest magnitude for $x^2 - ky^2 = 1$ and $x^2 - ky^2 = -1$ for $2 \leq k \leq 20$.

Pell Chart

(a) In this chart r and s are the least values of x and y that satisfy Pell's equation $x^2 - ky^2 = 1$

k	r	s	k	r	s
2	3	2	12	7	2
3	2	1	13	649	180
5	9	4	14	15	4
6	5	2	15	4	1
7	8	3	17	33	8
8	3	1	18	17	4
10	19	6	19	170	39
11	10	3	20	9	2

(b) In this chart r and s are the least values of x and y that satisfy Pell's equation $x^2 - ky^2 = -1$

k	r	s	k	r	s
2	1	1	13	18	5
5	2	1	17	4	1
10	3	1			

Observe a couple of things about the tables. First, notice there are not representatives for the Pell equation $x^2 - ky^2 = -1$ for most of the k. We have already seen that in the examples. Second, notice that the size of k does not necessarily dictate the size of r and s. The largest entry in this table is for $k = 13$. For $k < 100$, the largest least values r and s that solve the Pell equation $x^2 - ky^2 = 1$ occur for $k = 61$ and we have noted how big they are: $r = 1766319049$ and $s = 226153980$. For $k < 1000$ the largest least values r and s that solve the Pell equation $x^2 - ky^2 = 1$ occur for $k = 991$. For your information and amazement they are

$$r = 379,516,400,906,811,930,638,014,896,080$$
$$s = 12,055,735,790,331,359,447,442,538,767.$$

As you can see, r^2 has 59 digits.

Let us now switch to the Pythagorean equations. Unlike the Pell equations, the solutions to the Pythagorean equation $x^2 + y^2 = z^2$ were known long ago. It is believed that complete solutions to this equation were known more than three millenia ago by the Babylonians. The ancient Chinese and Indian mathematicians may have had the complete solutions as well. Certainly the Pythagoreans, for whom the Pythagorean theorem is named, had thoroughly analyzed this equation by 500 B.C. Perhaps the reason for the attention to this equation was its geometric significance; integral solutions represented right triangles with sides of integer length. Perhaps the numbers that formed the solutions were interesting in themselves. Very likely it was both; mathematics was not fragmented then, as it is now, into separate fields. The fact that this Diophantine equation occupied the mathematicians from different continents more than 2500 hundred years ago is reason enough to study it. Beyond this, however, the mathematics involved in an analysis of this equation is not at all ancient; it is as modern as Euler and Gauss and as fascinating as ever.

Let us begin with $x^2 + y^2 = c$. Our search is for nonzero integral solutions. The chart shows those numbers $c \leq 100$ that can be written as the sum of two nonzero squares. Notice that there are numbers for which there is no integral solution, only one integral solution, and more than one integral solution. For example, if $c = 3$ no integers work. The same is true for $c = 4$ because we are insisting on nonzero solutions. For $c = 5$, $x = 2$, $y = 1$ works. So does $x = 2$, $y = -1$; $x = -2$, $y = 1$ and $x = -2$, $y = -1$ So also does the reverse; that is, $x = \pm 1$, $y = \pm 2$ and $x = \mp 1$, $y = \pm 2$. We will not consider these as distinct, however. For $c = 65$ we have two distinct solutions: $x = 8$, $y = 1$ and $x = 7$, $y = 4$. It may surprise you that there are only 35 numbers, $c \le 100$, that can be written as the sum of two nonzero squares. That means 64 numbers cannot be represented this way. There doesn't appear to be a pattern for those numbers, c that made this list. Let's analyze them.

Pythagorean Chart

c	x	y	c	x	y	c	x	y
2	1	1	37	6	1	72	6	6
5	2	1	40	6	2	73	8	3
8	2	2	41	5	4	74	7	5
10	3	1	45	6	3	80	8	4
13	3	2	50	5	5	82	9	1
17	4	1		7	1	85	7	6
18	3	3	52	6	4		9	2
20	4	2	53	7	2	89	8	5
25	4	3	58	7	3	90	9	3
26	5	1	61	6	5	97	9	4
29	5	2	65	7	4	98	7	7
32	4	4		8	1	100	8	6
34	5	3	68	8	2			

The list contains primes: 2, 5, 13, 17, 29, 37, 41, 53, 61, 73, 89, and 97. The odd primes are of the form 1 (mod 4). This is no big surprise to us; the sum of two squares cannot be of the form 3 (mod 4) because $x^2 \equiv 0$ or 1 (mod 4). Doubling each such prime yields 10, 26, 34, 58, 74, 82, and they, too, are represented. Presumably 106, 122, 146, 178, and 194 would be on the list if it were extended. And this is true; for example, $106 = 9^2 + 5^2$, $122 = 11^2 + 1^2$, and $146 = 11^2 + 5^2$. Also, twice the perfect squares make the list. This is obvious because $2n^2 = n^2 + n^2$. This accounts for 8, 18, 32, 50, 72, and 98. The numbers 25 and 100 are the only squares on the list, and hence the only numbers, z^2, such that $x^2 + y^2 = z^2$. Notice that 25 and 100 are squares of 5 and 10, numbers already on the list. And 13^2 would be on an extended list: $13^2 = 12^2 + 5^2$. Also, 20, 40, 45, 52, 68, 80, 90, and 100 are on the list. Notice that they are a square times a number on the list; for example, $20 = (2^2)5$, $40 = (2^2)10$, $45 = (3^2)5$. Finally, 65 and 85, along with 50, are the only numbers on the list expressible in two different ways. They also are the products of numbers already on the list.

Synthesizing these observations, we may say that the primes on the list; that is, 2, and odd primes of the form 1 (mod 4), form a basis for the composites on the list. All such combinations of these primes whose product is ≤ 100 are on the list with the exception of even powers of 2. And most of the numbers on the list, when written in prime form, are composed of the primes on the list. There are three exceptions to that: $18 = 2 \times 3^2$, $72 = 2^3 \times 3^2$, and $98 = 2 \times 7^2$. Notice that these exceptions are all twice a perfect square.

Let us organize these observations formally. First we attend to the primes on the list. Other than the prime 2 they all must be of the form 1 (mod 4) because, as was pointed out in the proof of Theorem 2.3.8, x^2 (mod 4) can be only 0 or 1 (mod 4), and if $p = a^2 + b^2$, then p (mod 4) can be only 0, 1, or 2 (mod 4). Since $p \neq 2$, this leaves $p = 1$ (mod 4). The deeper result is that the converse is true; that is, if p is a prime of the form 1 (mod 4), then it can be written as the sum of two squares. This important result is credited to both Euler and Fermat. It was stated by Fermat in a letter to Mersenne in 1640. He said he had an irrefutable proof yet, like his famous last theorem, he didn't provide the proof. It was proved more than 100 years later by Euler in 1754. Its proof is slightly beyond the scope of our presentation, so, like Fermat, we will state it here without proof. A proof can be found in most elementary number theory books.

Theorem 2.3.9 (Fermat-Euler) *A prime of the form* 1 (mod 4) *can be expressed as the sum of two squares in exactly one way.*

Next let us look into the product of primes or, for that matter, the product of any two numbers on the list. The next theorem not only tells which products can be represented as sums of two squares but also how to construct the sums.

Theorem 2.3.10 *Suppose that m and n can both be expressed as the sum of two nonzero squares. Then mn can be expressed as the sum of two nonzero squares in*

1. *at least two different ways if the representations of m and n are distinct and neither representation is $k^2 + k^2$*

2. *at least one way if the representations m and n are the same and that representation is not $k^2 + k^2$*

3. *at least one way if one of m or n is expressed in the form: $k^2 + k^2$*

Proof Let $m = r^2 + s^2$ and $n = t^2 + u^2$. Now

$$mn = (r^2 + s^2)(t^2 + u^2) = r^2t^2 + r^2u^2 + s^2t^2 + s^2u^2$$

$$= (ru + st)^2 + (rt - su)^2 = (rt + su)^2 + (ru - st)^2.$$

(1) If these representations for m and n are distinct (that is, the pair r, s is different from the pair t, u), then the two representations of mn, $(ru + st)^2 + (rt - su)^2$ and $(rt + su)^2 + (ru - st)^2$, are distinct.

(2) If the representations for m and n are the same (that is, either $r = t$ and $s = u$ or $r = u$ and $s = t$), then only one of these two representations of mn is valid because either $rt - su$ or $ru - st = 0$.

(3) If m, for example, is $k^2 + k^2$, then $r = s$ and the two representations of mn are the same. □

Example 2.3.11

(a) Consider $17 \times 53 = 901$. The theorem tells us that there are at least two ways of expressing 901 as the sum of two squares. Using the Pythagorean chart, we see that we may form 17 out of 4 and 1 and 53 out of 7 and 2. So, letting $r = 4$, $s = 1$, $t = 7$, $u = 2$, we obtain $901 = (ru + st)^2 + (rt - su)^2 = 15^2 + 26^2$ and also $(rt + su)^2 + (ru - st)^2 = 30^2 + 1^2$.

(b) Consider $41 \times 98 = 4018$. The theorem tells us that there is at least one way of expressing 4018. Using the chart, we may let $r = 5$, $s = 4$, $t = u = 7$. Then $4018 = (ru + st)^2 + (rt - su)^2 = 63^2 + 7^2$.

(c) Consider $73 \times 73 = 5329$. The theorem tells us that there is at least one way of expressing 5329. Using the chart, we may let $r = 8$, $s = 3$, $t = 8$, $u = 3$. Then $5329 = (ru + st)^2 + (rt - su)^2 = 48^2 + 55^2$ and $(rt + su)^2 + (ru - st)^2 = 73^2 + 0^2$. ■

The following theorem tells the full story about what numbers can be written as the sum of two squares.

Theorem 2.3.12 *The number N can be written as the sum of two nonzero squares if and only if N is of the form uv, $2uv$, or $2v$, where u is a prime or a product of primes of the form $1 \pmod 4$ and v is a perfect square.*

Proof Suppose that n is of one of the forms mentioned above. Now if u is a prime or a product of primes of the form $1 \pmod 4$, Theorems 2.3.9 and 2.3.10 prove that it is representable as sum of two squares. Letting $u = a^2 + b^2$ and $v = k^2$, we have $uv = (ka)^2 + (kb)^2$. Also, $2u = (a + b)^2 + (a - b)^2$, $2uv = [k(a + b)]^2 + [k(a - b)]^2$. Finally, $2v = k^2 + k^2$.   □

This is half of the proof. The proof that if N can be written as the sum of nonzero squares it must be of one of the above forms is slightly beyond our treatment, but it can be found in most elementary number theory texts.

Corollary 2.3.13 *If the prime decomposition of N contains at least one prime of the form $3 \pmod 4$ raised to an odd power, then N cannot be written as the sum of two nonzero squares.*

The corollary provides a quick way of telling if a number N has a chance of being expressible as the sum of two nonzero squares. Of course, if primes of the form $3 \pmod 4$ are present and all raised to even powers, this is not enough to ensure that N can be written as the sum of nonzero squares. There still must be present either at least one prime of the form $1 \pmod 4$ or a 2 raised to an odd exponent.

As for multiple representations, we know at this point that if N has a particular prime form, then it can be represented as the sum of two nonzero squares. What those squares are we do not know. How many such representations there are, we are not sure. Theorem 2.3.10 indicates that if at least two different odd primes of the form $1 \pmod 4$ make up the prime factorization, there will be more than one representation.

Example 2.3.14

(a) Let $N = 637$. Putting this in prime form, we have $N = 7^2 \times 13$. Since $13 = 3^2 + 2^2$, we may write $7^2 \times 13$ as $7^2(3^2 + 2^2) = (7 \times 3)^2 + (7 \times 2)^2 = 21^2 + 14^2$.

(b) Let $N = 495$. In prime form $N = 3^2 \times 11 \times 5$. Since the prime 11 is raised to an odd power, 1, there is no way to write 495 as the sum of two nonzero squares.

(c) Let $N = 290 = 2 \times 5 \times 29$. We may proceed like this: $2 \times 5 = 10 = 3^2 + 1^2$; $29 = 5^2 + 2^2$. So, using Theorem 2.3.10, we have $10 \times 29 = 11^2 + 13^2 = 17^2 + 1^2$.

(d) Let $N = 15725 = 5^2 \times 17 \times 37$. Let's begin with 17×37. Using Theorem 2.3.10, we have $17 \times 37 = (4^2 + 1^2)(6^2 + 1^2) = 25^2 + 2^2 = 10^2 + 23^2$. So we have

$$5^2(25^2 + 2^2) = 125^2 + 10^2, \quad 5^2(10^2 + 23^2) = 50^2 + 115^2.$$

Also, we have, from Theorem 2.3.10,

$$(4^2 + 3^2)(25^2 + 2^2) = 106^2 + 67^2 = 83^2 + 94^2.$$
$$(4^2 + 3^2)(10^2 + 23^2) = 109^2 + 62^2 = 122^2 + 29^2$$

This makes a total of six ways of writing 15725 as the sum of two nonzero squares. ∎

We leave it as an exercise to discover a formula for determining the number of ways a number can be expressed as the sum of two squares.

Let us move on to arguably the most famous of the Diophantine equations: $x^2 + y^2 = z^2$. Notice from the Pythagorean chart that only two numbers c are perfect squares, 25 and 100: $3^2 + 4^2 = 5^2$ and $6^2 + 8^2 = 10^2$. Since we are interested not only in the numbers but also in the geometry of the situation, we notice that the triangles represented by these equations are right triangles. One has sides measuring 3 and 4 with a hypotenuse 5; the other has sides 6, 8, with hypotenuse 10. Notice also that these triangles are similar; that is, they have the same shape. This translates numerically in the observation that one triple of numbers is a multiple of the other triple of numbers; in this case the multiple is 2. Notice also that the numbers 3, 4, 5 are pairwise relatively prime; we will be most interested in those triangles whose sides are relatively prime.

Definition 2.3.15

(a) The natural numbers x, y, z are called a **Pythagorean triple** *if $x^2 + y^2 = z^2$.*

*(b) A Pythagorean triple is called **primitive** if x, y and z are pairwise relatively prime; that is, $\gcd(x, y) = \gcd(x, z) = \gcd(y, z) = 1$.*

*(c) A **Pythagorean triangle** is a right triangle with integer length
sides.*

*(d) A Pythagorean triangle is **primitive** if the lengths of the sides
are pairwise relatively prime.*

Notation: A Pythagorean triple will be denoted by an ordered triple;
for example, the Pythagorean triple defined above will be denoted by
(x, y, z). Note that the numbers in a Pythagorean triple are, respectively, the two sides and the hypotenuse of a Pythagorean triangle.
A Pythagorean triangle will also be denoted by a triple, but since
the ordering of the sides is not important we shall use a different
notation: $[x, y, z]$.

Recalling Theorem 2.3.12, we can say that if z is the hypotenuse
of a primitive right triangle, then z is a prime or product of primes
of the form 1 (mod 4). So we know what the hypotenuse can be.
Finding the sides of such triangles will be our next challenge.

Theorem 2.3.16 *Given any two distinct natural numbers a and b
with a > b, then (x, y, z) is a Pythagorean triple where $x = a^2 - b^2$,
$y = 2ab$, and $z = a^2 + b^2$. Furthermore, if (x, y, z) is a primitive
Pythagorean triple, then such an a and b exist. In this case a and b
are relatively prime and they are not both odd.*

Proof It is straightforward to show that if $x = a^2 - b^2$, $y = 2ab$,
and $z = a^2 + b^2$, then $x^2 + y^2 = z^2$. Simply look at the equations:

$$(a^2 - b^2)^2 + (2ab)^2 = a^4 - 2a^2b^2 + b^4 + 4a^2b^2 = (a^2 + b^2)^2.$$

Now suppose that (x, y, z) is a primitive Pythagorean triple. We
know that for any integer x, $x^2 \equiv 0$ or 1 (mod 4) because the square
of an even number is congruent to 0 (mod 4), the square of an odd
number is congruent to 1 (mod 4). Since $\gcd(x, y) = 1$, both cannot
be even so the sum is not congruent to 0 (mod 4). But $x^2 + y^2$
is a perfect square so it must be congruent to 1 (mod 4). Thus
they cannot both be odd. Let us then say that x is odd and y is
even. Letting $y = 2u$, we have the equation $x^2 + 4u^2 = z^2$. So
$4u^2 = z^2 - x^2 = (z + x)(z - x)$. Now x and z are odd so both $(z + x)$

and $(z - x)$ are even. Let $(z + x) = 2s$, $(z - x) = 2r$. Solving these two equations simultaneously, we get $z = r + s$, $x = s - r$. Also, $4u^2 = 2r2s$ or $u^2 = rs$. Now r and s are relatively prime because x and z are relatively prime. So $u^2 = rs$ requires that both r and s be perfect squares. Let $r = b^2$ and $s = a^2$. Then $u = ab$. And $x = a^2 - b^2$, $y = 2ab$, and $z = a^2 + b^2$.

Reviewing the restrictions on a and b, note that $a > b$ so that x will be positive. Note also that a and b must have no common factor or it would be shared by x, y, and z. Finally, a and b must not both be odd for then x and z would share the factor 2. Thus, if we let a and b take on all arbitrary values within these limitations, we get all primitive solutions. □

To get all the Pythagorean triples, we need only multiply the primitive Pythagorean triples by any natural number. We should point out that we do not intend to be finicky about the order of the x and y. Technically we cannot get the triple $(4, 3, 5)$ using the generators a and b, but this does not interest us. Our thinking is entirely geometric here. We are thinking of sides and the hypotenuse of a right triangle. We should also point out that some nonprimitive triangles can arise from the generators a and b, where both a and b are both odd or both even. The following theorem summarizes this.

Theorem 2.3.17 *Let* $\gcd(a, b) = d$. *Then*

1. *If* $(a/d)^2 + (b/d)^2 = n$, *then* $a^2 + b^2 = nd^2$.

2. *If* a *and* b *are both odd numbers and* $a > b$, *then* $a^2 + b^2 = 2n$, *where* $n = [(a + b)/2]^2 + [(a - b)/2]^2$.

Example 2.3.18

Letting $a = 2$ and $b = 1$, we get the famous $[3, 4, 5]$ triangle. Letting $a = 3$ and $b = 1$, we get $[8, 6, 10]$, a multiple of the 3, 4, 5 triangle. Another multiple of this triangle, $[9, 12, 15]$, cannot be obtained using the generators a and b. ∎

In the following chart we list all the primitive triangles resulting from letting $1 \le b < a < 10$.

Primitive Triangle Chart

a	b	x	y	z		a	b	x	y	z
2	1	3	4	5		8	3	55	48	73
4	1	15	8	17		5	4	9	40	41
6	1	35	12	37		7	4	33	56	65
8	1	63	16	65		9	4	65	72	97
3	2	5	12	13		6	5	11	60	61
5	2	21	20	29		8	5	39	80	89
7	2	45	28	53		7	6	13	84	85
9	2	77	36	85		8	7	15	112	113
4	3	7	24	25		9	8	17	144	145

In inspecting this chart there is a lot to notice about primitive Pythagorean triangles. For example, looking at all three sides, you may notice that one side is of even length, one side has length divisible by 3, and one side has length divisible by 5. Two of those conditions might be filled with a single side. For example, the side 15 in the triangle [15, 8, 17] is divisible by both 3 and 5; the side 12 in the triangle [5, 12, 13] is divisible by both 2 and 3.

Theorem 2.3.19 *In the Pythagorean triple* (x, y, z)*, one of* x *and* y *is divisible by 2, one is divisible by 3, and one of* x*,* y*, and* z *is divisible by 5.*

Proof In primitive triples we know that $y = 2ab$ for some a and b so it follows that y is even. Suppose y is not divisible by 3. So neither a nor $b \equiv 0 \pmod{3}$. It follows that both a^2 and $b^2 \equiv 1 \pmod{3}$. Therefore, $a^2 - b^2 \equiv 0 \pmod{3}$. It follows that x is divisible by 3. Since nonprimitive triples are multiples of primitive ones, we have established this part of the theorem. The proof that one of the sides is divisible by 5 is left as an exercise.   □

We finish this section on the Pythagorean equation with two investigations, both geometric. The first involves the search for an isosceles right triangle with integer length sides. The problem is that no such triangle exists. The reason is simple; if there were integers x, y, and z such that $x = y, x^2 + y^2 = z^2$, then $2x^2 = z^2$. This implies that $\sqrt{2} = z/x$, but we already know (and if we don't we will find out in Chapter 4) that $\sqrt{2}$ is not a fraction. So the best we can do is search for isosceles Pythagorean triangles whose legs are close to the same length; in particular, one unit apart, like the [3, 4, 5] and the [21, 20, 29] triangles that are listed on the Primitive triangle chart. It turns out that finding these triangles involves solving a Diophantine equation.

Theorem 2.3.20 *Suppose that* $x = a^2 - b^2$, $y = 2ab$, *and* $z = a^2 + b^2$. *If* a *and* b *satisfy the Diophantine equation* $a^2 - 2ab - b^2 = \pm 1$, *it follows that* $x^2 + y^2 = z^2$ *and* $|x - y| = 1$.

The proof is left as an exercise.

It turns out that we can find solutions to this Diophantine equation by noticing a recursive relationship between a_n and b_n. And it turns out further, as in Example 2.3.5, that quotient tables can be a handy way of producing solutions.

Theorem 2.3.21 *The equation* $a_n^2 - 2a_n b_n - b_n^2 = (-1)^n$ *is satisfied by the pair* a_n, b_n, *where* a_n, b_n *may be generated in two different ways.*

1. $a_1 = 2$, $b_1 = 1$; *and* $a_{n+1} = 2x_n + b_n$, $y_{n+1} = b_n$

2. $a_{-1} = 0$, $a_0 = 1$, $b_{-1} = 1$, $b_0 = 0$ *and*

    $a_{n+2} = 2a_{n+1} + a_n$ *and* $b_{n+2} = 2b_{n+1} + b_n$

The proof of this theorem is also left as an exercise.

Example 2.3.22

Let us generate some "near" isosceles Pythagorean triangles. From Theorem 2.3.21 (2) we may set up the following quotient table.

		2	2	2	2	2	
a_n	0	1	2	5	12	29	70
b_n	1	0	1	2	5	12	29

With these values for a and b we may generate the following Pythagorean triangles, where $x = a^2 - b^2$, $y = 2ab$, and $z = a^2 + b^2$:

[4, 3, 5], [21, 20, 29], [119, 120, 169], [697, 696, 985], and [4059, 4060, 5741]. ■

Our final investigation concerns a geometric interpretation of a different kind: geometry on the plane of integer lattice points. We defined integer lattice points in Exercise 10 of Section 2.2. The lattice plane is like an infinite geoboard. The **geoboard** is the tool that is used by school students to explore geometric concepts. It is made up of 25 pegs arranged in a 5-by-5 square array. The infinite geoboard can be modeled by the points in the Cartesian plane that have integer coordinates. Here is the formal definition. Both "integer" and "integral" are standard terminology.

Definition 2.3.23 *An* **integer (integral) lattice point** *is a point* (x, y), *where* x *and* y *are integers. The* **integer (integral) lattice plane** *is the set of all integer lattice points in the plane.*

There are many mathematical problems that can be posed on the integral lattice plane. One exploration we can conduct is the search for the number of lattice points that lie on a circle whose center is at the origin.

Example 2.3.24
Consider the three circles of Figure 2.1.

The circle of radius 1 has four integer lattice points: $(1, 0)$, $(0, 1)$, $(-1, 0)$, and $(0, -1)$. The circle of radius $\sqrt{2}$ has four also: $(1, 1)$, $(-1, 1)$, $(1, -1)$, and $(-1, -1)$. The circle of radius $\sqrt{3}$ has no integer lattice points on it. This makes sense; the numbers 1 and 2 can be expressed as the sum of two integer squares and 3 cannot.   ∎

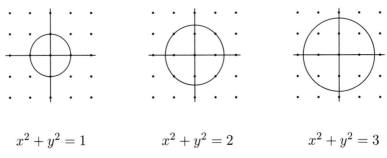

$$x^2 + y^2 = 1 \qquad x^2 + y^2 = 2 \qquad x^2 + y^2 = 3$$

Figure 2.1

Notice how this study ties in with our study of Pythagorean triangles. If a lattice point (a, b) is on a circle of radius r and it is not on either axis, we can see in Figure 2.2 that it forms one vertex of a right triangle with a second vertex at the origin and a third on the x-axis. The sides of the triangle have integer length and the length of the hypotenuse is the length of the radius. We know that the length of the radius is $\sqrt{a^2 + b^2}$, where a and b are nonzero integers, and we are very familiar with such numbers.

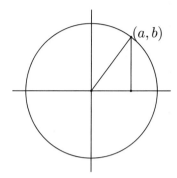

Figure 2.2

Example 2.3.25

(a) Consider the circle of radius 5; that is, $x^2 + y^2 = 5^2 = 25$. Not only does it pass through the points $(5, 0)$, $(0, 5)$, $(-5, 0)$, $(0, -5)$, but also the points $(3, 4)$, $(4, 3)$, $(-3, 4)$, $(-4, 3)$, $(-3, -4)$, $(-4, -3)$, $(3, -4)$, and $(-4, 3)$. That makes 12.

(b) Consider the circle $x^2 + y^2 = 5$. It passes through the points $(1, 2)$, $(2, 1)$, $(-1, 2)$, $(-2, 1)$, $(1, -2)$, $(2, -1)$, $(-1, -2)$, and $(-2, -1)$. That makes 8. The radius of this circle is $\sqrt{5}$.

(c) Consider $x^2 + y^2 = 65$. Its radius is $\sqrt{65}$. We know that $8^2 + 1^2$ and $7^2 + 4^2 = 65$. Let us count only those points in the first octant; that is, the points (a, b), where $a > 0$, $b > 0$ and $a \geq b$. For the grand total we will multiply this total by 8. So counting only those points in the first octant, we find $(8, 1)$ and $(7, 4)$. Thus there are 16 lattice points on the circle.

(d) Consider the circle of radius 65: $x^2 + y^2 = 65^2 = 4225$. Let us carefully count the lattice points in the first octant; they are all vertices of a right triangle, one vertex at $(0, 0)$, another at $(x, 0)$ and a third at (x, y), where $x > y$ and $x^2 + y^2 = 65^2$. So we look for Pythagorean triples (x, y, z), where $z = 65$.

We find that $8^2 + 1^2 = 65$ and $7^2 + 4^2 = 65$. So, letting $a = 8$, $b = 1$, we have

$$x = 8^2 - 1^2, \ y = 2 \cdot 8 \cdot 1, \ z = 8^2 + 1^2;$$

that is, the $(63, 16, 65)$ primitive Pythagorean triple. Also, we may let $a = 7$, $b = 4$, obtaining

$$x = 7^2 - 4^2, \, y = 2 \cdot 7 \cdot 4, \, z = 7^2 + 4^2;$$

that is, the $(33, 56, 65)$ primitive Pythagorean triple. We have found two lattice points in the first octant thus far: $(63, 16)$ and $(56, 33)$.

We may use Theorem 2.3.10 to find two more lattice points in the first octant:

$$(8^2 + 1^2)(7^2 + 4^2) = 60^2 + 25^2 = 39^2 + 52^2 = 65^2.$$

These yield the triples $(60, 25, 65)$ and $(39, 52, 65)$. These are not primitive; $(60, 25, 65)$ is an enlarged, 5-fold version of the primitive triple $(12, 5, 13)$ and $(39, 52, 65)$ is the 13-fold version of the primitive triple $(3, 4, 5)$. We could have expected that since $65 = 5 \times 13$.

We conclude that there are four lattice points in the first octant: $(63, 16)$, $(60, 25)$, $(56, 33)$, and $(52, 39)$. That makes 32 points excluding the points on the axes. There are four points on the axes $(\pm 65, 0)$ and $(0, \pm 65)$, so the total number of lattice points on the circle $x^2 + y^2 = 65^2$ is 36.

We can also find these representations of $65^2 = 5^2 \times 13^2$ strictly arithmetically. Here is how it looks:

$$5^2(12^2 + 5^2) = 60^2 + 25^2; \; 13^2(3^2 + 4^2) = 39^2 + 52^2$$

$$(3^2 + 4^2)(5^2 + 12^2) = 63^2 + 16^2 = 56^2 + 33^2. \qquad \blacksquare$$

Generally, the larger the circle the more possible lattice points it may have on it. We found that the circle with radius 25 has 12 lattice points and a circle with radius 65 has 36. On the other hand, relatively small circles can have lots of lattice points, too. For example, we find that a circle with radius $\sqrt{65} \approx 8.06226$ has 16 lattice points. It seems that the nonintegral radius, if chosen properly, may admit more lattice points for their radius size than their integral radii counterparts.

Remark At this point we have developed the tools to do some really neat mathematics. If we view the plane of integral lattice points as points in the complex plane, we open up a whole new mathematical system called the **Gaussian integers**. The set of Gaussian integers consists of complex numbers of the form $a + bi$, where a and b

are integers. The Gaussian integers, under complex arithmetic, form a system much like the integers, and the questions of divisibility we are exploring here arise naturally in the domain of Gaussian integers. The Fermat-Euler theorem has an elegent proof using Gaussian integers. This book does not include a chapter on complex numbers, so consider this treatment as a springboard to study more algebra and number theory. To get you started we encourage you to work on Project 5.10, "The Gaussian Integers."

EXERCISES

1. Suppose that r, s is a solution to $x^2 - ky^2 = -1$. Show that $r^2 + ks^2$, $2rs$ is a solution to $x^2 - ky^2 = 1$.

2. Suppose that r, s is a solution to $x^2 - ky^2 = n$ and t, u is a solution to $x^2 - ky^2 = 1$. Show that $rt + ksu$, $ru + st$ is also a solution to $x^2 - ky^2 = n$.

3. Find all the values for s for $|s| < k$ such that $x^2 - ky^2 = s$, where k is

   (a) 11
   (b) 12
   (c) 13

4. We know there are numbers that cannot be written as the sum of two squares. There are also numbers that cannot be written as the sum of three squares.

   (a) Show that a number of the form 7 (mod 8) cannot be the sum of three squares.
   (b) Show that a number of the form 28 (mod 32) cannot be the sum of three squares.
   (c) Can you find other types of numbers that cannot be written as sums of three squares?

5. Representing numbers as sums of cubes is a challenge.

   (a) Show that a number of the form 3 or 4 (mod 7) cannot be the sum of two cubes.

(b) Can you make a similar statement about the sums of three cubes? Prove your statement.

(c) What is the minimum number of cubes it takes to represent 23 as their sum?

6. Extend the Pell chart up to $k = 25$.

7. Prove, by induction, that if $x_n = 1$, $y_n = 1$ then the sequence of numbers defined by $x_{n+1} = 2x_n + x_{n-1}$ and $y_{n+1} = 2y_n + y_{n-1}$ is the same as the sequence of numbers defined by $x_{n+1} = x_n + 2y_n$ and $y_{n+1} = x_n + y_n$.

8. In the proof of Theorem 2.3.16 it is stated that $z = r + s$, $x = s - r$. Since x and z are relatively prime so, too, are r and s. Prove this.

9. Consider the Pell equations $x^2 - 5y^2 = \pm 1$.

   (a) Find five solutions to these equations.

   (b) Make up a "table of quotients" as we did for $x^2 - 2y^2 = \pm 1$.

   (c) Find a recurrence relationship between the entries.

   (d) Prove a theorem like Theorem 2.3.6.

10. Repeat Exercise 9 for the Pell equations $x^2 - 10y^2 = \pm 1$.

11. For the following numbers, state which can be written as the sum of two squares. If there is more than one way of doing this, demonstrate all the ways.

   (a) 75
   (b) 405
   (c) 125
   (d) 226
   (e) 196
   (f) 637

12. For the following numbers, state how many ways they can be written as the sums of two squares. List the ways.

   (a) $2^2 \times 3^2 \times 5$

    (b) $2 \times 5^2 \times 13$

    (c) $5 \times 13 \times 17$

    (d) 5^3

    (e) $5^3 \times 13^2$

13. In Theorem 2.3.10 (1), (2), and (3), the phrase "at least" is used. Can this be replaced by the term "exactly" in any of the instances? Which instances? Prove your answers.

14. Finish the following statements; use several examples to back up your statements; then try to prove them. If N can be written as the sum of two squares in k ways, then

    (a) $2N$ can be written as the sum of squares in . . . ways.

    (b) N^2 can be written as the sum of squares in . . . ways.

    (c) N^m can be written as the sum of squares in . . . ways.

15. Finish the following statements; use several examples to back up your statement; then try to prove them. If N can be written as the sum of two squares in k ways and p is a prime of the form 1 (mod 4) and p does not divide N, then

    (a) Np can be written as the sum of two squares in . . . ways.

    (b) Np^2 can be written as the sum of two squares in . . . ways.

    (c) Np^m can be written as the sum of two squares in . . . ways.

16. Find a formula for the number of ways that a number N can be written as the sum of two nonzero squares.

17. Complete the proof of Theorem 2.3.19; that is, in the Pythagorean triple (x, y, z), one of x, y, and z is divisible by 5.

18. Prove the following:

    (a) In the primitive Pythagorean triple (x, y, z), neither x nor y can be of the form 2 (mod 4).

    (b) In any Pythagorean triangle the hypotenuse, z, cannot be of the form 3 (mod 4).

19. Prove that in the primitive Pythagorean triple (x, y, z), both $x + y$ and $x - y$ are congruent to 1 or 7 (mod 8).

20. Recall the Primitive triangle chart.

    (a) List the next number after 20 that has multiple listings in the y column.

    (b) List the next number after 15 that has multiple listings in the x column.

    (c) Find a number that appears three times in the y column.

    (d) Find a number that appears three times in the x column.

21. List all of the Pythagorean triangles

    (a) with side of length 48. How many of them are primitive?

    (b) with hypotenuse of length 1105. How many are primitive?

22. Find how many lattice points the circle with radius r passes through where r is

    (a) $\sqrt{325}$

    (b) 325

    (c) $\sqrt{1105}$

    (d) 1105

23. Find the circle, centered at the origin, that passes through at least 100 lattice points with the smallest:

    (a) radius

    (b) integer radius

    (c) Compare answers in (a) and (b) and make comments.

24. Prove Theorem 2.3.20: If a and b are solutions to the Diophantine equation $a^2 - 2ab - b^2 = \pm 1$ then $x^2 + y^2 = z^2$ and $|x - y| = 1$, where $x = a^2 - b^2$, $y = 2ab$, and $z = a^2 + b^2$.

25. Prove Theorem 2.3.20 The equation $a_n^2 - 2a_n b_n - b_n^2 = (-1)^n$ is satisfied by the pair a_n, b_n, where a_n, b_n may be generated in two different ways.

i) $a_1 = 2$, $b_1 = 1$; and $a_{n+1} = 2x_n + b_n$, $y_{n+1} = b_n$

ii) $a_{-1} = 0$, $a_0 = 1$, $b_{-1} = 1$, $b_0 = 0$ and

$a_{n+2} = 2a_{n+1} + a_n$ and $b_{n+2} = 2b_{n+1} + b_n$

26. Find a "near" isosceles Pythagorean triangle whose sides are both longer than

    (a) $10,000$
    (b) $100,000$
    (c) $1,000,000$.

27. Do an analysis of Pythagorean triangles where one side is nearly twice the other. Explain why one cannot be exactly twice the other using (a) and (b).

    (a) Assume that $y = 2x \pm 1$. Form the Diophantine equation, find a sequence of solutions, and form a quotient table. Find solutions where both sides are larger than 1000.

    (b) Do the same as (a) for $x = 2y \pm 1$. Is there any difference in your answer here from your answer in (a)?

28. Repeat Exercise 27, where the ratio of the sides approaches 1 to 3.

29. Form triangles on the integral lattice plane, where the vertices are lattice points and the sides are of integer length. We are familiar with the Pythagorean triangles. They are right triangles and have two sides parallel to the coordinate axes. On the other hand, the triangle $(0,0)$, $(4,3)$, $(8,0)$ has one side parallel to a coordinate axis. This triangle is just the union of two 3, 4, 5 triangles and has sides of lengths 5, 5, and 8.

    (a) Find other examples of triangles with one side parallel to a coordinate axis.

    (b) Find examples of right triangles with one side parallel to a coordinate axis.

    (c) Find examples of triangles, none of whose sides is parallel to a coordinate axis.

(d) Find examples of right triangles, none of whose sides is parallel to a coordinate axis.

(e) Find a triangle that is "nearly" equilateral whose sides are
   (i) greater than length 10
   (ii) greater than length 100

30. Find all the Pythagorean triangles whose area and perimeter are the same number.

31. Prove:

(a) The diameter of the circumscribed circle of a Pythagorean triangle is an integer.

(b) The radius of the inscribed circle of a Pythagorean triangle is an integer.

   Hint: Prove these first for primitive Pythagorean triangles. Figure 2.3 shows a triangle X, Y, Z, with O representing the center of the inscribed circle and x, y, and z representing the sides opposite X, Y, and Z, respectively; z is the hypotenuse. Note that the area of $\triangle XYZ$ is the sum of the areas of triangles $\triangle XOZ$, $\triangle XOY$, and $\triangle YOZ$. Find r in terms of x, y, and z; then express x, y, and z in terms of integers a and b, thus giving r in terms of a and b.

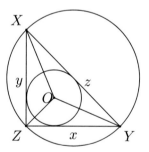

Figure 2.3

32. Suppose we have a circle of diameter d. Find all the Pythagorean triangles that can be inscribed in this circle. Find the diameters of their respective inscribed circles where d is

(a) 25

(b) 8

(c) 65

(d) 24

2.4 Factoring Large Numbers

Chapter 1 examined the set of natural numbers. If there were a theme to that chapter it would be that, as common and ordinary as they are, the natural numbers are still the most fascinating objects in all of mathematics. Much of this allure has to do with the multiplicative makeup of the numbers; in particular, the prime factorization of numbers. Certainly the allure captured the genius of Carl Friedrich Gauss, regarded by many as the greatest mathematician who ever lived. The assertion of Gauss in his *Disquisitiones Arithmeticae* that was quoted in Section 1.3 bears repeating here.

> The problem of distinguishing prime numbers from composite numbers and of resolving the latter into their prime factors is known to be one of the most important and useful in arithmetic. It has engaged the industry and wisdom of ancient and modern geometers to such an extent ... [that] the dignity of the science itself seems to require that every possible means be explored for the solution of a problem so elegant and so celebrated.

We have looked at the search for prime numbers in Section 1.4. In this section we will look into factoring composite numbers. As Gauss suggests, the search for large primes and the search for factors of large numbers go hand in hand. Coincidentally, we shall employ Gauss's invention, congruence arithmetic, to aid in our efforts. We shall approach factoring large numbers along three different avenues: the first avenue is theoretical; the second, practical, the third, applied. Fermat's little theorem is the vehicle for the first avenue, the Pollard ρ method is the vehicle for the second, cryptography and signature verification is the vehicle for the third.

We begin with Fermat's little theorem. Fermat first communicated his ideas to Frénicle in a letter on October 18, 1640. As with

his famous last theorem, Fermat did not produce a proof but did offer
conviction; in his letter he said "I would send you a demonstration if I
did not fear its being too long." It took nearly 100 years before Euler
published the first proof of this theorem in 1736, although the same
proof was given by Leibniz in 1683 in an unpublished manuscript.

Theorem 2.4.1 (Fermat's Little Theorem) *If p is a prime that
does not divide a, then $a^{p-1} \equiv 1 \pmod{p}$.*

Proof Consider the following numbers: $[a \pmod{p}]$, $[2a \pmod{p}]$,
$\ldots$, $[(p-1)a \pmod{p}]$. These are $p-1$ numbers all strictly between
0 and p and they are all different. This is because if two numbers
were the same —say, $[ja \pmod{p}] = [ka \pmod{p}]$ for different j
and k— then $ja \equiv ka \pmod{p}$. This implies that $p \mid j - k$, which
is impossible because $0 < j \neq k < p$. Therefore, these numbers
are simply $1, 2, 3, \ldots, p-1$ in some order. So their products are
equivalent mod p. That is,

$$1 \times 2 \times 3 \times \cdots \times (p-1)a^{p-1} \equiv 1 \times 2 \times 3 \times \cdots \times (p-1) \pmod{p}.$$

Since p does not divide the product $1 \times 2 \times 3 \times \cdots \times (p-1)$, we may
cancel this term from both sides. This yields $a^{p-1} \equiv 1 \pmod{p}$. $\square$

Corollary 2.4.2 *If p is a prime then, for all a, $a^p \equiv a \pmod{p}$.*

This theorem and its proof have a nice generalization due to
Euler. The idea behind Euler's generalization is the concept of the
number of numbers relatively prime to a given number. That is a
mouthful.

Definition 2.4.3 *For a given natural number N, define $\phi(N)$ to be
the number of positive numbers $n < N$ and relatively prime to N.
$\phi(N)$ is called the **Euler phi function** or the **Totient function**.*

Theorem 2.4.4 (Euler's Theorem) *If n is a natural number and
$\gcd(a, n) = 1$, then $a^{\phi(n)} \equiv 1 \pmod{n}$.*

We recognize that Fermat's little theorem is a special case of
Euler's theorem because if p is prime then $\phi(p) = p - 1$. We leave
the proof of Euler's theorem as an exercise.

Theorem 2.4.5 *If $a^t \equiv 1 \pmod{p}$ and if t is the least positive
exponent for which this is true, then $t \mid s$, where $a^s \equiv 1 \pmod{p}$.*

Proof Suppose $a^s \equiv 1 \pmod{p}$ and suppose that t is the least of the exponents for which this congruence is true. Thus $t \le s$. The division theorem tells us that $s = tq + r$ for some natural number q, where $0 \le r < t$. Now $a^s \equiv a^{tq+r} \equiv a^{tq}a^r \equiv a^r \equiv 1 \pmod{p}$. If $r > 0$, then $t \le r$. This is impossible so $r = 0$. Hence $t \mid s$. □

Corollary 2.4.6 *If $a^t \equiv 1 \pmod{p}$ and if t is the least exponent for which this is true, then $t \mid p - 1$.*

Corollary 2.4.7 *If $a^t \equiv 1 \pmod{n}$ and if t is the least exponent for which this is true, then $t \mid \phi(n)$.*

Fermat tells us that if p is a prime and if $0 < a < p$, then $a^p \equiv a \pmod{p}$. It strikes number theorists that this might be a good way to test whether a number, m, is prime. Just take all numbers, a, less than m and raise them to the mth power. If you get $[a \pmod{m}]$ then you have a prime. Well, not quite. This would be true if the converse of Fermat's theorem were true. But it is not. This leads to some ancient history and some modern definitions.

The ancient Chinese believed that if $m \mid 2^m - 2$, then m is a prime. This is the converse of this special case of Fermat's theorem: If p is a prime, then $p \mid 2^p - 2$; that is, $2^p \equiv 2 \pmod{p}$. This belief is understandable because the smallest number for which the converse fails is $m = 341$; 341 is not prime, it is equal to 11×31. As mentioned in Chapter 1, the number $2^{341} - 2$ has 103 digits and it would have been next to impossible to show that $341 \mid 2^{341} - 2$; that is, if we didn't have the power of modular arithmetic. We will display this power in Theorem 2.4.10. Numbers such as these are important enough to be given a name.

Definition 2.4.8 *Composite numbers, m, for which $m \mid 2^m - 2$ are called* **base 2 pseudoprimes**.

It has been shown that there are infinitely many base 2 pseudoprimes; in fact, there are infinitely many even base 2 pseudoprimes, the smallest being $161038 = 2 \times 73 \times 1103$. A way of generating base 2 pseudoprimes is to work out the numbers of the form $n = (4^p - 1)/3$ for primes $p \ge 5$. These are all base 2 pseudoprimes. The first four of these are 341, 5461, 1398101, and 22369621.

We may still ask about the converse of Fermat's theorem in the general setting; that is, if m is such that $m \mid a^m - a$ for all $a < m$, then m is prime. This statement is not true either. There are indeed composite numbers such that $m \mid a^m - a$ for all $a < m$. They are rarer than base 2 pseudoprimes and also merit a special name.

Definition 2.4.9 *Composite numbers for which* $m \mid a^m - a$ *for all* $a < m$ *are called* **absolute pseudoprimes** *or* **Carmichael numbers**.

Carmichael numbers were named for the number theorist Robert Carmichael (1879−1966). The smallest Carmichael number is 561 = $3 \times 11 \times 17$; the next few are 1105, 1729, 2465, 2821, 6601, 8911, and 10585. Carmichael numbers are very rare; there are only 1547 of them less than 10^{10}. Nevertheless, it was finally proved in 1992 that there are infinitely many of them.

We now prove that 341 is a base 2 pseudoprime. The methods used in proving this give an indication of how powerful congruence methods can be in number theory.

Theorem 2.4.10 $341 \mid 2^{341} - 2$

Proof Notice that $341 = 11 \times 31$. Fermat's little theorem tells us that $2^{10} \equiv 1 \pmod{11}$. Therefore,

$$2^{340} \equiv (2^{10})^{34} \equiv (1)^{34} \equiv 1 \pmod{11}.$$

Multiplying both sides by 2 we get $2^{341} \equiv 2 \pmod{11}$. So $11 \mid 2^{341} - 2$.
Also, $2^5 \equiv 32 \equiv 1 \pmod{31}$ so

$$2^{340} \equiv (2^5)^{68} \equiv (1)^{68} \equiv 1 \pmod{31}.$$

Multiplying both sides by 2 we get $2^{341} \equiv 2 \pmod{31}$. Therefore, $31 \mid 2^{341} - 2$. Since 11 and 31 are relatively prime, their product 341 also divides $2^{341} - 2$.   □

While the converse of Fermat's theorem is not true it is almost true. As you go further and further out in the number universe, the chances of finding base 2 pseudoprimes and Carmichael numbers diminish so dramatically that it is a near guarantee that a number for which $m \mid 2^m - 2$ is a prime. Of course, a near guarantee, while good enough for cryptographers, is really not good enough for mathematicians.

Next we introduce an actual procedure that can be used to factor large composite numbers. The size of the numbers we can factor depends on the speed of our computer and how much computer time we have. We shall factor small numbers here as a demonstration and leave a fuller investigation for a project. This method of factoring is called the **Pollard ρ method**, named after J. M. Pollard, an Englishman who works for the phone company and whose hobby is number theory. Here is how it works.

1. Take a number we would like to factor, call it N.

2. Beginning with the number, 1, successively form numbers of the form $[n^2 + 1 \pmod{N}]$. Label this sequence $\{u_n\}$. So $u_0 = 1$ and $u_{n+1} \equiv u_n^2 + 1 \pmod{N}$.

3. Form another sequence that is designed to produce factors of N. This sequence, call it d_n is defined as follows:

$$d_n = \gcd(|u_{2n} - u_n|, N).$$

The idea here is that the numbers in the sequence $\{u_n\}$ offer a random sample of numbers $< N$. There will be a repetition in this sequence, so the sequence is periodic and the number of distinct numbers is relatively small. From this set of numbers another periodic set is generated: numbers of the form $|u_{2n} - u_n|$. This second periodic set should produce two numbers equivalent mod p for some prime factor of N. Thus, there will be a term k such that d_k is a prime. Words and phrases such as "random," "relatively small," and " should" indicate that this procedure comes from the field of probability. If the likelihood is high enough and if k is small enough compared to N, then the procedure is useful. Like any iterative process, it must produce results efficiently to be useful.

Definition 2.4.11 *Let the natural number N be given. Then the sequence $\{u_n\}$ is called the **Pollard sequence mod** N, if $u_0 = 1$, and $u_{n+1} \equiv u_n^2 + 1 \pmod{N}$.*

Example 2.4.12
    (a) Let $N = 15$. Beginning with 1, we successively get $1^2 + 1 \equiv 2 \pmod{15}$, $2^2 + 1 \equiv 5 \pmod{15}$, $5^2 + 1 \equiv 11 \pmod{15}$, $11^2 + 1 \equiv 2 \pmod{15}$, and the repetition begins. The Pollard sequence for 15 is 1, 2, 5, 11, 2, 5, 11, ... Notice that after the initial "1" it is cyclic of period 3. The sequence $\{d_n\}$ looks like this.

n	u_n	$\lvert u_{2n} - u_n \rvert$	d_n
0	1		
1	2	3	3
2	5	3	3
3	11	0	
4	2	3	3
5	5		
6	11		
7	2		
8	5		

We carried this table on well beyond what was necessary. In this example we can stop when we reach a value for d_n that is not 1. In this case we have found a prime factor, 3. Dividing 3 into 15 completes the search for the prime factors.

(b) Let $N = 1883$. The Pollard sequence looks like this: 1, 2, 5, 26, 677, 761, 1041, 957, 712, 418, 1489, 831, 1384, 446, 1202, 544, 306, 1370, 1433, 1020, 985, 481, 1636, 754, 1734, 1489. This is a repeat, $u_{10} = u_{25} = 1489$. Notice that this sequence can be drawn in the form of the greek letter ρ with the tail consisting of the non-repeating beginning $u_1, \ldots u_9$ and the head consisting of the cycle of length 15 consisting of $u_{10}, \ldots u_{24}$. Here is the important part of the chart with the Pollard sequence and the values for the sequence $\{d_n\}$.

n	u_n	$\lvert u_{2n} - u_n \rvert$	d_n
0	1		
1	2	3	1
2	5	672	7
3	26		
4	677		

We stop at the prime 7 in the d_n column. Dividing 7 into 1883 yields 269, which happens to be prime. So the factorization of 1883 is the product: 7×269.

(c) Let $N = 5183$.

n	u_n	$|u_{2n} - u_n|$	d_n
0	1		
1	2	3	1
2	5	672	1
3	26	103	1
4	677	1883	1
5	2226	2466	1
6	129	1241	73
7	1093		
8	2560		
9	2289		
10	4692		
11	2664		
12	1370		

Dividing 73 into 5183 yields 71; so $5183 = 73 \times 71$. ■

The example worked fine for the composite numbers. If you try to find factors of a prime number with the Pollard ρ method, it will not work well. We leave it as an exercise to discover what happens. The following theorem sums up what our examples have demonstrated; we leave its proof for the project on the Pollard ρ method (Project 5.7).

Theorem 2.4.13 *Let $\{u_n\}$ be the Pollard sequence (mod N) and define the sequence $\{d_n\}$ as $\{\gcd(|u_{2n} - u_n|, N)\}$. If N is composite, then there will eventually be an n such that $d_n > 1$ and this term will be a factor of N.*

Our final foray into factoring takes us into the real world: the world of personal security and national security. With the advent of computers the worry over privacy and secrecy has risen, in some circles, to the level of paranoia. And building protection has become a field of study that has involved mathematics. The field of cryptography, in particular, has become extremely active. We will take a peek into this field with the following very simple version of signature verification. The important formula in our discussion comes from Euler's theorem; it is $\phi(pq) = (p-1)(q-1)$, where p and q are prime numbers. This formula is a special case of a general formula for $\phi(N)$. You are invited to find this general formula as an exercise.

Here is the scenario. I am a wealthy man and everyone knows it. Lots of people would like a piece of my wealth. When money is withdrawn from my bank account, it is imperative that the bank can identify me from an impostor. While my friendly face, my good name, even my mother's maiden name all can help identify me, we want a secure system that electronically identifies me. This means numbers. Here is how it works. I have an identification number, call it N. The bank, upon receiving N, will offer a number questioning my identity, call it Q. I will then send back a response based on the bank's question, call it R. The bank will check to see if my response affirms my identity; it will if my response yields a certain target number, call it T. If it does, I get the withdrawal; if not, I don't. Now I might as well make all of these numbers public since someone will find them out anyway. So N, Q, R, and T will all be known. Furthermore, the bank's method of checking my response will be known. The only thing secret in all of this is my method of creating R. It looks like a real challenge. Let's look at an example of how this might work.

Example 2.4.14

Suppose my identification number is 365; the bank tests my identity by sending me a testing number, say 1876. In order to affirm my identity I must offer the bank a response that will yield 1 after it performs its checking process. And I know how the bank's checking process works; it multiplies my response by the checking number and calculates this modulo my ID number. So we have $N = 365$, $Q = 1876$, $T = 1$. My job is to find a response number, R, so that $[1876R \bmod(365)] = 1$. I will find R by solving the Diophantine equation $1876R - 365y = 1$. I can do this because I have read Section 2.1; in fact, this very equation was solved in that section; it is $(1876 \times 136) - (365 \times 699) = 1$. So my response R is 136. And I get my money.   ■

Now this signature verification procedure was given strictly for demonstration purposes. It is obviously not useable. For starters, if the bank gives me a number, Q, that is not relatively prime to 365, I cannot respond with any R that will affirm my identity. This is minor, however; I can always ask for a different Q. The procedure is fatally flawed because it is not secure. All mathematicians since Euclid are familiar with the Euclidean algorithm and they would catch on quickly as to what I was doing to create my response, R. Then they would steal my money.

Let us examine a serious procedure that contains the elements of a verification procedure that is actually used. Here is how it will work. As before, N will be my ID number, Q will be the bank's query, R will be my response. This time the bank's procedure will be to calculate $[R^Q \pmod{(N)}]$ and the target number, T, will be 2. In this procedure I will accept only numbers, Q, from the bank if $\gcd(\phi(N), Q) = 1$. I calculate my response, R, by computing the number $[2^S \pmod{N}]$, where S is my secret number. It is calculated by solving the congruence $QS \equiv 1 \pmod{\phi(N)}$.

Example 2.4.15

Let us use my same ID number, 365. The bank offers the same Q, 1876, and I reject it. The bank then gives me $Q = 1877$ and I go to work, find a secret number S, and then produce a public reply, $R = 77$. The bank checks this out and finds me to be the right person and I collect my withdrawal. Here is what I did and here is what the bank did to affirm my identity.

(i) I know that $N = 365 = 5 \times 73$ so I know that $\phi(365) = \phi(5 \times 73) = 4 \times 72 = 288$. I reject 1876 because $\gcd(288, 1876) \neq 1$.

(ii) I accept $Q = 1877$ because $\gcd(1877, 288) = 1$. The Euclidean algorithm may be used to check this. In fact, $1877(29) - 288(189) = 1$ so $1877(29) \equiv 1 \pmod{288}$.

(iii) I take the 29 from (ii) and call it my secret number, S. I then calculate $[2^{29} \pmod{365}]$. This is my response number R. $R = 77$.

(iv) The bank then calculates $[77^{1877} \pmod{365}]$ and gets 2.

Here is why my response, R, works.

$$
\begin{aligned}
[R^Q \pmod{N}] &= [77^{1877} \pmod{365}] & (1) \\
&= [2^{29(1877)} \pmod{365}] & (2) \\
&= [2^{288k+1} \pmod{365}] & (3) \\
&= [(1)2^1 \pmod{365}] & (4) \\
&= [2 \pmod{365}] & (5)
\end{aligned}
$$

Note that (2) is true because $2^{29} \equiv 77 \pmod{365}$; (3) is true because $29(1877) \equiv 1 \pmod{288}$; and (4) is true because Euler's theorem tells us that $a^{288} \equiv 1 \pmod{365}$ for all a that are devoid of 5s and 73s in their prime factorization. Clearly 2 is such a number. ∎

Is my money secure in this example? In a word, no. The reason is that everyone knows the bank's procedure and nearly everyone knows how to factor 365 into its prime form. Fewer people know that $2^{(p-1)(q-1)} \equiv 1 \pmod{pq}$, but we know this because we know Euler's theorem and we know that $\phi(pq) = (p-1)(q-1)$. But mathematicians would put the word out on that, too. So my secret number of 29 is really no secret. And calculators and computers can easily find numbers of the form $[a^b \pmod c]$ even for large a, b, and c. So, while the whole procedure looks daunting it is, in fact, child's play for factorable numbers like 365. But what if my ID number were really large? Suppose it was made up of the product of two primes and I was the only one who knew the two primes, And the product was so large that even the most mathematically inclined thieves could not factor it. Then my situation is good. My number is N, $N = pq$, and no one but I knows p and q. So when the bank gives me Q, I can find my secret number S, such that $QS \equiv 1 \pmod{(p-1)(q-1)}$. Since the bank doesn't know p and q it would not know $(p-1)(q-1)$. My secret number, S, remains secret as I produce my public response, R. Although the public knows Q and N and it knows I must come up with an R so that $R^Q \equiv 2 \pmod N$, the numbers are too large for thieves to find such an R. After I produce it, they can't work backward to figure out how I did it.

Example 2.4.16

Let us do an example with fairly large numbers and see what might happen. I will choose my identification number to be $N = 31417$. Only I know that it is the product of two primes. Well, not quite. It turns out that a person who knows the Pollard ρ method of factoring knows my number too. He enters my number into his computer that is programmed with Pollard ρ method and, after 28 iterations it tells him that $\gcd(|u_{28} - u_{14}|) = 8099$ and $\gcd(8099, 31417) = 89$. Thus 89 is a factor. It then provides the other prime factor, 353. Since he knows Euler's theorem he knows that $\phi(89 \cdot 353) = 88 \times 352 = 30976$. He enters my number; the bank responds with $Q = 6651$. He checks to see if $\gcd(6651, 30976) = 1$. Here's his work.

$$
\begin{aligned}
30976 &= 6651(4) + 4372 \\
6651 &= 4372(1) + 2279 \\
4372 &= 2279(1) + 2093 \\
2279 &= 2093(1) + 186 \\
2093 &= 186(11) + 47 \\
186 &= 47(3) + 45 \\
47 &= 45(1) + 2 \\
45 &= 2(22) + 1 \\
2 &= 1(2) + 0
\end{aligned}
$$

Since $\gcd(30976, 6651) = 1$ he accepts the number and moves on to the quotient table.

	4	1	1	1	11	3	1	22
30976	1	0	1	1	2	3	35	108	143	3254
6651	0	1	4	5	9	14	163	503	666	15155

So $6651(15155) - 30976(3254) = 1$. And so $S = 15155$ because $6651S \equiv 1 \pmod{30976}$.

Next he must calculate $[2^S \pmod{31417}]$ to come up with his response, R. Calculating $[2^{15155} \pmod{31417}]$ appears daunting, but his calculator could make short work of it. Nevertheless he prefers to show off his mathematical skills. He knows $31417 = 89 \times 353$; he knows from Fermat's little theorem that $2^{88} \equiv 1 \pmod{89}$ and $2^{352} \equiv 1 \pmod{353}$; he knows that if $2^n \equiv 1 \pmod{p}$ and $2^m \equiv 1 \pmod{q}$, then $2^l \equiv 1 \pmod{pq}$, where $l = \mathrm{lcm}(m, n)$. This is true by Euler's theorem, he leaves the details as an exercise for us. Applying this knowledge to the heist, he obtains

$$l = \mathrm{lcm}(88, 352) = 352 \text{ so } 2^{352} \equiv 1 \pmod{31417}.$$

Actually 352 is not the smallest exponent for which $2^{88} \equiv 1 \pmod{31417}$, but that is beside the point here. He is anxious to find $2^{15155} \pmod{31417}$. So he calculates the remainder of 15155 upon division by 352. It is 19. In fact, $15155 = 43(352) + 19$. So $2^{15155} = 2^{43(352)} \times 2^{19}$ and therefore

$$2^{15155} \equiv 2^{43(352)} \times 2^{19} \equiv 1(2^{19}) \pmod{31417}.$$

Now 2^{19} is not so daunting; in fact, $2^{19} = 524288$. Long division yields $524288 = 16(31417) + 21616$, so $[2^{19} \pmod{31417}] = 21616$. His response, R, is 21616.

The bank takes his response and calculates (with its fancy computer) $[21616^{6651} \pmod{31417}]$. Not surprisingly, it gets 2 because

$$21616^{6651} \equiv 2^{(19)6651} \equiv 2^{[(19+43(352)]6651} \equiv 2^{(15155)6651} \equiv 2^{30976(3254)+1}$$
$$\equiv 1 \times 2 \pmod{31417}.$$

So he gets my money.   ∎

Clearly my number is too small. Beyond this there are other mathematical considerations that would make me pause. Here is an important one: There can be many secret numbers that work. My number S was 15155, but we see that $S = 19$ works also. This is true because there may exist several numbers d such that $2^d \equiv 1 \pmod{31417}$ and $d < 30976$. In this example we used $d = 352$. Any multiple of 352 would have worked, and even some divisors of 352 might work. In fact, $2^{88} \equiv 1 \pmod{31417}$. This state of affairs occurred because we know $2^{88} \equiv 1 \pmod{89}$ and $2^d \equiv 1 \pmod{353}$, where $d \mid 352$. Since $88 \mid 352$, there is a chance that $2^{88} \equiv 1 \pmod{353}$ also. And this is true. So any multiple of 88 would serve in this context. Let's face it, my choice of an ID number was not a good one.

A final word about the size of numbers that are secure in this day and age. Current factoring algorithms can factor numbers up to 130 digits, but to be on the safe side you choose your primes p and q to have about 100 digits each. That gives an ID number of about 200 digits. One of the most important aspects of predicting whether a number will remain secure or not is understanding the speed of calculations. For example, we must know how many steps the Euclidean algorithm will take and how long each step takes, we must know the speed of calculating numbers like $ab \pmod{c}$, and we must guess what new methods might arise to factor large numbers more quickly. Generally, it is believed that every three extra digits added to the size of a number will double the time it takes to factor the number. So, for me, at the end of this millennium, I must come up with two 100-digit primes. Probably two primes of 80 or so digits would be good enough. I'll just hire that job out; I'm wealthy — whoops— there goes the security.

These signature verification systems come under the heading of cryptography. The system we have outlined here is based on the Rivest-Shamir-Adleman (RSA) cryptosystem. RSA is a public key system; that is, a system where the exchange numbers are public and the security comes from a difficult mathematical problem. Our example is based on signature verification but, generally, this system is used to transmit secret messages. As you can imagine, this is of vital importance to governments.

EXERCISES

1. What can you say about n and m if

   (a) $n^{96} \equiv m \pmod{17}$?

   (b) $n^9 \equiv m \pmod{19}$?

   (c) $n^{216} \equiv m \pmod{91}$?

   (d) $n^{25} \equiv m \pmod{20}$?

2. Show:

   (a) If 7 does not divide n, then 7 does divide $n^{12} - 1$.

   (b) $n^{13} - n$ is divisible by 2, 3, 5, 7, and 13 for all numbers n.

3. Find the prime factorization for the following numbers.

   (a) $2^{13} - 2$

   (b) $3^{13} - 3$

   (c) $2^{37} - 2$

   (d) $3^{37} - 3$

4. Prove that a number of the form $n^4 + 4$ for $n > 1$ cannot be prime.

5. Prove the following theorem: Suppose that p and q are primes and $a < p$. If $p \mid a^q - 1$, then p is of the form $qk + 1$.

6. Using the theorem of Exercise 5, the table of primes from Section 1.3, and Theorem 1.3.1, given the following numbers, (i) list the primes that could be their factors, (ii) write down their prime factorizations.

(a) $2^{23} - 1 = 8388607$

(b) $3^{13} - 1 = 1594322$

(c) $2^{30} - 1 = 1073741823$

(d) $5^{12} - 1 = 244140624$.

7. (a) Using algebra, write down the various ways that $a^n - 1$ can be factored where

   (i) $n = 12$

   (ii) $n = 30$

   (iii) $n = 20$

   (b) Using (a) find the prime form of

   (i) $5^{12} - 1$

   (ii) $2^{30} - 1$

   (iii) $2^{20} - 1$

8. Find prime form for the following numbers:

   (a) $7^7 - 1$

   (b) $3^{15} - 1$

   (c) $2^{19} - 1$

   (d) $9^{12} - 1$

9. Find the following sums mod 11 for each number $a < 11$.

   (a) $1 + a + a^2 + a^3 + a^4 + a^5 + a^6 + a^7 + a^8 + a^9$

   (b) Explain what is happening.

   (c) Try to make a general rule about such things and prove it.

10. For primes $p < 50$

    (a) Find the smallest n such that $2^n \equiv 1 \pmod{p}$. Call your answer n_p.

    (b) For which p is $(p - 1)/n_p$ the largest? The smallest?

11. Show that there is a multiple of a prime

    (a) except for 2 and 5 that is made up of digits that are all 9s

(b) except for 2, 3, and 5 that is made up of all 1s

12. Let $N = 111\ldots1$, where N is made up of p 1s and where p is a prime other than 3. Show that $N \equiv 1 \pmod{p}$.

13. Show that the following are base 2 pseudoprimes.

 (a) 561
 (b) 645
 (c) 5461

14. Show that 561 is a Carmichael number. In particular, use the factoring of 561 to show that

 (a) $2^{561} \equiv 2 \pmod{561}$
 (b) $3^{561} \equiv 3 \pmod{561}$
 (c) $5^{561} \equiv 5 \pmod{561}$
 (d) $a^{561} \equiv a \pmod{561}$ for any $a < 561$

15. Show that the following are Carmichael numbers.

 (a) 1105
 (b) 1229
 (c) 2465

16. Test the primality of the following:

 (a) $9! - 1,\ 10! + 1$
 (b) $11! - 1,\ 12! + 1$
 (c) $15! - 1,\ 16! + 1$

17. Find possible numbers n and m for the following congruences. Try to prove your assertions.

 (a) $m! \equiv 1 \pmod{n}$
 (b) $m! \equiv -1 \pmod{n}$

18. Make a list of $\phi(n)$ for $n \leq 25$. Using this list, find a formula for $\phi(n)$ where

(a) $n = p$ where p is a prime number

(b) $n = pq$ where p and q are primes

(c) $n = p^2$, $n = p^3$, $n = p^k$

(d) n is any natural number

19. (a) Prove the following: Let $S = \{n : n$ is relatively prime to $N\}$; let $T = \{[a^n \pmod{N})]\}$ where a is a fixed number relatively prime to N. Then $S = T$.

   (b) Using (a) prove Euler's theorem, 2.4.4.

20. Use the Pollard ρ method on several different prime numbers and describes what happens.

21. Continue the chart for Example 2.4.12 (c) and find out

   (a) when $\{u_n\}$ begins repeating

   (b) what the repetition pattern is for $\{d_n\}$

22. Factor the following numbers using the Pollard ρ method.

   (a) $N = 8,131$

   (b) $N = 7,807$

   (c) $N = 13,081$

   (d) $N = 16,019$

   (e) $N = 36,287$

   (f) $N = 75,007$

   (g) $N = 551,987$

   (h) $N = 23,356,727$

23. Let $N = 4444^{4444}$; let a be the sum of the digits of N, b be the sum of the digits of a, c be the sum of the digits of b. Find c.

24. Given the ID number $N = 221$ and the response from the bank, $Q = 23$,

   (a) Find R.

   (b) What is S? Can you find another S that will work? If so, what is it?

25. Let $N = 1891 = 31 \times 61$. Let $Q = 323$.

    (a) Find the smallest n and m such that $2^n \equiv 1 \pmod{31}$ and $2^m \equiv 1 \pmod{61}$.

    (b) Find the smallest l such that $2^l \equiv 1 \pmod{1891}$.

    (c) Find R.

26. Let $N = 16637 = 131 \times 127$. Let $Q = 7979$.

    (a) Find the smallest n and m such that $2^n \equiv 1 \pmod{131}$ and $2^m \equiv 1 \pmod{127}$.

    (b) Find the smallest l such that $2^l \equiv 1 \pmod{16637}$.

    (c) Find R.

27. Using the Pollard ρ Method factor each of my ID numbers, N, and give a response, R, when the bank offers Q that will allow you to steal my money.

    (a) $N = 551,987$; $Q = 441,847$

    (b) $N = 269,997,421$; $Q = 97,578,799$

    (c) $N = 163,276,871$; $Q = 68,652,139$

    (d) $N = 30,796,045,833$; $Q = 3,885,039,731$

28. Prove that if a person knows N and also the product $p-1 \times q-1$, then she can figure out pq.

29. Given that p and q are primes, prove the following:

    (a) If $2^n \equiv 1 \pmod{p}$ and $2^n \equiv 1 \pmod{q}$, then $2^n \equiv 1 \pmod{pq}$.

    (b) If $2^n \equiv 1 \pmod{p}$, $2^m \equiv 1 \pmod{q}$, and l is the lcm(m, n), then $2^l \equiv 1 \pmod{pq}$.

    (c) If $2^n \equiv 1 \pmod{pq}$, $2^m \equiv 1 \pmod{pq}$, and d is gcd(n, m), then $2^d \equiv 1 \pmod{pq}$.

Chapter 3

The Rational Numbers

The rational numbers have the more familiar name of fractions; they are expressed as m/n, where m and n are integers and $n \neq 0$. These numbers belong to four worlds: the worlds of number theory, geometry, algebra, and analysis. If we include the world of common sense, it makes five; after all, fractions are as much a part of everyday life and experience as the counting numbers and whole numbers. We look into two of these worlds at some length in this chapter (number theory and algebra) and only touch on the worlds of geometry and analysis. We return to the worlds of algebra, geometry, and analysis in greater depth in Chapter 4.

Ancient people surely were familiar with the idea of a part of a whole and understood the concept of a half, a third, a fourth, and so on. But it was quite another thing for them to develop a way of dealing with these ideas mathematically. The ancient Egyptians tended to write all fractions as sums of unit fractions. A **unit fraction** is a fraction with a numerator of 1. *The Rhind Papyrus*, one of the oldest mathematical documents dating back about 4000 years, showed representations of the following fractions: $2/11 = 1/6 + 1/66$ and $2/97 = 1/56 + 1/679 + 1/776$. Apparently the Egyptian mathematicians of the time were loathe to use the same unit fraction twice; it seems that $2/11 = 1/11 + 1/11$ might be easier to work with. You would expect that today we have especially neat ways of working with fractions. This is not true, however. In fact, it is fair to say that the first three sections of this chapter deal with the problems of expressing fractions in the "best" possible way. Furthermore, there are four projects in Chapter 5 that deal with these subjects.

Fractions fit naturally into the world of geometry. When we look at sizes we naturally come up with the ideas of comparison, similarity, and ratio. The ancient Greek geometers were fascinated by ratios of lengths of different line segments. They established theorems concerning ratios of lengths and similarity of figures. Two segments are **commensurable** if a unit fraction exists that can measure both segments in whole number lengths. It was devastating to their philosophy of harmony in the universe when they discovered that the length of the side of a square and the length of its diagonal are not commensurable. In today's language we say that $\sqrt{2}$ is irrational.

Fractions are algebraic objects too; they are the solutions to the linear equations $ax + b = 0$, where a and b are integers. The set of rational numbers under addition and multiplication form an algebraic structure called a field. This means that the basic arithmetic operations of addition, subtraction, multiplication, and division (except division by 0) can be carried out with the guarantee that you will get an answer. This was not true for subtraction in the set of natural numbers; for example, $5 - 7$, has no natural number answer. Of course it does have an answer within the integers. The situation is similar for division within the integers; for example, $5 \div 7$ has no integral answer. Obviously it has a rational solution: $5/7$. We will examine the abstract structure of the set of rationals in Chapter 4. In Section 3.4 we shall use fractions to help solve Diophantine equations. While solutions to Diophantine equations are limited to integers, it turns out that by expanding our range of solutions to rational numbers we can develop methods for finding integral solutions.

Beyond number-theoretic, geometric, and algebraic considerations, the rational numbers present an opportunity to examine the field of mathematical analysis. The study of analysis concerns itself with concepts of distance, nearness, and proximity. The rational numbers are important in the study of analysis because they are ubiquitous on the real line. If we want to refer to a point on the real line we can count on there being a fraction close by, *really* close by. Our notation for real numbers, the decimal notation, depends upon this fact. A decimal expansion of a real number is simply the sum (albeit infinite) of fractions. Just as the primes form the building blocks and multiplication forms the cement for the natural numbers, the rationals form the building blocks and addition the cement for the real numbers. We begin this chapter with a study of the decimal representation of rational numbers.

3.1 Rational Numbers as Decimals

In Chapters 1 and 2 we represented whole numbers in base 10 notation. This was exact; after all, every whole number can be represented as a sum of powers of 10. A fraction can also be represented in base 10 notation. Since a fraction represents a part of a whole, the representation will consist of sums of powers of $\frac{1}{10}$. These are familiar to us as **decimals**.

Definition 3.1.1 *The number N, written in* **decimal form,** *looks like this:*

$$N = a_n a_{n-1} \ldots a_1 a_0 . a_{-1} \ldots a_{-k} \ldots .$$

This is shorthand for

$$a_n 10^n + a_{n-1} 10^{n-1} + \cdots + a_1 10 + a_0 + a_{-1} 10^{-1} + \cdots + a_{-k} 10^{-k} + \cdots .$$

We also say that $a_n a_{n-1} \ldots a_1 a_0 . a_{-1} \ldots a_{-k} \ldots$ is the **decimal expansion** *of N or the* **decimal representation** *of N.*

Remark We shall not use decimal expansions that end in all 9s. The reason for this is that the decimal expansion

$$a_n a_{n-1} \ldots a_1 a_0 . a_{-1} a_{-2} \ldots a_{-k},$$

where $a_{-k} \neq 0$, represents the same number as the expansion

$$a_n a_{n-1} \ldots a_1 a_0 . a_{-1} a_{-2} \ldots (a_{-k} - 1) \, 9 \, 9 \, 9 \ldots .$$

Since uniqueness of notation is important, we shall not use such expansions.

While natural numbers can get arbitrarily large, the base 10 representation of any specific whole number is finite. But for non-integers the situation is different; the decimal can continue forever. Numerical expressions with infinite expansions can pose problems both conceptually and philosophically, but we will sidestep that here with an axiom and definition. The axiom is an informal version of what is called the completeness axiom. It is given in Section 4.3.

A Completeness Axiom A decimal expansion represents an existing number.

These numbers, whose existence is guaranteed by the completeness axiom, are called real numbers.

Definition 3.1.2 *A **real number** is a number expressible in decimal form.*

Notation: For simplicity we will not use negative subscripts; we shall express a real number like this: $N.a_1a_2\ldots a_k\ldots$, where N represents an integer.

So we can say that a real number *is* its decimal expansion. Since natural numbers and integers have decimal representations, they are real numbers. Their decimal representations are uninteresting; they consist entirely of zeros on the righthand side of the decimal point. Rational numbers are also real numbers, but their decimal expansions are quite interesting.

Example 3.1.3

(a) Let us write 3/8 in decimal notation. Using long division, or a hand calculator, we find $3/8 = 0.375$. So $3/8 = (3)10^{-1}+(7)10^{-2}+(5)10^{-3} = 3/10 + 7/100 + 5/1000$.

(b) The fraction $9/20 = 0.45$. That is, $9/20 = (4)10^{-1}+(5)10^{-2} = 4/10 + 5/100$. This, too, is simple.

(c) The fraction 5/7 appears as 0.7142857 on the calculator. So $5/7 = (7)10-1 + (1)10^{-2} + (4)10^{-3} + (2)10^{-4} + (8)10^{-5} + (5)10^{-6} + (7)10^{-7}$. Or does it? The fact is that 5/7 does not equal this sum. Not only is the decimal representation longer, but it extends infinitely far out. Through long division, we can find a pattern. If we carry out the division process of dividing 7 into 5, we get the successive remainders of 5, 1, 3, 2, 6, 4, and 5 again. So the quotient of 0.714285 repeats these same six numbers in the same order over and over again. The decimal representation never becomes exact but it does get exceedingly close; in fact, as close as you wish. And we have discovered a pattern: The sequence of digits 714285 repeats in cycles of length six. So it is a bad news, good news situation: The bad news is that the decimal representation is infinite; the good news is that we know exactly which digits go where in the sequence. Here is the time-honored long division process that shows the pattern.

$$
\begin{array}{r}
0.7\ 1\ 4\ 2\ 8\ 5 \\
7\ \overline{\smash{)}5.0\ 0\ 0\ 0\ 0\ 0} \\
\underline{0\ 0} \\
5\ 0 \\
\underline{4\ 9} \\
1\ 0 \\
\underline{0\ 7} \\
3\ 0 \\
\underline{2\ 8} \\
2\ 0 \\
\underline{1\ 4} \\
6\ 0 \\
\underline{5\ 6} \\
4\ 0 \\
\underline{3\ 5} \\
5
\end{array}
$$

(d) The fraction $47/72$ appears as 0.6527778 on the calculator. This is $(6)10^{-1} + (5)10^{-2} + (2)10^{-3} + (7)10^{-4} + (7)10^{-5} + (7)10^{-6} + (8)10^{-7}$. This sum probably does not represent $47/72$ exactly, but we can see a pattern developing. It looks as if the 7s may continue forever. Long division will confirm this.

(e) The fraction $11/49$ appears as 0.2244898 on the calculator. This is $(2)10^{-1} + (2)10^{-2} + (4)10^{-3} + (4)10^{-4} + (8)10^{-5} + (9)10^{-6} + (8)10^{-7}$. We do not know if this is exact, but our suspicion is that it is not. We can do the long division on $11/49$ and hope eventually that the remainders will repeat, but we might take awhile. ∎

Let us try to sort out what the decimal representations of fractions must look like.

Theorem 3.1.4 *The decimal expansion of a fraction m/n must either be finite or eventually repeat.*

Proof As we divide n into m with the long division process, one of the following two things must occur. A remainder of 0 occurs, bringing the process to an end, or a remainder of 0 does not occur. If 0 does not occur, then the remainders must eventually repeat. This is because the remainders upon division by n must necessarily be less than n. The division theorem ensures this. Since 0 does not occur, there are only $n - 1$ different remainders and each new division in the process produces a remainder. So after at most $n - 1$ successive divisions a remainder must repeat. Once a remainder comes around a second time, the long division process repeats and the cycle begins again.   □

Example 3.1.5

(a) Consider the fraction $1/6$. As a decimal, this reads $0.1666\ldots$. The 6 repeats forever. The long division process of $6\sqrt{1.0000\ldots}$ yields the remainders: 1, 4, 4, 4, and 4 again and forever more.

(b) The fraction $9/25 = 0.36$. The remainders in the long division process here are 9, 15, and 0. Then the process ends.

(c) The fraction $3/11 = 0.272727\ldots$. The remainders are 3, 8, and 3 again. This second 3 signals the repeat.   ∎

These three examples display three different behaviors. For $1/6$ the decimal begins with a 1 and then repeats forever with a 6. For $9/25$ the decimal terminates after two numbers: 36. For $3/11$ the decimal repeats with 27 forever. These three behaviors have names.

Definition 3.1.6

*(a) A decimal expansion is **terminating** if it is finite. The number of places after the decimal point is called the **length** of the expansion. We assume that the last digit of the expansion is not 0.*

*(b) A decimal expansion is **periodic** if, beginning just after the decimal point, it begins a repetitive cycle. The length of one cycle is called the **period**.*

*(c) A decimal expansion is **mixed** if it has a cycle of repetition but the cycle does not begin immediately after the decimal point. If the cycle begins in the $(d + 1)$st place after the decimal point, it has **delay** d.*

Notation: Here is the notation for a decimal that has a repeating part beginning after the dth place. The number a represents an integer.

$$a.a_1 a_2 \ldots a_d \overline{a_{d+1} \ldots a_{d+k}}$$

It turns out that the three behaviors of decimal expansions can be nicely translated into their fractional equivalents and conversely. A fraction has a terminating decimal expansion when the fraction is equivalent to the fraction $N/10^k$. A fraction has a periodic decimal expansion when it is equivalent to $N/(10^k - 1)$ (that is a fraction with denominator composed entirely of nines). A fraction has a mixed expansion when it is equivalent to $N/(10^k - 1)(10^l)$ (that is a fraction beginning with nines and ending with zeros). In other words all fractions are equivalent to either a fraction of the form $N/(10\ldots0)$, $N/(99\ldots9)$, or $N/(9\ldots90\ldots0)$.

Example 3.1.7

(a) The fraction $1/6$ has mixed behavior. It is equivalent to $15/90$.

(b) The fraction $9/25$ has a terminating expansion. It is equivalent to $36/100$.

(c) The fraction $3/11$ has a repeating expansion. It is equivalent to $27/99$. ∎

We now present our observations formally. While all that follows is valid for all kinds of fractions, both proper and improper, both positive and negative, we shall present the theorems and the proofs with the understanding that the fractions m/n are positive and proper; that is, $0 < m < n$.

Theorem 3.1.8

1. *The fraction m/n has a terminating decimal expansion if and only if $m/n = N/(10^k)$. The length of the expansion is the least k for which this is true.*

2. *The fraction m/n has a repeating decimal expansion if and only if $m/n = N/(10^k - 1)$. The period of the expansion is the least k for which this is true.*

3. *The fraction m/n has a mixed decimal expansion if and only if $m/n = N/(10^k - 1)(10^l)$. The delay in the expansion is the least l for which this is true, the period is the least k for which this is true.*

Proof

(1) It is clear that if m/n has a terminating expansion of length k then $m/n = 0.\,a_1a_2\ldots a_k$ and $a_k \neq 0$. This is equivalent to a fraction $N/(10^k)$, where $N = a_1a_2\ldots a_k$ Conversely, if $m/n = N/(10^k)$ and k is the least number for which this is true, then the numerator does not end in a zero and the fraction has a terminating expansion of length k.

(2) Suppose that m/n has a periodic expansion of period k. Then

$$m/n = 0.\overline{a_1a_2\ldots a_k}.$$

Now $10^k(m/n) = a_1a_2\ldots a_k.\overline{a_1a_2\ldots a_k}$ and so

$$10^k(m/n) - (m/n) = a_1a_2\ldots a_k = N.$$

Therefore $m/n = N/(10^k - 1)$. Since the period is k, this is the least such number that can represent m/n with an equivalent fraction whose denominator is comprised of all nines.

Conversely suppose that $m/n = N/(10^k-1)$, where $N = a_1a_2\ldots a_k$. The long division process of

$$99\ldots9\sqrt{a_1a_2\ldots a_k.0\,0\,0\,0\ldots}$$

yields the remainder: $a_1a_2\ldots a_k$ after k divisions because

$$10^k(a_1a_2\ldots a_k) = (10^k - 1)(a_1a_2\ldots + a_k) + (a_1a_2\ldots a_k).$$

So $m/n = 0.\overline{a_1a_2\ldots a_k}$. Now the remainder $a_1a_2\ldots a_k$ may occur before k divisions, but no matter where it occurs the fraction m/n will be equivalent to a fraction with denominator composed of nines. The number that counts the number of nines before the first repetition is the period of the expansion.

(3) The proof of this is left as an exercise.

$\square$

Since we know there are only three types of behaviors and we know that all fractions are equivalent to fractions with denominators of the form $10^k, 10^k - 1$, or a combination of these, it might seem that there is no challenge left. It is surely true that if we know the decimal expansion of m/n, then we know something about the prime form of n.

Theorem 3.1.9

1. If $m/n = 0.a_1a_2\ldots a_k$, then the prime form of n is composed entirely of 2s and/or 5s. Furthermore, $n = 2^j5^l$ where $k = \max(j, l)$.

2. If $m/n = 0.\overline{a_1a_2\ldots a_k}$, then the prime form of n is composed only of primes that factor $10^k - 1$.

3. If $m/n = a.a_1a_2\ldots a_d\overline{a_{d+1}\ldots a_{d+k}}$, then n has 2s and/or 5s as well as at least one other prime in its prime form. The maximum number of 2s or 5s in n is d, the other primes in its prime form must divide $10^k - 1$.

The proof of this theorem is left as an exercise.

Example 3.1.10

(a) Consider the number $N = 0.1172$. So $N = 1172/10000$. Whatever this fraction reduces down to, it will have at least one of the primes 2 and/or 5 represented to the 4th power. And indeed it does. The fraction reduces to $293/2500$ and 2500 has the prime decomposition 2^25^4.

(b) Let us find denominators d that yield terminating expansions of length two. So the fractions $m/d = 0.ab$ where $b \neq 0$. Since the decimal expansion is of length 2, then d must be one of the following: 2^2, 2^25, 2^25^2, 5^22, or 5^2; that is, $4, 20, 100, 50$, or 25.

(c) Let $N = 0.\overline{358974}$. Then $N = 358974/999999$. Letting m/n be the reduced fraction equivalent to N, we know that the prime form of n contains divisors of 999999. There are many such divisors and it is not at all clear which ones will be in the reduced fraction. It turns out that the reduced fraction is $14/39$ and both 3 and 13 divide 39.

(d) Let us find denominators d that yield periodic expansions of length two. So the fractions $m/d = 0.\overline{ab}$. Since the period is 2, we must find all factors of 99. But we should not include 3 or 9 because those denominators have expansions of period 1. That leaves 11, 33, and 99.

(e) Consider the number $N = 0.72\overline{297}$. Letting $N = m/n$, we can say that n contains a 2 and/or a 5 to the second power along with at least one prime that divides 999. Let's work it out.

$$10^5 N - 10^2 N = 72297.\overline{297} - 72.\overline{297} = 72225$$

So $N = 72225/(10^2(10^3 - 1)) = 72225/99900$. This reduces to $107/148$.
Notice that $148 = 2^2 \times 37$.

(f) Let us find denominators d that yield mixed expansions with
delay one and period one. So the fractions $m/d = 0.a\overline{b}$. We wish to
find numbers that are the product of single powers of 2s and 5s and
also divisors of 9; that is, 3 and 9. Putting it another way, we want
to find divisors of 90 that are not single primes nor powers of single
primes. These are 2×3, 2×9, 5×3, 5×9, 10×3, and 10×9; that
is, 6, 18, 15, 45, 30, and 90, respectively.   ∎

Let us reverse our study. Suppose that a fraction is given us and
we are asked to examine the nature of the decimal expansion. For
example, if we are handed the reduced fraction $5/7$, we might be hard
pressed to come up with the fact that it has period 6. The challenge
of analyzing the behavior of ordinary fractions, especially those with
denominators devoid of 2s and 5s, is exciting. Let us gather some
data and examine the decimal expansions of reduced fractions.

The chart that follows lists the behavior of the unit fractions with
denominators d for $2 \leq D \leq 20$. The type of expansion is indicated
by the letters T, P, and M. The letter T stands for terminates, the
length of the expansion follows; P stands for periodic, the period
follows;   M, stands for mixed, the delay and period follow. As
an aside you should note that the numerators play no part in the
behavior of the fraction other than to reduce it. This follows from our
preceding analysis. So the behavior of unit fractions $1/n$ is identical
to the behavior all fractions m/n, where $\gcd(m, n) = 1$.

Look for the following clues in the chart. The denominators of
the fractions with terminating representations are 2, 4, 5, 8, 10, 16,
and 20, all numbers whose prime form is composed of 2s and/or 5s.
The length of the expansion is the greater of the exponents for 2 and
5. This we expected. The denominators of fractions with periodic
representations are 3, 7, 9, 11, 13, 17, and 19. With the exception of
9, these numbers are prime; and 9 is a prime squared. Again, this is
expected; denominators that are devoid of 2s and 5s in their prime
form are periodic. As for the period, it is not so clear what is going
on. The period of $1/3$ is 1, of $1/7$ is 6, of $1/11$ is 2, of $1/13$ is 6,
of $1/17$ is 16, and of $1/19$ is 18. The period is certainly less than

the prime and in two cases it is just one less. The denominators of fractions with mixed decimal representation are 6, 12, 14, 15, and 18, numbers whose prime form is composed of combinations of 2s and/or 5s and other primes. The delay matches the exponent of the 2 or the 5, while the period matches the period of the prime. This is not unexpected. Perhaps an extension of this chart is in order because we really don't know about 1/21. Our calculator shows 0.047619, but we do not know where it is going.

Decimal Expansion Chart

D	Expansion	Type of expansion
2	$1/2 = 0.5$	T: 1
3	$1/3 = 0.\overline{3}$	P: 1
4	$1/4 = 0.25$	T: 2
5	$1/5 = 0.2$	T: 1
6	$1/6 = 0.1\overline{6}$	M: 1, 1
7	$1/7 = 0.\overline{142857}$	P: 6
8	$1/8 = 0.125$	T: 3
9	$1/9 = 0.\overline{1}$	P: 1
10	$1/10 = 0.1$	T: 1
11	$1/11 = 0.\overline{09}$	P: 2
12	$1/12 = 0.08\overline{3}$	M: 2, 1
13	$1/13 = 0.\overline{076923}$	P: 6
14	$1/14 = 0.0\overline{714285}$	M: 1, 6
15	$1/15 = 0.0\overline{6}$	M: 1, 1
16	$1/16 = 0.0625$	T: 4
17	$1/17 = 0.\overline{0588235294117647}$	P: 16
18	$1/18 = 0.0\overline{5}$	M: 1, 1
19	$1/19 = 0.\overline{052631578947368421}$	P: 18
20	$1/20 = 0.05$	T: 2

Theorem 3.1.11 acknowledges what we know. Its statements are the converses of Theorem 3.1.9. Its proof is left as an exercise.

Theorem 3.1.11

1. *Let the reduced fraction m/n be given. If the prime form of n is composed entirely of 2s and/or 5s, then $m/n = 0.a_1a_2 \ldots a_k$, where $n = 2^j 5^l$ and $k = \max(j, l)$.*

2. Let the reduced fraction m/n be given. If the prime form of n is composed only of primes that factor $10^k - 1$, then $m/n = 0.\overline{a_1 \ldots a_l}$, where $l \leq k$.

3. Let the reduced fraction m/n be given. If n has 2s and/or 5s as well as at least one other prime in its prime form, then $m/n = a.a_1 a_2 \ldots a_d \overline{a_{d+1} \ldots a_{d+k}}$, where the maximum number of 2s or 5s it has is d and the other primes in its prime form must divide $10^k - 1$.

Example 3.1.12

(a) Consider the fraction $17/40$. Notice that $40 = 2^3 5^1$, so we know that the decimal expansion of $17/40$ is terminating after three terms. And indeed it is: $17/40 = 0.425$.

(b) Consider the fraction $11/27$. Notice that $27 \mid 10^3 - 1$. I am not sure how you would notice this, but notice it anyway. So $11/27$ has a periodic expansion of period 3. Indeed it does: $11/27 = 0.\overline{407}$.

(c) Consider the fraction $23/28$. The denominator is $2^2 \times 7$. So we expect the decimal expansion to be mixed. The delay will be 2; the period will be 6. Using a calculator, we get the answer $0.82\overline{142857}$. Using a bit of algebra and no calculator, we can get the answer like this:

$$23/28 = 23/(7 \times 2^2) = (23 \times 5^2)/(7 \times 2^2 \times 5^2)$$
$$= (1/100) \times (575/7) = (1/100) \times (82 + (1/7)). \qquad \blacksquare$$

Let us turn our attention to fractions with denominators devoid of 2s and 5s. In the chart these are the fractions $1/n$, where $n = 3, 7, 9, 11, 13, 17,$ and 19. All of these fractions have periodic expansions. But what is the period of the expansion? Euler's theorem provides a partial answer.

Theorem 3.1.13 *If n is a number having no 2 or 5 in its prime form, then the proper fraction m/n has a periodic expansion of period k, where $k \mid \phi(n)$.*

Proof Since n is not divisible by 2 or 5, we know from Euler's theorem that $10^{\phi(n)} \equiv 1 \pmod{n}$. This means that $n \mid 10^{\phi(n)} - 1$. It follows from Theorem 3.1.8 (2) and Corollary 2.4.7 that m/n has an expansion of period k, where $k \mid \phi(n)$. $\qquad \square$

Corollary 3.1.14 *If p is a prime other than 2 or 5, the fraction m/p has a periodic expansion of period k, where $k \mid p - 1$.*

Example 3.1.15

(a) Consider $1/21$. Our chart does not cover this but the theorem helps. Since $\phi(21) = 6$, we know that our calculator can give us the answer: It is $0.\overline{047619}$.

(b) Consider $32/41$. Since 41 is prime, this fraction should be periodic with a period that divides 40. In fact, $23/41 = 0.\overline{56097}$; the period is 5.

(c) Consider $26/37$. Since 37 is prime, this fraction should be periodic with a period that divides 36. In fact, $26/37 = 0.\overline{702}$; the period is 3.

(d) The fraction $722/1313 = 0.\overline{549885757806}$. Since

$$\phi(1313) = \phi(101 \times 13) = \phi(101) \times \phi(13) = 100 \times 12 = 1200$$

the period must divide 1200. It does; the period is 12. ∎

We can refine our analysis of the period of a periodic expansion. We cannot improve on Corollary 3.1.14, but if we know the periods of the expansions of m/p and m/q for primes p and q, we can say something about the period of the expansion for m/pq. Let us look at examples.

Example 3.1.16

(a) The fraction $1/21 = 0.\overline{047619}$. It has a periodic expansion with period 6. Notice that $21 = 3 \times 7$ and that the periods of $m/3$ and $m/7$ are 1 and 6, respectively.

(b) The fraction $41/91 = 0.\overline{450549}$. This is also a periodic expansion with period 6. Notice $91 = 7 \times 13$ and that the periods of $m/7$ and $m/13$ are both 6.

(c) The fraction $288/407 = 0.\overline{707616}$. This is also a periodic expansion with period 6. Notice that $407 = 11 \times 37$ and that the periods of the expansions of $m/11$ and $m/37$ are 2 and 3, respectively.

(d) The fraction $722/1313 = 0.\overline{549885757806}$. This is periodic of period 12. We know $1313 = 13 \times 101$ and the periods of the expansions of $m/13$ and $m/101$ are 6 and 4, respectively. ∎

From these examples we may guess that the period of the expansion of m/pq is related to the periods of expansions of m/p and m/q. And it is.

Lemma 3.1.17 *Let k and n be natural numbers such that $k \mid n$. Then $(x^k - 1) \mid (x^n - 1)$.*

Proof Since $k \mid n$, it follows that $n = kt$ for some t. So the conclusion follows from the algebraic equality:

$$x^n - 1 = (x^k - 1)(x^{k(t-1)} + x^{k(t-2)} + \cdots + x^k + 1). \qquad \square$$

Corollary 3.1.18 *If $k \mid n$, then $10^k - 1 \mid 10^n - 1$.*

Theorem 3.1.19 *Let p and q be distinct primes. If m/pq has a periodic expansion of period l, then $l = \operatorname{lcm}(k, j)$, where k is the period of m/p and j is the period of m/q.*

Proof Since k is the period of the expansion of m/p, it follows that $p \mid 10^k - 1$. Similarly, $q \mid 10^j - 1$. Since $l = \operatorname{lcm}(k, j)$, we obtain $k \mid l$ and $j \mid l$. So, by Corollary 3.1.17, it follows that $10^k - 1 \mid 10^l - 1$ and $10^j - 1 \mid 10^l - 1$. Hence $pq \mid 10^l - 1$. Now l is the smallest such exponent for which $pq \mid 10^l - 1$ by the following reasoning. Suppose $pq \mid 10^{l'} - 1$. Now $l' = ks + r$ for some r, where $0 \le r < k$. So we have $10^{l'} \equiv 10^{ks}10^r \equiv 1 \cdot 10^r \pmod{p}$. But $p \mid 10^{l'} - 1$ so $10^{l'} \equiv 1 \pmod{p}$. It follows that $10^r \equiv 1 \pmod{p}$. Since k is the period and $r < k$, it follows that $r = 0$. Thus l' is a multiple of k. Similarly, l' is a multiple of j. So l' is a common multiple. Since l is the least common multiple of k and j, it is the smallest such exponent. $\quad\square$

At this point we have settled nearly every question that might arise concerning the decimal behavior of fractions. Here are some answers.

1. There are exactly three behaviors of decimal expansions of fractions.

2. Any decimal expansion exhibiting one of the three behaviors automatically represents a fraction.

3. If n is composed only of 2s and/or 5s, then m/n has a terminating expansion. If n is devoid of 2s and 5s, then m/n has a periodic representation. If n has a prime form containing primes both of the form 2 and/or 5 and other primes, then m/n has a mixed expansion.

4. While the length of a terminating expansion can be found knowing the prime form of the denominator, the period of a periodic expansion cannot be pinned down. What can be said is that if the expansion of m/n is periodic, then it has period k, where $k \mid \phi(n)$.

5. There is a formula for the period of fractions m/n that works for most primes p, where $n = p^k$. Unfortunately, it fails for $p = 3$ and another prime $p < 1000$. We leave it as an exercise to find the formula and the prime for which it fails.

EXERCISES

1. Find the reduced fractions corresponding to the following decimal expansions:

   (a) $0.45\overline{425}$

   (b) $0.3\overline{780219}$

   (c) $0.2556\overline{2330}$

2. Find the unit fraction with the smallest denominator d such that

   (a) $1/d$ is mixed with delay of 3 and repeat of 3

   (b) $1/d$ terminates after 6 places

   (c) $1/d$ repeats with period 5

3. Find all denominators d such that m/d has a terminating decimal representation and its length is

   (a) 1

   (b) 2

   (c) 3

(d) 4

4. Find all the denominators d such that m/d has a periodic decimal expansion and its period is

(a) 1

(b) 2

(c) 3

(d) 4

5. Find all the denominators, n, such that m/n is of the form:

(a) $0.ab\bar{c}$

(b) $0.a\overline{bc}$

(c) $0.ab\overline{cde}$

(d) $0.\overline{abcdefgh}$

6. Find the periods of the following reduced proper fractions. Explain why your answer is true.

(a) $m/707$

(b) $m/119$

(c) $m/1919$

7. Prove that if m/p has period k, where p is a prime, then p is of the form $nk + 1$ for some number n.

8. Prove Theorem 3.1.8 (3). The fraction m/n has a mixed decimal expansion if and only if $m/n = N/(10^k - 1)(10^l)$. The delay in the expansion is the least l for which this is true; the period is the least k for which this is true.

9. Prove Theorem 3.1.9.

(a) If $m/n = 0.a_1 a_2 \ldots a_k$, then the prime form of n is composed entirely of 2s and/or 5s. Furthermore, $n = 2^j 5^l$, where $k = \max(j, l)$.

(b) If $m/n = 0.\overline{a_1 a_2 \ldots a_k}$ then the prime form of n is composed only of primes that factor $10^k - 1$.

(c) If $m/n = a.a_1 a_2 \ldots a_d \overline{a_{d+1} \ldots a_{d+k}}$, then n has 2s and/or 5s as well as at least one other prime in its prime form. The maximum number of 2s or 5s in n is d; the other primes in its prime form must divide $10^k - 1$.

10. Prove Theorem 3.1.11.

(a) Let the reduced fraction m/n be given. If the prime form of n is composed entirely of 2s and/or 5s, then $m/n = 0.a_1 a_2 \ldots a_k$, where $n = 2^j 5^l$ and $k = \max(j, l)$.

(b) Let the reduced fraction m/n be given. If the prime form of n is composed only of primes that factor $10^k - 1$, then $m/n = 0.\overline{a_1 \ldots a_l}$, where $l \le k$.

(c) Let the reduced fraction m/n be given. If n has 2s and/or 5s as well as at least one other prime in its prime form, then $m/n = a.a_1 a_2 \ldots a_d \overline{a_{d+1} \ldots a_{d+k}}$, where the maximum number of 2s or 5s it has is d and the other primes in its prime form must divide $10^k - 1$.

11. A **repunit** number is a natural number whose digits are all 1s. The **order** of the repunit number is the number of 1s. Prove: If p is a prime other than 2, 3, or 5, then p divides a repunit number of order n if and only if the decimal expansion of m/p has period k, where $k \mid n$.

12. Factor the following repunit numbers; then list all the primes p such that m/p has the periodic expansion of the order of the repunit number.

(a) 11111 (five ones)

(b) 11111111 (eight ones)

(c) 1111111111 (ten ones)

(d) 111111111111 (twelve ones)

(e) 111111111 (nine ones)

(f) 1111111 (seven ones)

(g) Test the truth of the following statement: Let p be a prime other than 3 and suppose that $p \mid 11 \ldots 1$ (n ones). Suppose further that n is the least such number of ones for which this is true. Then $p \equiv 1 \pmod{n}$.

13. Explore the periods of m/p^k, where p is a prime; in particular, find the periods of the following:

    (a) $m/3$

    (b) $m/9$

    (c) $m/27$

    (d) $m/81$

    (e) $m/243$

    (f) Finish the thought: The period of $(m/3)^n$ is . . ..

14. If you have a calculator or computer capability of finding decimal expansions with long periods, continue your exploration. Find the periods of

    (a) $m/121$ and $m/1331$

    (b) $m/49$ and $m/343$

    (c) $m/169$ and $m/2197$

    (d) If p is prime and $p \neq 3$, then the period of $(m/p)^n$ is . . ..

15. You may have found from the previous two exercises that there is a formula for the period of fractions m/n that works for most primes p, where $n = p^k$. Unfortunately, it fails for $p = 3$. There is another prime $p < 1000$ for which it also fails. Find that prime.

16. Examine the behavior of the following reduced fractions; that is, tell whether they are terminating, periodic, or mixed, and give the details of the length, the period, and the delay.

    (a) $m/333$

    (b) $m/4444$

    (c) $m/55555$

    (d) $m/666666$

    (e) $m/7777777$

    (f) $m/88888888$

    (g) $m/999999999$

17. A natural number $N = a_n a_{n-1} \ldots a_3 a_2 a_1$ is **cyclic** if for all $k < n$ the number kN consists of the very same digits as N in a "cyclic" permutation; that is, $kN = a_j \ldots a_2 a_1 a_n \ldots a_{j+1}$, for some $j < n$. Show that the following numbers are cyclic:

   (a) The integer part of $10^6(1/7)$; denoted by $[10^6(1/7)]$

   (b) The integer $[10^{16}(1/17)]$

   (c) The integer $[10^{18}(1/19)]$

   (d) What can you say about $[10^6(1/13)]$? Is it cyclic? How about $[10^6(2/13)]$? Try to make a definition of a semi-cyclic number that fits the fractions $1/13$ and $2/13$. Find other examples of semicyclic numbers.

18. Examine the digits that make up the decimal expansions of periodic fractions m/p.

   (a) What can you say about the sum of the digits of one period of the expansion of m/p?

   (b) If m/p has an even period, $2k$, can you say anything about the relationship between the first k digits of the expansion and the last k digits of the expansion?

19. From Fermat's little theorem, (Section 2.4), we know that $a^{p-1} \equiv 1 \pmod{p}$ if a is not a multiple of p. If $p - 1$ is the first power of a for which this congruence is true, then a is called a **primitive root** of p. So, for example, 10 (or 3) is a primitive root of 7 and 10 is not a primitive root of 13.

   (a) Explain how this fact translates into the fact that $m/7$ has period 6 when written in base 10 notation; $m/13$ does not have period 12 in base 10 notation.

   (b) Find all the primes less than 50 for which 10 is a primitive root of p.

3.2   Decimals as Rational Numbers

In Section 3.1 we studied the decimal expansions of rational numbers. Here we do the reverse; we examine the rational equivalents of decimal expansions. Unfortunately, this is not always possible. As we learned from 3.1, there are irrational numbers; they are the real numbers whose decimal expansions neither terminate nor eventually repeat. So our task here and in the next section will be to find fractions that get close to a real number. Consider, for example, $\sqrt{2}$. As you surely know, it is irrational; you may have seen a proof of this. On a calculator $\sqrt{2}$ is represented by 1.414213562. This is the fraction 1414213562/1000000000 which reduces to 707106781/500000000. Of course, the decimal expansion for $\sqrt{2}$ goes beyond what the calculator shows and it does not have a repeating pattern, or, for that matter, any pattern that we can capture.

Theoretically, there are rational numbers with denominators of the form 10^n that can approximate $\sqrt{2}$ to any accuracy you like. But in practice, we cannot go much beyond 12 place accuracy with fractions of the form $k/10^n$. In our quest to express real numbers with approximating rationals, we will find fractions with denominators much smaller than those of the form 10^n. In fact, we will find that certain fractions seem to be tailor-made for fitting particular decimals. This is true of $\sqrt{2}$.

Example 3.2.1

The decimal expansion of $\sqrt{2}$ begins 1.414213562. So $14/10 = 7/5$ agrees with $\sqrt{2}$ in one place, 141/100 agrees in two places, $1414/1000 = 707/500$ agrees in three places, and we may continue the approximations with fractions of the form $k/10^n$ all the way up to $n = 9$. As promised, we can find fractions with smaller denominators that do even an better job of approximating this decimal. Let us look at fractions that are close to $\sqrt{2}$ with denominators $2, 3, \ldots 10$. The closest fractions with ascending denominators are $3/2, 4/3, 6/4 = 3/2, 7/5, 8/6 = 4/3, 10/7, 11/8, 13/9, 14/10 = 7/5$. They differ from $\sqrt{2}$ by approximately

$$
\begin{aligned}
|3/2 - \sqrt{2}| &\approx .086 \\
|4/3 - \sqrt{2}| &\approx .081 \\
|7/5 - \sqrt{2}| &\approx .0142 \\
|10/7 - \sqrt{2}| &\approx .0144 \\
|11/8 - \sqrt{2}| &\approx .04 \\
|13/9 - \sqrt{2}| &\approx .03.
\end{aligned}
$$

Of these 7/5 is the closest approximation. Even a denominator of 11 does not offer an improvement on 7/5; $|16/11 - \sqrt{2}| \approx .04$. The next improvement comes with denominator 12; $|17/12 - \sqrt{2}| \approx .002$.

In fact, 17/12 is a better approximation to $\sqrt{2}$ than 141/100 and rivals 1414/1000. ∎

Let us discuss what a "good approximation" should mean. In general terms, the larger the denominator, the more accurate the fraction can be. To understand this, think of the various fractions, k/n, as points along the number line. They partition the line into segments of length $1/n$. Now the decimal falls within one of these segments, and the furthest it could be from an endpoint would be $1/2n$. So, in order to gauge the quality of an approximation, k/n, to a decimal r, we should relate the closeness of the approximation, d, with the size of the interval. That is, we should calculate the ratio of d to $1/n$, where $|k/n - r| = d$. Thus we shall be interested in $d/(1/n) = dn = |k - nr|$.

Definition 3.2.2 *Let r be a given real number and suppose that s/t and u/v are approximations of r. Then s/t is a **better approximation** than u/v if $|s - tr| < |u - vr|$.*

Example 3.2.3

Let us revisit our approximations of $\sqrt{2}$.

$$|3 - 2\sqrt{2}| \approx .172$$
$$|4 - 3\sqrt{2}| \approx .243$$
$$|7 - 5\sqrt{2}| \approx .071$$
$$|10 - 7\sqrt{2}| \approx .1$$
$$|11 - 8\sqrt{2}| \approx .314$$
$$|13 - 9\sqrt{2}| \approx .272$$
$$|16 - 11\sqrt{2}| \approx .444$$
$$|17 - 12\sqrt{2}| \approx .029$$

So the ordering of our approximations from best to worst is

$$17/12, 7/5, 10/7, 3/2, 4/3, 13/9, 11/8, 16/11.$$ ∎

Let us revisit our approximations using an organized technique.

Example 3.2.4

(a) Here is one more look at the approximation of $\sqrt{2}$. This time we shall list the multiples of $\sqrt{2}$ and moniter how close they are to a whole number.

$$n\sqrt{2} \;=\; n(1.414\ldots) \;=\; [n(1.414\ldots] + e_n \qquad d_n \qquad k/n$$

$n\sqrt{2}$		$n(1.414\ldots)$		$[n(1.414\ldots]+e_n$	d_n	k/n
$\sqrt{2}$	=	$1.414\ldots$	=	$1 + 0.414\ldots$	0.414	$1/1$
$2\sqrt{2}$	=	$2.828\ldots$	=	$2 + 0.828\ldots$	0.172	$3/2$
$3\sqrt{2}$	=	$4.242\ldots$	=	$4 + 0.242\ldots$	0.243	$4/3$
$4\sqrt{2}$	=	$5.656\ldots$	=	$5 + 0.656\ldots$	0.343	$6/4$
$5\sqrt{2}$	=	$7.071\ldots$	=	$7 + 0.071\ldots$	0.071	$7/5$
$6\sqrt{2}$	=	$8.485\ldots$	=	$8 + 0.485\ldots$	0.485	$8/6$
$7\sqrt{2}$	=	$9.899\ldots$	=	$9 + 0.899\ldots$	0.1	$10/7$
$8\sqrt{2}$	=	$11.313\ldots$	=	$11 + 0.313\ldots$	0.314	$11/8$
$9\sqrt{2}$	=	$12.727\ldots$	=	$12 + 0.727\ldots$	0.272	$13/9$
$10\sqrt{2}$	=	$14.142\ldots$	=	$14 + 0.142\ldots$	0.142	$14/10$
$11\sqrt{2}$	=	$15.556\ldots$	=	$15 + 0.556\ldots$	0.444	$16/11$
$12\sqrt{2}$	=	$16.970\ldots$	=	$16 + 0.970\ldots$	0.029	$17/12$

Here the brackets mean the "greatest integer less than," e_n is the decimal part left over, d_n is the rounded off difference between $[n\sqrt{2}]$ and its nearest integer neighbor, and k is the nearest integer neighbor.

(b) Here is a look at the number π.

$$n\pi \;=\; n(3.141\ldots) \;=\; [n(3.141\ldots] + e_n \qquad d_n \qquad k/n$$

$n\pi$		$n(3.141\ldots)$		$[n(3.141\ldots]+e_n$	d_n	k/n
π	=	$3.141\ldots$	=	$3 + 0.141\ldots$	0.142	$3/1$
2π	=	$6.283\ldots$	=	$6 + 0.283\ldots$	0.283	$6/2$
3π	=	$9.424\ldots$	=	$9 + 0.424\ldots$	0.425	$9/3$
4π	=	$12.566\ldots$	=	$12 + 0.566\ldots$	0.434	$13/4$
5π	=	$15.707\ldots$	=	$15 + 0.707\ldots$	0.292	$16/5$
6π	=	$18.849\ldots$	=	$18 + 0.849\ldots$	0.15	$19/6$
7π	=	$21.991\ldots$	=	$21 + 0.991\ldots$	0.009	$22/7$
8π	=	$25.132\ldots$	=	$25 + 0.132\ldots$	0.133	$25/8$
9π	=	$28.274\ldots$	=	$28 + 0.274\ldots$	0.274	$28/9$
10π	=	$31.415\ldots$	=	$31 + 0.415\ldots$	0.416	$31/10$
11π	=	$34.557\ldots$	=	$34 + 0.557\ldots$	0.442	$35/11$
12π	=	$37.699\ldots$	=	$37 + 0.699\ldots$	0.301	$38/12$

So the ordering of these approximations (using only reduced fractions) from best to worst is

$$22/7, 25/8, 3/1, 19/6, 28/9, 16/5, 31/10, 13/4, 35/11. \qquad \blacksquare$$

This technique for finding good approximations serves as a basis for the proof of a theorem that gives is a handle on approximations in general.

Theorem 3.2.5 *Suppose that a natural number N and a real number r are given. There exists a rational number k/n, where $n < N$ such that $|k - nr| \leq 1/N$.*

Proof Let N and r be given. Consider the sequence of numbers r, $2r$, ..., $(N-1)r$. Let $e_n = nr - [nr]$ for $1 \leq n < N$. We have four cases:

(1) If $e_n = 0$ for some n, then $[nr] - nr = 0 < 1/N$.

(2) If $e_j = e_k$ for $j < k$, then we have $([kr] - [jr]) - (k-j)r = 0 < 1/N$ and $(k-j) < N$.

If all $e_n > 0$ are distinct, we can arrange them in order of magnitude along the open interval between 0 and 1. Here is that ordering: $0 < e_{n_1} < e_{n_2} < \ldots < e_{N-1} < 1$. These numbers partition the interval into N intervals.

(3) If all the intervals were of length $1/N$, then, for some $i < N$, $e_i = 1/N$ so $ir - [ir] = 1/N$.

(4) If the intervals are of different length, then at least one of the intervals, call it the interval from e_a to e_b, is less than $1/N$ in length. Let $a < b$. Since

$$|e_a - e_b| = |ar - [ar] - (br - [br])| = |([br] - [ar]) - (b-a)r| < 1/N$$

and $|b - a| < N$ we have our result. If $a > b$ we still have our result because

$$|([br] - [ar]) - (b-a)r| = |([ar] - [br]) - (a-b)r|.   \qquad \square$$

Corollary 3.2.6 *Given a real number r, there are infinitely many fractions k/n such that $|k/n - r| < 1/n^2$.*

Let us see if we can find these special fractions that the corollary guarantees. We give them a name so we can refer to them in the discussion that follows.

Definition 3.2.7 *Let r be a real number and q be a rational number, where $q = k/n$. Then q is an r-**approximant** if $|q - r| < 1/n^2$.*

Example 3.2.8

(a) We know that $7/5$ and $17/12$ are $\sqrt{2}$-approximants. In fact, not only is $|7/5 - \sqrt{2}| < 1/5^2 = 1/25$, but $|7/5 - \sqrt{2}| \approx 0.0142 < 1/70$. Also, not only is $|17/12 - \sqrt{2}| < 1/12^2 = 1/144$, but $|17/12 - \sqrt{2}| < 1/407$.

(b) We know that $22/7$ is a π-approximant because $|22/7 - \pi| \approx 0.00126 < 1/7^2$. In fact, $|22/7 - \pi| < 1/791$.   ■

Let us search for more r-approximants. Keep in mind that our goal is to find the fraction with smallest denominator that lies within a small region about a given real number. Putting it another way; given $\epsilon > 0$, find the smallest n such that $|k/n - r| < \epsilon$.

Example 3.2.9

(a) Let $r = \sqrt{2}$ and $\epsilon = 0.005$. We are looking for a $\sqrt{2}$-approximant that is closer than ϵ from $\sqrt{2}$. Since $17/12$ is a $\sqrt{2}$-approximant, it is within $1/144$ of $\sqrt{2}$ but $1/144 > 0.005$. Actually, $|17/12 - \sqrt{2}| < 1/407$ and $1/407 < 0.005$ so $17/12$ fills the bill. Incidentally, the approximant $7/5$ does not work because $|7/5 - \sqrt{2}| > 0.005$.

(b) Let $r = \sqrt{2}$ and $\epsilon = 0.001$. Now the approximant $17/12$ does not work because $|17/12 - \sqrt{2}| > 1/408 > 0.001$. So where shall we search for a closer next approximant? Theorem 3.2.5 gives us a clue. It says that, given N, there exists $n < N$ such that $|17/12 - \sqrt{2}| < 1/nN$. Substituting 407 for nN and 12 for n, we find $N \approx 33.97$. So, if $N = 33$, then $n = 12$ satisfies the theorem; if $N = 34$, then $n = 12$ does not. Thus we should look for denominators ≤ 33 for a closer $\sqrt{2}$-approximant. Actually, we already know an answer to this. It comes from our discussion of Pell equations in Section 2.3, specifically Example 2.3.5. A closer $\sqrt{2}$-approximant is $41/29$; $|41/29 - \sqrt{2}| \approx 0.0004$.

(c) Let $r = \pi$ and $\epsilon = 0.001$. We know that $22/7$ is a π-approximant because $|22/7 - \pi| < 1/49$. In fact, $1/792 < |22/7 - \pi| < 1/791$. Using Theorem 3.2.5 as in part (b), we find that $|22/7 - \pi| < 1/(7 \cdot 113)$. So if $N = 113$, $n = 7$ satisfies the theorem; if $N = 114$, $n = 7$ does not. Thus it makes sense to look for denominators ≤ 113 for a closer π-approximant. In fact, it turns out that we should look at 113 because $|355/113 - \pi| \approx 0.00000027$.   ■

In Example 3.2.9 (b) we recalled solutions to Pell's equation to find a $\sqrt{2}$-approximant. In fact, solutions to Pell equations do yield $\sqrt{k}$-approximants.

Theorem 3.2.10 *If x, y is a solution to $|x^2 - ky^2| = 1$, then x/y is a $\sqrt{k}$-approximant.*

Proof Suppose that $|x^2 - ky^2| = 1$. Then $|x^2/y^2 - k| = 1/y^2$. Factoring gives us

$$|x^2/y^2 - k| = |x/y + \sqrt{k}||x/y - \sqrt{k}| < |x/y - \sqrt{k}| = 1/y^2. \quad \square$$

Theorem 2.3.6 shows a way of generating infinitely many $\sqrt{2}$-approximants. Using the recurrence relation in that theorem, we find that another approximant after $41/29$ is $(41+2(29))/(41+29) = 99/70$. Notice that $|99/70 - \sqrt{2}| \approx 0.00007 < 1/13859 < 1/4900$.

In the next section we develop a way of finding r-approximants generally but first we shall take a fun detour into a related topic, Farey sequences.

Since our goal in this section is to seek fractions with the smallest possible denominator to approximate a real number, it makes sense to examine the nature of fractions with small denominators. These fractions are not only of interest in this context but they are a fascinating study in their own right. We temporarily restrict our study to reduced proper fractions; that is, fractions that lie on the number line between 0 and 1 that are reduced to lowest terms. Let us begin with the fraction of smallest denominator, denominator 2. That would be $1/2$. Moving to denominator 3, there are two such fractions: $1/3$ and $2/3$. With denominator 4 there is $1/4$ and $3/4$; $2/4$ has already been accounted for since $2/4 = 1/2$. If we set a limit on the size of the denominator, we can list all of the fractions with denominators up to that size, and we can list them in order of size. Here is what we mean with denominators ≤ 7. The numbers are

$$\frac{1}{7}, \frac{1}{6}, \frac{1}{5}, \frac{1}{4}, \frac{2}{7}, \frac{1}{3}, \frac{2}{5}, \frac{3}{7}, \frac{1}{2}, \frac{4}{7}, \frac{3}{5}, \frac{2}{3}, \frac{5}{7}, \frac{3}{4}, \frac{4}{5}, \frac{5}{6}, \frac{6}{7}.$$

The series of numbers listed here is an example of a Farey sequence. It is named after John Farey, a sometime mathematician, who, in 1816, wrote about some interesting properties of this sequence. The Farey sequence is sometimes called the Farey series.

Definition 3.2.11 *The* **Farey sequence** *or* **Farey series** *of order n is an ascending sequence of reduced fractions r/s, where $0 < r < s \leq n$. We denote this sequence by F_n.*

Let us examine the nature of F_n. Looking at F_7 may give us some clues; for example, consider the successive terms

$$1/3, \ 2/5, \ 3/7, \ 1/2, \ 4/7.$$

Notice that the sums of two successive terms > 7, the denominators of successive terms are different, the difference between successive terms is small, and, for any three successive terms, the mediant of the outer terms is the middle term.

Definition 3.2.12 *The **mediant** of a/b and c/d is $(a + c)/(b + d)$.*

Notation: We denote the mediant of a/b and c/d as $a/b \oplus c/d$.

Note that the mediant is defined on any pair of rationals, not just those in a Farey sequence.

While this is a very limited sample, here is the general formulation of these observations for F_n. It turns out that all of them are true.

1. The sum of denominators of successive terms always exceeds n.

2. The denominators of two successive terms are different.

3. The difference between successive terms r/s and t/u is $1/su$.

4. If r/s, t/u, and v/w are three successive terms, then $t/u = r/s \oplus v/w$.

We prove two of these and leave the other two for the student.

Theorem 3.2.13

1. *If r/s and t/u are successive terms in F_n, then $s + u > n$.*

2. *If $n > 1$, then no two successive terms of F_n have the same denominator.*

The proof is left as an exercise.

Theorem 3.2.14 *If r/s and t/u are successive terms in F_n, then $st - ru = 1$.*

Proof Given r/s let us examine what its successor, t/u, must look like. Since r/s is in lowest terms and hence $\gcd(r, s) = 1$ we may use our results from linear Diophantine equations to obtain an x and y such that $sx - ry = 1$. Labeling one such solution x_0, y_0, the other solutions are of the form $x = x_0 + kr$, $y = y_0 + ks$ for integer k.

Because $s < n$, we can find a solution where $n - s < y \le n$. Letting y_1 be that solution and x_1 be its partner, we have found our choice of the successor of r/s; $t = x_1$, $u = y_1$. Notice that $st - ru = 1$.

We still must check to see that t/u, as we have defined it previously, really is the successor of r/s. Suppose that there were a fraction v/w such that $r/s < v/w < t/u$. We have

$$v/w - r/s = (vs - wr)/(sw) \ge 1/(sw)$$

and

$$t/u - v/w = (tw - uv)/(uw) \ge 1/(uw).$$

It follows that

$$1/(su) = t/u - r/s = (t/u - v/w) + (v/w - r/s) \ge 1/(uw) + 1/(sw)$$
$$= (u + s)/(suw) > n/(suw) \ge 1/(su).$$

Now $(u + s)/(suw) > n/(suw)$ because $s + u > n$, and $n/(suw) \ge 1/(su)$ because $w \le n$. Thus we have that $1/(su) > 1/(su)$, a contradiction. So we conclude that there is no such v/w. $\square$

Theorem 3.2.15 *If $r/s < t/u$ and $st - ru = 1$, then there is no fraction v/w between them such that $w < s + u$.*

Proof Suppose that $r/s < v/w < t/u$, and $st - ru = 1$. We know that either $v/w < (r + t)/(s + u)$ or $v/w > (r + t)/(s + u)$. Suppose the former. Then the distance from v/w to r/s is closer than the distance from $(r + t)/(s + u)$ to r/s; that is, $(vs - wr)/(ws) < [s(r + t) - r(s + u)]/(s(s + u)) = 1/(s(s + u))$. Since $(vs - wr) \ge 1$, we conclude that $w \ge s + u$. The latter case is proved similarly. $\square$

It follows immediately from theorems 3.2.14 and 3.2.15 that the mediant is the rational of smallest denominator that lies between two successive Farey fractions.

Corollary 3.2.16 *If f_1 and f_2 are successive members of F_n for some n, then $f_1 \oplus f_2$ is the next fraction of smallest denominator that will fit in between.*

Let us see how Farey sequences and r-approximants fit together. Since we interested in fractions with small denominators, we shall use the corollary to search for r-approximants. First we should note that while Farey sequences are sequences made up of proper fractions, we can carry out our analysis away from the interval of numbers between 0 and 1. The following theorem states this.

Theorem 3.2.17 *Let $a/b = n + r/s$ and $c/d = n + t/u$, where r/s and t/u are proper fractions. Then $a/b \oplus c/d = n + (r/s \oplus t/u)$.*

The proof is left as an exercise.

Example 3.2.18

Let us search for $\sqrt{2}$-approximants again. We know that $7/5$ and $17/12$ are approximants. We can also find that $\sqrt{2}$ lies between these two fractions. By the corollary, the fraction with smallest denominator that lies in the interval between $7/5$ and $17/12$ is $7/5 \oplus 17/12 = 24/17$. This is, indeed, a $\sqrt{2}$-approximant because $|24/17 - \sqrt{2}| < 1/17^2$. Proceeding in this manner, we note that $\sqrt{2}$ is between $24/17$ and $17/12$, so, we take the next mediant and get $41/29$ a $\sqrt{2}$-approximant we are familiar with. Now $\sqrt{2}$ is between $41/29$ and $17/12$ so we obtain the mediant $58/41$. It, too, is a $\sqrt{2}$-approximant because $|58/41 - \sqrt{2}| < 1/41^2$. Finally, from $41/29$ and $58/41$ we get the mediant $99/70$, another $\sqrt{2}$-approximant we have encountered.   ∎

We have seen from this example that the mediants of $\sqrt{2}$-approximants are, themselves, $\sqrt{2}$-approximants. Notice that we have not stated a general theorem to this effect; that is that the mediant of two r-approximants is an r-approximant. We examine this proposition in the exercises.

As we search for r-approximants, q, we notice that $q = [q] + k/m$ where k/m is in a Farey sequence. So in order to assess the magnitude of our task in finding r-approximants in F_n, we should count how many members F_n has. It goes without saying that as n increases, the number of members of F_n gets larger. There are 17 fractions in F_7. For F_8, we would have to add $1/8$, $3/8$, $5/8$, and $7/8$ to the mix. This would give us 21. And F_9 would include the new fractions $1/9$, $2/9$, $4/9$, $5/9$, $7/9$, and $8/9$. This makes 27 fractions in all. Clearly the new members in F_{n+1} are those reduced fractions with denominator $n+1$. The number of those fraction is the number of numbers less than and relatively prime to $n+1$. We recognize this as the Euler phi function $\phi(n+1)$ from Definition 2.4.3.

Theorem 3.2.19 *The number of members of F_n is the sum $\phi(2) +$ $\phi(3) + \phi(4) + \cdots + \phi(n) = \sum_{k=2}^{n} \phi(k)$.*

The proof follows from our preceding observations.

Now this sum is not all that easy to calculate. The chart helps us out.

Farey Chart

n	$\phi(n)$	$\sum \phi(k)$	$3n^2/\pi^2$	$(3n^2/\pi^2)/(\sum \phi(k))$
2	1	1		
3	2	3		
4	2	5		
5	4	9		
6	2	11		
7	6	17		
8	4	21		
9	6	27		
10	4	31	30.40	0.981
15	8	71	68.39	0.963
20	8	127	121.59	0.957
25	20	199	190.00	0.948
50	20	773	759.91	0.983
100	40	3043	3039.64	0.999
200	80	12231	12158.54	0.994
300	80	27397	27356.72	0.998
400	160	48677	48634.17	0.999
500	200	76115	75990.89	0.998

Notice that there is a formula for approximating this sum: $3n^2/\pi^2$ gives an approximation of the number of members in F_n. The approximation gets better as n increases in the sense $(3n^2/\pi^2)/(\sum \phi(n))$ gets closer and closer to 1 as n gets larger and larger. An advanced number theory course will provide a proof of this. Since the number of members of F_n is proportional to n^2, sifting through these to find a good r-approximant would be a challenge without computer help or without a mathematician's help.

While the decimal representation for numbers is preferable for expressing numbers in general, we have made a case for using Farey sequences along the number line for locating real numbers. Let's compare the two approaches with $\sqrt{2}$ once more. Suppose we wish to locate fractions that agree with $\sqrt{2}$ on two decimal places. The decimal representations supply 99 benchmarks beginning with 1.01 and ending with 1.99. A comparable, but fewer, number of Farey fractions are in F_{17} which has 95 members. The closest decimal approximation is 1.41. There are several $\sqrt{2}$-approximants in F_{17}; they are $3/2, 4/3, 7/5, 10/7, 17/12$, and $24/17$. The best two $\sqrt{2}$-approximants are $17/12$ and $24/17$; both differ from $\sqrt{2}$ by about 0.0025. This is almost twice as good an approximation as 1.41, which differs from $\sqrt{2}$ by about .0042. The effectiveness of the Farey sequence is much more striking if greater accuracy is wanted. For example, there are 9999 decimal fractions between 1 and 2 that are separated by 0.0001. There are fewer than one-third as many Farey fractions in F_{100}, yet the $\sqrt{2}$-approximant from F_{100} that best approximates $\sqrt{2}$ is almost three times as good an approximation as the comparable decimal fraction. You can check it out: $|1.41421\ldots - \sqrt{2}| \approx 0.00021$ while $|99/70 - \sqrt{2}| \approx 0.00007$.

We conclude this section with one of many interesting problems that can be approached with Farey sequences. Other problems are given in the exercises.

Example 3.2.20

Suppose we randomly select two natural numbers p and q from the infinity of natural numbers in the universe. What is the probability these two numbers are relatively prime? Would you guess that it is $1/2$, greater than $1/2$, or less than $1/2$? Before reading on you should think about this a while. Think of it this way: If you and a friend each pick a number, what are the chances that the numbers share at least one prime in their decompositions?

We may proceed as follows: If the two numbers are $\leq n$, then we are dealing with n^2 possible pairs of numbers. If we look at only those pairs where the first number is less than the second, we are looking at approximately $n^2/2$ numbers; $n(n-1)/2$ to be exact. Letting $p < q$, we know that the fraction p/q is in F_n if p and q are relatively prime. And we know that approximately $3n^2/\pi^2$ pairs are relatively prime. So out of the approximately $n^2/2$ pairs, approximately $3n^2/\pi^2$ are relatively prime. Thus the probability that two numbers are relatively prime, if they are both are $\leq n$, is approximately $(3n^2/\pi^2)/(n^2/2) = 6/\pi^2$. As n grows larger, this approximation gets relatively more and more accurate. So this is the number that the approximations are tending toward. Since $6/\pi^2 \approx$

0.6079, the chances are $> 1/2$ that the numbers are relatively prime. Putting it another way, the chances that the two numbers that you and your friend chose share at least one prime are less than 50%, in fact; they are $< 40\%$. ∎

EXERCISES

1. Find all the r-approximants, k/n, for $n \leq 25$ and pick out the best one where $r =$

   (a) $2344/733$
   (b) $25567/35557$
   (c) 0.8642
   (d) 1.23456

2. Find all the r-approximants, k/n, for $n \leq 25$ and pick out the best one where $r =$

   (a) $\sqrt{3}$
   (b) $\sqrt{19}$
   (c) $\sqrt[3]{2}$
   (d) $\sqrt{\pi}$

3. Prove Corollary 3.2.6: Given a real number r there are infinitely many fractions k/n such that $|k/n - r| < 1/n^2$.

4. In Example 3.2.8 we saw that our best $\sqrt{2}$-approximants k/n not only are within $1/n^2$ accuracy of $\sqrt{2}$ but they are within $1/K$ as well, where K is the largest such denominator. What can you say about the ratio of n^2 to K? Cite several of our best $\sqrt{2}$-approximants to make your point.

5. Prove or disprove the converse of Theorem 3.2.10: If x/y is a $\sqrt{k}$-approximant, then x, y is a solution to the Pell equation $|x^2 - ky^2| = 1$.

6. Prove that if $|x^2 - 2y^2| = 1$, then $|x/y - \sqrt{2}| < 1/2y^2$.

7. See if you can find a number k such that $|p/q - \sqrt{2}| > 1/kq^2$ for all natural numbers q.

8. Prove Theorem 3.3.13.

   (a) If r/s and t/u are successive terms in F_n, then $s + u > n$.

   (b) If $n > 1$, then no two successive terms of F_n have the same denominator.

9. Prove or disprove this converse of Theorem 3.2.15: If $r/s < t/u$ and there is no fraction, v/w, between them such that $w < s+u$, then $st - ru = 1$.

10. Prove Theorem 3.2.17: Let $a/b = n + r/s$ and $c/d = n + t/u$, where r/s and t/u are proper fractions. Then $a/b \oplus c/d = n + (r/s \oplus t/u)$.

11. Prove or disprove:

   (a) The mediants of $\sqrt{2}$-approximants are, themselves, $\sqrt{2}$-approximants.

   (b) The mediants of r-approximants are, themselves, r-approximants.

12. In the following Farey sequences find the immediate predecessor and the immediate successor to the given fraction.

   (a) $17/24$ in F_{25}

   (b) $17/24$ in F_{57}

   (c) $61/79$ in F_{100}

13. This exercise carries on the comparison of Farey sequence accuracy to decimal accuracy. We are looking for fractions that are closer to $\sqrt{2}$ than 0.001.

   (a) Find the n such that F_n has the most members fewer than 999.

   (b) For the F_n of part (a) find the member of that Farey sequence that is closest to $\sqrt{2}$.

   (c) How does the accuracy of your answer in part (b) compare to 0.001.

   (d) Find the smallest n such that F_n contains a fraction that is closer to $\sqrt{2}$ than 0.001.

(e) How many members are there in your answer to (d) as compared to 999?

(f) Compare your answers in (c) and (e).

14. This exercise carries further the comparison of Farey sequence accuracy to decimal accuracy. We are looking for fractions that are closer to $\sqrt{2}$ than 0.00001.

   (a) Find the n such that F_n has the most members fewer than 99999.

   (b) For the F_n of part (a) find the member of that Farey sequence that is closest to $\sqrt{2}$.

   (c) How does the accuracy of your answer in part (b) compare to 0.00001.

   (d) Find the smallest n such that F_n contains a fraction that is closer to $\sqrt{2}$ than 0.00001.

   (e) How many members are there in your answer to (d) as compared to 99999?

   (f) Compare your answers in (c) and (e).

15. Carry out an accuracy study comparing the fractions F_{17} with the fractions $1/100, \ldots 99/100$. Use the following real numbers (these are the numbers from exercises 1 and 2) for your study: $2344/733$, $25567/35557$, 0.8642, 1.23456, $\sqrt{3}$, $\sqrt{19}$, $\sqrt[3]{2}$, $\sqrt{\pi}$.

   (a) Find the fraction $q = [q] + k/n$ where k/n is in F_{17} that best approximates each of the preceding numbers.

   (b) Compare your answers in (a) to the accuracy of the fraction $r = [r] + l/100$ that best approximates each of the preceding numbers. What do you find?

16. Recall the plane of integer lattice points from Section 2.3. Suppose that I live at the origin $(0,0)$. As I look out into the universe I see points of the integer lattice in all directions. I see the point $(1,1)$, for example, but I do not see $(2,2)$, $(3,3)$, or (n,n) for $n > 1$ because these are hidden from my view by $(1,1)$. If the universe contains points from the integral lattice such that

(a) $-1000 \le x \le 1000$, $-1000 \le y \le 1000$, how many points are there? How many points are visible from $(0,0)$? What is the ratio of visible points to all points?

(b) $x^2 + y^2 \le 100^2$, approximate (an exact count would be nice but it is difficult), how many points are there? Approximate how many are visible from $(0,0)$. What is the approximate ratio of visible points to all points?

17. Recall, again, the plane of integer lattice points from Section 2.3. We are interested in doing some basic geometry in this plane; for example, finding areas of polygons and finding distances between points and lines. The following theorem is helpful in this regard. It is called Pick's theorem.

Theorem 3.2.21 Pick's Theorem   *In the plane of lattice points the area of a polygon is $A = \frac{1}{2}b + i - 1$, where b represents the number of boundary points, i the number of interior points of the polygon.*

For example letting A, B, C, D be the points $(0,0)$, $(3,1)$, $(4,3)$, $(2,4)$, respectively we see in Figure 3.1 that quadrilateral $ABCD$ has 5 boundary points and 6 interior points, so its area is $7\frac{1}{2}$. The triangle ACD has 4 boundary points and 4 interior points, so its area is 5. The triangle ABC has 3 boundary points and 2 interior points, so its area is $2\frac{1}{2}$.

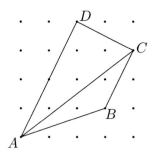

Figure 3.1

18. Notice that the smallest area a polygon can have is $1/2$. This occurs for triangles that have no interior points, such as OXY in Figure 3.2.

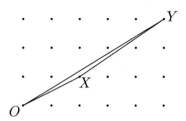

Figure 3.2

(a) Prove the following theorem: Let a, b, c, d be such that $0 \le a/b < c/d \le 1$. Then the condition $bc - ad = 1$ is true if and only if the triangle OXY, with coordinates $O = (0,0)$, $X = (b,a)$, $Y = (b+d, a+c)$, has area $1/2$.

(b) How many different-shaped triangles of area $1/2$ are there in the plane of points (x,y), where $0 \le x \le 7$ and $0 \le y \le 7$?

(c) Approximately how many different shaped triangles of area $1/2$ are there in the plane of points (x,y), where $0 \le x \le 100$ and $0 \le y \le 100$?

19. Consider the triangles of area $1/2$ that can be constructed on the lattice integer lattice plane, where $0 \le x \le 10$, and $0 \le y \le 10$.

(a) How many different shaped triangles are there?

(b) Which has the longest perimeter? How long is it?

(c) Which has the shortest height? How high is it?

20. Here is one more exercise dealing with the integer lattice plane. For each of the following lines find points in the finite integral lattice of dimension $0 \le x \le 100$ and $0 \le y \le 100$ that are closest without being on the line itself. Also give that closest distance.

(a) $y = (19/24)x$

(b) $y = (31/57)x$

(c) $y = (21/97)x$

3.3   Continued Fractions

In this section we introduce a method for finding the fractions with smallest denominators that approximate reals numbers. It consists of writing numbers as a series of quotients. This new mathematical object is called a continued fraction, and while it looks strange, it is really quite natural. The process that generates the continued fraction was known by the Greeks; they called it "antanairesis" and, as you will see, it is based on the Euclidean algorithm.

Definition 3.3.1 *A **continued fraction** is a number of the form*

$$a_0 + \cfrac{1}{a_1 + \cfrac{1}{a_2 + \cfrac{1}{a_3 + \cfrac{1}{\ddots \atop a_{n-1} + \cfrac{1}{a_n} \atop \ddots}}}}$$

where $a_i > 0$ for all $i > 0$, $a_0 \geq 0$.

If the series of quotients is finite, then we call this representation a **finite continued fraction**, *if it is not finite, then it is called an* **infinite continued fraction**.

Notation: A finite continued fraction is written $[a_0; a_1, a_2, \ldots, a_n]$; an infinite continued fraction is denoted by $[a_0; a_1, a_2, \ldots, a_n, \ldots]$.

Definition 3.3.2 *If r is a number of the form $[a_0; a_1, a_2, \ldots, a_n \ldots]$, then $[a_0; a_1, a_2, \ldots, a_n \ldots]$ is called the* **continued fraction expansion** *of r. The expansion $[a_0; a_1, a_2, \ldots, a_i] = p_i/q_i$ is called the ith* **convergent** *of r.*

Notation: We denote the ith convergent by p_i/q_i.

Finite continued fractions represent fractions. Infinite continued fractions are not so easy to understand. You might say they are "infinite quotients." Whether they represent numbers, as do their finite counterparts, is something that needs to be clarified. Generally, the concept of an "infinite process," whether it be an infinite sum, an infinite product, or an infinite quotient, is a concept that needs to be examined with care. In this chapter we have already examined a specific case of an infinite sum, an infinite decimal. We learned that infinite decimal expansions that eventually repeat or terminate represent rational numbers; the other decimal expansions represent nonrational real numbers. We will touch on infinite continued fractions in this section and we will discuss the general topic of infinite processes at length in Section 4.3.

Example 3.3.3

(a) Consider the continued fraction $[0; 1, 6, 9]$ This is

$$0 + \cfrac{1}{1 + \cfrac{1}{6 + \cfrac{1}{9}}} = 0 + \cfrac{1}{1 + \cfrac{1}{55/9}} = 0 + \frac{1}{64/55} = 55/64.$$

(b) Consider the continued fraction $[1; 2, 3, 4]$. This is

$$1 + \cfrac{1}{2 + \cfrac{1}{3 + \cfrac{1}{4}}} = 1 + \cfrac{1}{2 + \cfrac{1}{13/4}} = 0 + \frac{1}{30/13} = 43/30.$$

■

Let us reverse the process and find the continued fraction associated with a fraction.

Example 3.3.4

(a) Consider the number $3/8$.

$$3/8 = 0 + \frac{1}{8/3} = 0 + \cfrac{1}{2 + \cfrac{1}{3/2}} = 0 + \cfrac{1}{2 + \cfrac{1}{1 + \cfrac{1}{2}}} = [0; 2, 1, 2]$$

(b) Consider 0.123.

$$0.123 = 0 + \cfrac{1}{1000/123} = 0 + \cfrac{1}{8 + \cfrac{1}{123/16}} = 0 + \cfrac{1}{8 + \cfrac{1}{7 + \cfrac{1}{16/11}}}$$

$$= 0 + \cfrac{1}{8 + \cfrac{1}{7 + \cfrac{1}{1 + \cfrac{1}{11/5}}}} = 0 + \cfrac{1}{8 + \cfrac{1}{7 + \cfrac{1}{1 + \cfrac{1}{2 + \cfrac{1}{5}}}}}$$

$$= [0; 8, 7, 1, 2, 5] \qquad\qquad\qquad \blacksquare$$

Notice that with each of the expansions we could carry the expansion out one more step and retain the equality with the original fraction. This could be accomplished by making the last fraction $1/1$. For example, with $3/8$ the continued fraction could be $[0; 2, 1, 1, 1]$, and 0.123 could be $[0; 8, 7, 1, 2, 4, 1]$. But we did not allow this with our condition that $a_n > 1$. We have this condition so that the continued fraction expansion for each number is unique. Notice also that this procedure of finding continued fractions terminates for ordinary fractions. Since every number that appears on our calculator is a terminating decimal and, hence, a fraction, every decimal on our calculator can be represented as a finite continued continued fraction.

We can use our calculator to find continued fraction expansions. Here is how it works. You need only use the keys that perform division, inversion (or taking reciprocal), and subtraction. We will represent these keys by $\div$, x^{-1}, and $-$, respectively. Inside the angled brackets we show what the calculator screen displays.

Example 3.3.5

(a) Consider $3/8$. Here is the procedure with a calculator:

$3 \div 8 = \langle\ 0.375\ \rangle\ -0 = \langle\ 0.375\ \rangle\ x^{-1}\ \langle\ 2.6666667\ \rangle\ -2 =$
$\langle\ 0.6666667\ \rangle\ x^{-1}\ \langle\ 1.5\ \rangle\ -1 = \langle\ 0.5\ \rangle\ x^{-1}\ \langle\ 2\ \rangle.$

Here is the procedure in words. We perform three operations: (i) divide, then (ii) subtract off the integer portion of the result, and then (iii) invert. We continue this until we reach an integer. In this example we subtracted 0, then 2, then 1, and we stopped with the integer 2. Collecting those numbers in order, we have [0; 2, 1, 2]. That is, in fact, the continued fraction expansion of 3/8 we saw in Example 3.3.4 (a).

(b) Consider 0.123.

$0.123 = \langle\ 0.123\ \rangle\ -0 = \langle\ 0.123\ \rangle\ x^{-1}\ \langle\ 8.1300813\ \rangle\ -8 =$
$\langle\ 0.1300813\ \rangle\ x^{-1}\ \langle\ 7.6875\ \rangle\ -7 = \langle\ 0.6875\ \rangle\ x^{-1}$
$\langle\ 1.4545455\ \rangle\ -1 = \langle\ 0.4545455\ \rangle\ x^{-1}\ \langle\ 2.2000000\ \rangle\ -2 =$
$\langle\ 0.2000000\ \rangle\ x^{-1}\ \langle\ 5.0000000\ \rangle$

We will stop here. The numbers we have subtracted, along with the final number, 5, make up our continued fraction [0; 8, 7, 1, 2, 5]. This coincides with our findings in Example 3.3.4 (b). You should know that sometimes we do not get a whole number in our final steps. In fact, in this example if our calculator displayed nine-place accuracy instead of seven, we would have seen 0.499999995 at the final step instead of 5. This is a problem that stems from the round-off error that the calculator necessarily imposes. We will have to be on the lookout for this.

(c) Consider 1876/365.

$1876 \div 365 = \langle\ 5.139726\ \rangle\ -5 = \langle\ 0.139726\ \rangle\ x^{-1}\ \langle\ 7.156828\ \rangle$
$-7 = \langle\ 0.156828\ \rangle\ x^{-1}\ \langle\ 6.3749998\ \rangle\ -6 = \langle\ 0.3749998\ \rangle\ x^{-1}$
$\langle\ 2.6666681\ \rangle\ -2 = \langle\ 0.6666681\ \rangle\ x^{-1}\ \langle\ 1.4999968\ \rangle\ -1 =$
$\langle\ 0.4999968\ \rangle\ x^{-1}\ \langle\ 2.0000128\ \rangle$

Here we will stop. Notice that the round-off is showing. But, in any case, our expansion is [5; 7, 6, 2, 1, 2]. ∎

This process of representing a fraction by a continued fraction will always stop because it is based on the Euclidean algorithm and the Euclidean algorithm is finite. Let us examine the relationship with our old friends 1876 and 365.

$$
\begin{aligned}
1876 &= 365(5) + 51 \\
365 &= 51(7) + 8 \\
51 &= 8(6) + 3 \\
8 &= 3(2) + 2 \\
3 &= 2(1) + 1 \\
2 &= 1(2)
\end{aligned}
$$

The quotients found in this process are, in order, 5, 7, 6, 2, 1, 2. This is the very continued fraction expansion of $1876/365$.

Here we see, side by side, the Euclidean algorithm and the continued fraction algorithm.

$$
\begin{array}{rcl}
1876 &=& 365(5) + 51 \\
365 &=& 51(7) + 8 \\
51 &=& 8(6) + 3 \\
8 &=& 3(2) + 2 \\
3 &=& 2(1) + 1 \\
2 &=& 1(2)
\end{array}
\quad\longrightarrow\quad
\begin{array}{rcl}
1876/365 &=& 5 + 51/365 \\
365/51 &=& 7 + 8/51 \\
51/8 &=& 6 + 3/8 \\
8/3 &=& 2 + 2/3 \\
3/2 &=& 1 + 1/2 \\
2/1 &=& 2
\end{array}
$$

Recall the table of quotients from Section 2.1 which helped us solve the linear Diophantine equation, $1876x - 365y = 1$.

	5	7	6	2	1	2		
1876	1	0	1	7	43	93	136	365
365	0	1	5	36	221	478	699	1876

Using quotient tables, we can see an easy way of finding fractional equivalents of continued fractions and the partial convergents of continued fractions.

Example 3.3.6

The continued fraction $[5; 7, 6, 2, 1, 2]$ represents $1876/365$. Its convergents are $[5]$, $[5; 7]$, $[5; 7, 6]$, $[5; 7, 6, 2]$, $[5; 7, 6, 2, 1]$, and, finally, the number itself: $[5; 7, 6, 2, 1, 2]$. The table of quotients we know.

	5	7	6	2	1	2		
1876	1	0	1	7	43	93	136	365
365	0	1	5	36	221	478	699	1876

The respective continued fraction convergents can be formed from this chart; they are $5/1$, $36/7$, $221/43$, $478/93$, $699/136$, and the number itself, $1876/365$. ∎

Notice that the convergents $5/1, 36/7, \ldots$ in this example can be read off the table of quotients, but they are upside down. For cosmetic reasons, we may simply invert the table like this:

	5	7	6	2	1	2		
p_k	0	1	5	36	221	478	699	1876
q_k	1	0	1	7	43	93	136	365

Now the convergents $5/1, 36/5, \ldots$ can be read off the table right side up. We call this inverted quotient table a **convergents table**.

Let us formally state and prove the relationship between continued fraction convergents and the convergents table.

Theorem 3.3.7 *Let $p/q = [a_0; a_1, a_2, \ldots, a_n]$ and let the kth convergent $p_k/q_k = [a_0; a_1, a_2, \ldots, a_k]$. Then $p_k = a_k p_{k-1} + p_{k-2}$ and $q_k = a_k q_{k-1} + q_{k-2}$, where we shall let $p_{-2} = 0, p_{-1} = 1$, and $q_{-2} = 1, q_{-1} = 0$.*

Proof We proceed by mathematical induction on k. Let $\mathcal{P}(k)$ be the following statement: If $p_k/q_k = [a_0; a_1, a_2, \ldots, a_k]$, then $p_k = a_k p_{k-1} + p_{k-2}$ and $q_k = a_k q_{k-1} + q_{k-2}$. Let us consider $\mathcal{P}(0)$. We know that $p_0/q_0 = a_0/1$. Now $p_0 = a_0 p_{-1} + p_{-2} = a_0$ because of our initial conditions that $p_{-2} = 0$, and $p_{-1} = 1$. Similarly, $q_0 = a_0 q_{-1} + q - 2 = 1$ because of the initial conditions that $q_{-2} = 1$, and $q_{-1} = 0$.

Suppose that $\mathcal{P}(k)$ is true and consider $\mathcal{P}(k + 1)$: Notice that p_{k+1}/q_{k+1} which is $[a_0; a_1, a_2, \ldots a_k, a_{k+1}]$, can be represented in our notation like this:

$$[a_0; a_1, a_2, \ldots, a_{k-1}, (a_k + 1/a_{k+1})].$$

So we have
$$
\begin{aligned}
p_{k+1}/q_{k+1} &= \frac{(a_k + 1/a_{k+1})p_{k-1} + p_{k-2}}{(a_k + 1/a_{k+1})q_{k-1} + q_{k-2}} \\
&= \frac{[(a_k a_{k+1} + 1)/a_{k+1}]p_{k-1} + p_{k-2}}{[(a_k a_{k+1} + 1)/a_{k+1}]q_{k-1} + q_{k-2}} \\
&= \frac{a_{k+1}(a_k p_{k-1} + p_{k-2}) + p_{k-1}}{a_{k+1}(a_k q_{k-1} + q_{k-2}) + q_{k-1}} \\
&= \frac{(a_{k+1}p_k + p_{k-1})}{(a_{k+1}q_k + q_{k-1})}.
\end{aligned}
$$

The final equality is true because of the induction hypothesis. So $p_{k+1} = a_{k+1}p_k + p_{k-1}$ and $q_{k+1} = a_{k+1}q_k + q_{k-1}$. This completes the induction proof. $\square$

Let us now consider the continued fraction expansion of an irrational number. The $\sqrt{2}$ is quite familiar to us, having been analyzed in Section 3.2. On a calculator with nine-decimal-place accuracy it looks like this: 1.414213562. Of course, as it reads, this is a rational number so we can use our calculator to find its finite continued fraction. Here is what happens.

Example 3.3.8
$$\sqrt{2} = \langle\, 1.414213562 \,\rangle\, -1 = \langle\, 0.414213562 \,\rangle\, x^{-1}$$
$$\langle\, 2.414213562 \,\rangle\, -2 = \langle\, 0.414213562 \,\rangle\, x^{-1}$$
$$\langle\, 2.414213562 \,\rangle\, -2 = \langle\, 0.414213562 \,\rangle$$

It appears as if the expansion is repeating with 2s. One more iteration yields
$$x^{-1} \,\langle\, 2.414213562 \,\rangle\, -2 = \langle\, 0.414213563 \,\rangle.$$

Though the repetition is not perfect because of round-off error, it does appear that the continued fraction expansion of $\sqrt{2}$ is an infinite expansion: $[1; 2, 2, 2, \ldots]$. Here is an informal proof that $[1; 2, 2, 2, \ldots]$ represents $\sqrt{2}$.

Let $x = [1; 2, 2, 2, \ldots]$; so

$$x = 1 + \cfrac{1}{2 + \cfrac{1}{2 + \cfrac{1}{2 + \ddots}}} \;\; ; \;\; x - 1 = \cfrac{1}{2 + \cfrac{1}{2 + \cfrac{1}{2 + \ddots}}}$$

Thus

$$x - 1 = \frac{1}{2 + (x - 1)}.$$

Therefore, $x - 1 = 1/(x + 1)$ and it follows that $x^2 = 2$. ■

Let us look at the convergents table for $[1; 2, 2, 2, \ldots]$.

Example 3.3.9

	1	2	2	2	2	2	2	...
p_k	0	1	1	3	7	17	41	99	239	...
q_k	1	0	1	2	5	12	29	70	169	...

The convergents, p_k/q_k, beginning with $k = 0$ are, respectively,

$$1, \quad 3/2, \quad 7/5, \quad 17/12, \quad 41/29, \quad 99/70, \quad 239/169.$$

Notice that these are $\sqrt{2}$-approximants that we found in 3.2. ■

Lest you think that infinite continued fractions all repeat consider the expansion for π. Using our calculator technique, we find the expansion.

Example 3.3.10

$\pi = \langle\, 3.141592654\,\rangle\; -3 = \langle\, 0.141592654\,\rangle\; x^{-1}$
$\langle\, 7.052513306\,\rangle\; -7 = \langle\, 0.052513306\,\rangle\; x^{-1}$
$\langle\, 15.39659441\,\rangle\; -15 = \langle\, 0.39659441\rangle\; x^{-1}$
$\langle\, 1.003417228\,\rangle\; -1 = \langle\, 0.003417228\,\rangle\; x^{-1}\; \langle\, 292.6348337\,\rangle$

Let us stop here. We have no idea where this expansion is going, or even if it is accurate considering the round-off error. The convergents table looks like this.

		3	7	15	1	292	...
p_k	0	1	3	22	333	355	103993	...
q_k	1	0	1	7	106	113	33102	...

The convergents, p_k/q_k, beginning with $k = 0$ are, respectively:

$$3, \quad 22/7, \quad 333/106, \quad 355/113, \quad 103993/33102.$$

Notice that these are π-approximants; the first few we found in 3.2. ∎

In fact, the convergents of the continued fraction expansion of a number r are, indeed, r-approximants. Let us see why.

Example 3.3.11

(a) Look at $3/8$ again. Its expansion is $[0; 2, 1, 2]$. Its convergents are, respectively, $[0]$, $[0; 2]$, $[0; 2, 1]$, $[0; 2, 1, 2]$.

		0	2	1	2
p_k	0	1	0	1	1	3
q_k	1	0	1	2	3	8

We can simply look at the convergents table and see that $3/8$ is approximated initially by 0, then by $1/2$, then by $1/3$, and then by $3/8$ itself. Notice something curious about these convergents: $1/2 - 0 = 1/2; 1/3 - 1/2 = -1/6$, and $3/8 - 1/3 = 1/24 = 1/(8 \cdot 3)$.

(b) Consider 0.123. Its expansion is $[0; 8, 7, 1, 2, 5]$.

		0	8	7	1	2	5
p_k	0	1	0	1	7	8	23	123
q_k	1	0	1	8	57	65	187	1000

Its convergents are 0, $1/8$, $7/57$, $8/65$, $23/187$, and $123/1000$. Notice that $1/8 - 0 = 1/8$; $7/57 - 1/8 = -1/456 = -1/(57 \cdot 8)$,

$8/65-7/57 = 1/(65 \cdot 57)$, $23/187-8/65 = -(187 \cdot 65)$, and $123/1000 - 23/187 = 1/(187 \cdot 1000)$. ∎

These examples lead us to guess that the successive convergents differ by $1/N$, where N is the product of their denominators. This is true.

Lemma 3.3.12 *Suppose we have three fractions: r/s, t/u, and v/w such that $v = at+r$, $w = au+s$, and $st-ru = x$. Then $uv-tw = -x$.*

Proof $uv - tw = u(at + r) - t(au + s) = ru - st = -x$. □

Theorem 3.3.13 *Suppose that $[a_0; a_1, a_2, \ldots, a_n, \ldots] = r$ and the kth convergent of r is $[a_0; a_1, a_2, \ldots, a_k] = p_k/q_k$. Then*

$$(p_k/q_k) - (p_{k-1}/q_{k-1}) = (-1)^{k-1}/q_k q_{k-1}.$$

Proof We proceed by induction. Suppose that r and p_k/q_k are as before and let $\mathcal{P}(k)$ be the following statement:

$$(p_k/q_k) - (p_{k-1}/q_{k-1}) = (-1)^{k-1}/q_k q_{k-1}.$$

Now $\mathcal{P}(1)$ is true because

$$(p_1/q_1) - (p_0/q_0) = a_0 + 1/a_1 - a_0 = 1/a_1 = 1/q_1 q_0$$

since $q_0 = 1$ and $q_1 = a_1$.

Suppose that $\mathcal{P}(k)$ is true. Therefore, $(p_{k-1}q_k) - (p_k q_{k-1}) = (-1)^k$. Consider $\mathcal{P}(k + 1)$. Since the three fractions p_{k-1}/q_{k-1}, p_k/q_k, p_{k+1}/q_{k+1} satisfy the hypotheses of the lemma, we have that $(p_k q_{k+1}) - (p_{k+1}q_k) = (-1)^{k+1}$. Thus

$$(p_{k+1}/q_{k+1}) - (p_k/q_k) = (p_{k+1}q_k - p_k q_{k+1})/q_k q_{k+1} = (-1)^k/q_{k+1}q_k.$$

So $\mathcal{P}(k + 1)$ is true and the theorem follows. □

This theorem leads to several results that bear upon the discussions in Section 3.2. The proofs are straight forward and left as exercises.

Corollary 3.3.14 *Let $p/q = [a_0; a_1, a_2, \ldots, a_n]$ and let $k < n$. If k is even, then $p_k/q_k < p/q$; if k is odd, then $p_k/q_k > p/q$.*

Corollary 3.3.15 *Let* $r = [a_0; a_1, a_2, \ldots, a_n, \ldots]$ *Then* $|(p_k/q_k) - r| < 1/q_k q_{k+1}$.

Corollary 3.3.16 *The convergents of a continued fraction expansion of the number r are r-approximants.*

Corollary 3.3.17 *The successive convergents of a continued fraction expansion are successive fractions in a Farey sequence.*

Now that we know that continued fractions convergents are r-approximants, let us examine just how accurate these approximants are. In Section 3.2 we did a brief survey comparing the accuracy of decimal fractions to the accuracy of fractions in a Farey sequence. Now let us be more explicit and compare the accuracy of a continued fraction expansion with respect to the size of the denominator of the convergent.

Example 3.3.18

Consider $1876/365$. With decimals this begins 5.139726. So its first few decimal convergents are 5, 5.1, 5.14, 5.140, and 5.1397. They differ from the fraction by about 0.14, 0.04, 0.0003, 0.0003, and 0.00003. The continued fraction expansion is $[5; 7, 6, 2, 1, 2]$. The convergents table looks like this.

The continued fraction convergents are 5, $36/7$, $221/43$, $478/93$, and $699/136$. They differ from the $1876/365$ by about 0.14, 0.003, 0.0002, 0.00008, and 0.00002, respectively. This last approximation is $1876/365 - 699/136 = 1/49640 = 1/(365 \cdot 136)$. So, with a denominator of 136 we have found a fraction closer to the original than the decimal expansion with a denominator of $10,000$. There are 9999 fractions with denominators less than 10000 and there are fewer than 5700 fractions with denominator ≤ 136, as we can deduce from the Farey chart in 3.2. ∎

Now let us do an in-depth study of continued fraction accuracy.

Example 3.3.19

Consider the expansion for $\sqrt{2}$; that is, $[1; 2, 2, 2, \ldots]$. Let us see if we can find the digit that comes after the 11th digit in the decimal representation on our fancy calculator. Our calculator tells us that $\sqrt{2} = 1.41421356237$.

	1	2	2	2	2	2	2	2	2	2
p_k	0	1	1	3	7	17	41	99	239	577	1393	3363
q_k	1	0	1	2	5	12	29	70	169	408	985	2378

	2	2	2	2	2	2	2
p_k	8119	19601	473211	114243	275807	665857	1607521
q_k	5741	13860	33461	80782	195025	470832	1136689

Here we list the convergents along with their accuracy.

k	p_k/q_k	$1/(q_{k+1}q_k)$	decimal accuracy
0	1/1	.5	0
1	3/2	.1	0
2	7/5	.02	1
3	17/12	3×10^{-3}	2
4	41/29	4.9×10^{-4}	3
5	99/70	8×10^{-5}	3
6	239/169	1.4×10^{-5}	4
7	577/408	2.4×10^{-6}	5
8	1393/985	4.3×10^{-7}	6
9	3363/2378	7.3×10^{-8}	6
10	8119/5741	1.3×10^{-8}	7
11	19601/13860	2.2×10^{-9}	8
12	47321/33461	3.7×10^{-10}	9
13	114243/80782	6.3×10^{-11}	9
14	275807/195025	1.1×10^{-11}	10
15	665857/470832	1.9×10^{-12}	11
16	1607521/1136689	3.2×10^{-13}	12

By **decimal accuracy** we mean that y matches x to an accuracy of n decimal places if $|x - y| < 5 \times 10^{-(n+1)}$. We see that p_{16}/q_{16} matches $\sqrt{2}$ to 12 place accuracy. We may work out p_{16}/q_{16} by hand, if necessary, and find out the 12th digit of $\sqrt{2}$. It is 3; that is, $p_{16}/q_{16} = 1.414213562373\ldots$.

It is interesting to check and see how good our $1/(q^{k+1}q_k)$ bound is on the accuracy. We can do that because we know $\sqrt{2}$ to 11 places. Let's check the actual difference between $\sqrt{2}$ and two convergents:

$$|\sqrt{2} - p_4/q_4| = |\sqrt{2} - 41/29| \approx 4.2 \times 10^{-4} \text{ while } 1/q_4q_5 \approx 4.9 \times 10^{-4}$$
$$|\sqrt{2} - p_9/q_9| = |\sqrt{2} - 3363/3789| \approx 6.3 \times 10^{-8} \text{ while } 1/q_9q_{10} \approx$$

7.3×10^{-8}. So, as you can see, our bound is pretty close to the true difference. ∎

We finish this section with an observation that you may have already made: The convergents of a continued fraction are in lowest terms. This comes as no surprise because convergents are fractions with the smallest denominator possible for a given approximation. But it takes on an interesting twist if we consider the convergent for the fraction itself. For example, suppose you are given a fraction p/q and asked if it is in lowest terms. If it is not, then you are asked to reduce it. Here is what you do. Use your calculator to whip out its continued fraction expansion, $[a_0; a_1, a_2, a_3, \ldots, a_n]$, form its convergents table, and note its final convergent p_n/q_n. This is the reduced fraction equivalent to p/q.

Example 3.3.20

Consider the fraction $226576/126497$. Its continued fraction expansion is $[1; 1, 3, 1, 3, 1, 2, 1, 1, 1, 1, 2]$ The convergents table looks like this

		1	1	3	1	3	1	2	1	1	1	1	2
p_k	0	1	1	2	7	9	34	43	120	163	283	446	729	1904
q_k	1	0	1	1	4	5	19	24	67	91	158	249	407	1063

So $226576/126497$ reduces to $1904/1063$, and this is in lowest terms. ∎

EXERCISES

1. Find the continued fraction expansions for

   (a) $7/11$
   (b) $81/35$
   (c) $45/73$
   (d) $21/68$
   (e) $91/67$

2. Find the continued fraction expansions for

   (a) 0.8642
   (b) 1.2345
   (c) 0.12345679
   (d) 1.618033989

(e) 3.16227766

3. Find the first eight terms of the continued fraction expansion for the following numbers. If you see a pattern in the expansion, you can extend it.

(a) $\sqrt{3}$

(b) $\sqrt{5}$

(c) $\sqrt{6}$

(d) $\sqrt[3]{2}$

(e) e (the base of the natural logarithm)

4. Find fractions with denominator ≤ 10 and ≤ 100 that most closely approximate each of the following numbers. Also tell how close the approximations are.

(a) 669/563

(b) 2344/733

(c) 0.150493827

(d) $\sqrt{7}$

(e) π^2

5. Reduce the following fractions using continued fraction expansions.

(a) 25567/35557

(b) 45765339/81972912

(c) 1331931458/4426364715

6. Given r find the least d such that $|c/d - r| < 0.0000005$ for an appropriate c, where r is

(a) 23456/123457

(b) 554057/874442

(c) 0.707106781

(d) 0.618033989

(e) $\sqrt{13}$

(f) $1/\pi$

7. Let $n/m = [a_0; a_1, a_2, \ldots, a_n]$. What is the continued fraction expansion of m/n? Explain why your answer is true.

8. If $p_n/q_n = [a_0; a_1, a_2, \ldots, a_n]$ where $a_0 > 0$ (we allow $a_n = 1$) what fraction is represented by the continued fraction

   (a) $[a_n; a_{n-1}, a_{n-2}, \ldots, a_0]$?

   (b) $[a_n; a_{n-1}, a_{n-2}, \ldots, a_1]$?

   (c) Look at numerical examples to help you in (a) and (b) and then prove your results.

9. Find which fraction is smaller, given their continued fraction expansions. Check to see if you are right by finding the fractions that these expansions represent.

   (a) $[0; 1, 6, 5, 3, 2, 2, 6, 7]$, $[0; 1, 6, 5, 3, 2, 1, 10]$

   (b) $[1; 4, 3, 7, 7, 8, 3, 6]$, $[1; 4, 3, 7, 7, 8, 3, 5, 36, 1, 9]$

   (c) $[2; 5, 5, 4, 2, 1, 300, 10]$, $[2; 5, 5, 4, 2, 1, 300]$

10. Let $r = [a_0; a_1, a_2, \ldots, a_n]$ and $s = [a_0; a_1, a_2, \ldots, a_n, a_{n+1}]$.

    (a) Complete this statement: $r < s$ if and only if ....

    (b) Complete this statement: $r > s$ if and only if ....

11. Given $[a_0; a_1, a_2, \ldots, a_n]$ and $[b_0; b_1, b_2, \ldots, b_m]$, how can you tell which represents a larger fraction? Explain your answer.

12. For each of the following pairs of continued fractions find the mediant of the fractions they represent.

    (a) $[0; 1, 2]$ and $[0; 1, 2, 2]$

    (b) $[1; 1, 2, 3]$ and $[1; 1, 2]$

    (c) $[0; 1, 2, 1, 1, 2]$ and $[0; 1, 2, 1, 1, 6]$

    (d) $[2; 1, 4, 1, 7, 2]$ and $[2; 1, 4, 1, 9]$

    (e) $[0; 1, 5, 1, 4]$ and $[0; 1, 5, 1, 1, 5]$

13. Prove: Let p_{n-1}/q_{n-1} and p_n/q_n be the successive convergents: $[a_0; a_1, \ldots, a_{n-1}]$ and $[a_0; a_1, \ldots, a_{n-1}, a_n]$, respectively. Then $p_{n-1}/q_{n-1} \oplus p_n/q_n = [a_0; a_1, \ldots, a_n, 1] = [a_0; a_1, \ldots, a_n + 1]$.

14. Given $[a_0; a_1, a_2, \ldots, a_n] = p/q$ and $[b_0; b_1, b_2, \ldots, b_m] = r/s$, for which cases can you give a formula for the continued fraction expansion of $p/q \oplus r/s$?

15. Prove Corollary 3.3.14: Let $p/q = [a_0; a_1, a_2, \ldots, a_n]$ and let $k < n$. If k is even, then $p_k/q_k < p/q$; if k is odd, then $p_k/q_k > p/q$.

16. Prove Corollary 3.3.15: Let $r = [a_0; a_1, a_2, \ldots, a_n, \ldots]$. Then $|(p_k/q_k) - r| < 1/q_k q_{k+1}$.

17. Prove Corollary 3.3.16: The convergents of a continued fraction expansion of the number r are r-approximants.

18. Prove Corollary 3.3.17: The successive convergents of a continued fraction expansion are successive fractions in a Farey sequence.

19. Show that the partial convergents of a continued fraction are fractions that are reduced to lowest terms.

20. Show that $p_k/q_k - p_{k-2}/q_{k-2} = (-1)^k a_k/q_k q_{k-2}$ for $2 \leq k \leq n$.

21. Given the following points a and b, find the fraction with smallest denominator that lies in the interval (a, b): Try to figure out a method of finding this answer using continued fraction expansions.

    (a) 7/32, 5/23
    (b) 31/71, 69/158
    (c) 128/279, 67/146
    (d) 0.53, 0.54
    (e) 0.628, 0.6289
    (f) 0.13972, 0.13973

22. Our calculator shows that $\sqrt{5}$ is 2.2360679775. Find the expansion of $\sqrt{5}$ to 11 decimal places.

23. For which numbers n does it appear as if $\sqrt{n}$ has an infinite continued fraction expansion of $[[n]; k, k, \ldots]$. Try to prove your observation.

3.4 Solving Equations on the Rational Plane

In Chapter 2 we introduced Diophantine equations. These are polynomial equations that call for integer solutions only. In this section we examine a method of solving these equations that uses rational numbers and analytic geometry. First we observe that finding rational solutions to one type of Diophantine equation allows us to find integer solutions to a related equation. Here is how it works.

Example 3.4.1

Consider the Diophantine equation $2x^2 + 4y^2 = 3z^2$. It has infinitely many solutions; here are three of them; we list them as triples (x, y, z): $(2, 1, 2)$, $(2, 5, 6)$, and $(22, 29, 38)$. The first one can be found by inspection, the second one might come that way too, but the third is definitely too big to find without some magic procedure. We will develop that magic procedure soon. The first thing we should establish is that if a rational solution is found for the Diophantine equation $2x^2 + 4y^2 = 3$, then an integer solution can be found for $2x^2 + 4y^2 = 3z^2$. Here is why: Suppose $x = p/q$, $y = r/s$ represents a rational solution to $2x^2 + 4y^2 = 3$. Then, getting a common denominator we can see that $2(ps)^2 + 4(rq)^2 = 3(qs)^2$. So (ps, rq, qs) is an integral solution to $2x^2 + 4y^2 = 3z^2$. ■

The technique used in this example works for quadratic Diophantine equations generally; that is, the integer solutions to $ax^2 + by^2 = cz^2$ can be discovered by solving for rational solutions of $ax^2 + by^2 = c$. Next we turn to the magic. Once we have found one rational solution we may generate others using analytic geometry on the rational lattice plane. The rational plane is analogous to the integral lattice plane introduced in Section 2.3.

Definition 3.4.2 *A **rational lattice point** is a point of the form (r, s), where r and s are rational numbers. The plane made up of rational points is called the **rational lattice plane**. A **rational circle***

is the collection of rational lattice points that lie a fixed distance from a given rational lattice point.

Example 3.4.3

(a) Consider the circle $x^2 + y^2 = 1$. This circle has lots of rational lattice points. We can figure them out from all our work with Pythagorean triples in Section 2.3. For example, $(\pm 1, 0)$, $(0, \pm 1)$ are four such points. And $(\pm 4/5, \pm 3/5)$ are four more. And $(\pm 3/5, \pm 4/5)$ are four more. These eight points are based on the $(3, 4, 5)$ Pythagorean triple. And then there are the eight points based on the $(5, 12, 13)$ Pythagorean triple as well. And on and on it goes.

(b) Consider the circle $x^2 + y^2 = 2$. This circle has lots of rational lattice points too but they are not so easy to come by: $(\pm 1, \pm 1)$ are four such points. And $(\pm 7/5, \pm 1/5)$ and $(\pm 1/5, \pm 7/5)$ provide eight more.

(c) Consider the circle $x^2 + y^2 = 3$. This circle has no rational lattice points. If $(a/b)^2 + (c/d)^2 = 3$, then $(ad)^2 + (bc)^2 = 3(bd)^2$. Corollary 2.3.13 tells us that the sum of perfect squares cannot be of the form $3 \times t^2$, where t is an integer.   ∎

Theorem 3.4.4 *If the integer,k, is not the sum of two squares then $x^2 + y^2 = k$ has no rational lattice points on it.*

Proof The argument follows that in Example 3.4.3 (c).   □

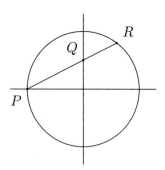

Figure 3.3

We shall examine carefully the nature and arrangement of rational lattice points on the unit circle. Consider the unit circle in Figure 3.3. Here P is the point $(-1, 0)$, Q is a point $(0, q)$ on the interval along the y-axis from 0 to 1, and R is the point of intersection of the line PQ and the circle. It turns out that R is a rational point on the circle.

Example 3.4.5

Suppose that $Q = (0, 5/8)$. Now $y = (5/8)x + 5/8$ is the equation for the line joining P to Q. Substituting this into the equation $x^2 + y^2 = 1$, we get

$$x^2 + (25/64)(x^2 + 2x + 1) = (89/64)x^2 + (25/32)x + 25/64 = 1.$$

Thus we have the quadratic equation $(1/64)(89x^2 + 50x - 39) = 0$. Using the quadratic formula we get:

$$x = (1/178)(-50 \pm \sqrt{16384}) = (1/178)(-50 \pm 128).$$

These two numbers are -1 and $39/89$. We recognize $x = -1$ as a solution. Substituting in $39/89$ for x in the linear equation, we obtain

$$y = (5/8)(x + 1) = (5/8)(39/89 + 1) = 80/89.$$

So we have point R: $R = (r, s) = (39/89, 80/89)$. Notice that this also gives us a solution to the Diophantine equation $x^2 + y^2 = z^2$: $x = 39$, $y = 80$, $z = 89$. ∎

Theorem 3.4.6 *If P, Q, R are described as above and if $q = b/a$ then R is the point (r, s), where $r = (a^2 - b^2)/(a^2 + b^2)$ and $s = 2ab/(a^2 + b^2)$.*

Proof The line PQ has the equation $y = qx + q$. So $x^2 + y^2 = (1 + q^2)x^2 + 2q^2x + q^2 = 1$. Using the quadratic formula, we find that $x = (1 - q^2)/(1 + q^2)$ and $(-1 - q^2)/1 + q^2)$. Substituting these into the linear equation, $y = 2q/(1 + q^2)$ and 0. The first answer gives R, the second, P. So $r = (1 - q^2)/(1 + q^2)$ and $s = 2q/(1 + q^2)$. Substituting in b/a for q, we get our answer above. □

Notice that the rational point on the circle has coordinates that have the same denominator. This must always be the case.

Corollary 3.4.7 *The coordinates of all rational points (x, y) on the unit circle have the same denominator.*

Proof Let $R = (x, y)$ be a rational point on the circle in the first quadrant and form the line joining P to R. It is of the form $y = mx + b$, where m and b are rational. So $Q = (0, b)$ and it follows from Theorem 3.4.6 that x and y have the same denominator.   □

Notice also that there is a close relationship between the coordinates of R: (r, s) given Q: $(0, b/a)$ and the primitive triangle chart of Section 2.3. Given a and b on the chart, we see that $r = x/z$ and $s = y/z$.

Theorem 3.4.8 *If $R = (r, s)$ is a rational point on the unit circle and $r = b/a$, $s = c/a$, then the line PR, where P is $(-1, 0)$ has a y-intercept of $c/(a + b)$; that is, $Q = (0, c/(a + b))$.*

The proof is left as an exercise.

Example 3.4.9

Referring to Figure 3.3,

(a) Let $Q = (0, 1/2)$. Then, using the formula from Theorem 3.4.6, $R = (3/5, 4/5)$.

b) Let $Q = (0, 4/9)$. Then $R = (65/97, 72/97)$.

(c) Let $R = (12/13, 5/13)$. Then, using the formula from Theorem 3.4.8, $Q = (0, 1/5)$.

(d) Let $R = (5/13, 12/13)$. Then $Q = (0, 2/3)$.   ■

Notice that the first quadrant of the unit circle is packed with rational points. The density of rational points seems to be the same as the density of the rational points on the interval on the y-axis between 0 and 1. And it is. In Section 3.3 we searched for fractions that lay between two numbers on the real line. Here we can do the same; that is, search for rational points on arcs of the unit circle.

Example 3.4.10

Let us find rational points on arcs of the circle.

(a) Let $R = (3/5, 4/5)$ and $R' = (65/97, 72/97)$. Let us find a rational point on the arc joining R and R'. Letting Q and Q' correspond to R and R', respectively, we know that $Q = (0, 1/2)$, $Q' = (0, 4/9)$ so we may take the point $Q'' = (0, 5/11)$ between them. Notice that 5/11 is the mediant of 1/2 and 4/9 and it lies between them. Then $R'' = (48/73, 55/73)$.

(b) Let's make it harder. Using the notation in (a), let's find a point on the arc joining R' to R''. Again we use the formulas: $Q''' = 9/20$. So $R''' = (319/481, 360/481)$. ∎

We leave it as an exercise to examine whether these rational points have the smallest denominator among all the rational points that lay on these arcs.

Example 3.4.11

Let us see how close we can come to finding a rational point (r, s) on the circle where $r = s$. We begin with $R_1 = (3/5, 4/5)$ and $R_2 = (4/5, 3/5)$. The respective points, Q, are $Q_1 = (0, 1/2)$, $Q_2 = (0, 1/3)$. Let $Q_2 = (0, 2/5)$. Then $R_2 = (21/20, 20/29)$. Letting $R_3 = (20/29, 21/29)$ we proceed, and find $Q_3 = (0, 3/7)$. Let $Q_4 = (0, 5/12)$ and so $R_4 = (119/169, 120/169)$. That's pretty close. We will pursue this more in the exercises. ∎

Had we begun with a search for rational solutions of $x^2 + y^2 = 1$, then we could have derived the very theory for integral solutions of $x^2 + y^2 = z^2$ that we did in Section 2.3. Let us now turn to the Diophantine equation $x^2 + y^2 = 2z^2$. We have not studied this Diophantine equation yet so we can develop its integral solutions by first studying the rational solutions of the equation $x^2 + y^2 = 2$. The picture is a bit different for this problem; the point $(-1, 0)$ is not on the circle.

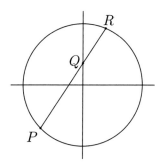

Figure 3.4

In Figure 3.4 we show the point $P = (-1, -1)$ on the circle. The point $Q = (0, q)$ ranges on the y-axis between 0 and $\sqrt{2}$.

Theorem 3.4.12 *Let P, Q, R be points as described in Figure 3.4 and let $q = b/a$. It follows that R is the point (r, s), where*

$$r = (2a^2 - b^2)/(2a^2 + 2ab + b^2), \quad s = (2a^2 + 4ab + b^2)/(2a^2 + 2ab + b^2).$$

Proof Note that b need not be smaller than a. The line PQ has the equation $y = (q + 1)x + q$. So

$$x^2 + y^2 = (q^2 + 2q + 2)x^2 + (2q^2 + 2q)x + q^2 = 1.$$

Using the quadratic formula, $r = (2 - q^2)/(q^2 + 2q + 2)$ and, substituting this into the linear equation, $s = (q^2 + 4q + 2)/(q^2 + 2q + 2)$. Substituting in b/a for q, we get our answer. $\square$

Example 3.4.13

Referring to Figure 3.4,

(a) Let $Q = (0, 1)$. Then, using the formula from Theorem 3.4.12, $R = (1/5, 7/5)$.

(b) Let $Q = (0, 1/2)$. Then $R = (7/13, 17/13)$.

(c) Let $Q = (0, 2/3)$ Then $R = (7/17, 23/17)$.

Solutions to the Diophantine equation $x^2 + y^2 = 2z^2$ that derive from these are $x = 1, y = 7, z = 5$; $x = 7, y = 17, z = 13$; and $x = 7, y = 23, z = 17$.   ∎

Let us examine one more Diophantine equation: $x^2 - 2y^2 = z^2$. We have already studied the equation $x^2 - 2y^2 = 1$. It is a Pell equation from Section 2.3. The picture of $x^2 - 2y^2 = 1$ is not a circle but a hyperbola. We show it in Figure 3.5.

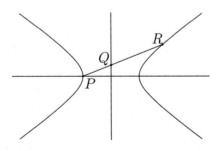

Figure 3.5

As before, P is the point $(-1, 0)$, Q is a point $(0, q)$ along the y-axis, and R is the point of intersection of the line PQ and the hyperbola. It turns out that R is a rational point on the hyperbola. Notice that q must be less than $\sqrt{2}/2$ because otherwise the line PQ will not meet the hyperbola but be parallel to the asymptotes.

Theorem 3.4.14 *If P, Q, R are described as above and if $q = b/a$, then R is the point (r, s), where $r = (a^2 + 2b^2)/(a^2 - 2b^2)$ and $s = 2ab/(a^2 - 2b^2)$.*

The proof is left as an exercise.

Example 3.4.15

(a) Let $Q = (0, 1/2)$. Then, using the formula from Theorem 3.4.14, $R = (3, 2)$.

(b) Let $Q = (0, 1/3)$. Then $R = (11/7, 6/7)$.

(c) Let $Q = (0, 1/4)$. Then $R = (9/7, 4/7)$.

These translate into solutions for the Diophantine equation $x^2 - 2y^2 = z^2$ as follows: $(3, 2, 1)$, $(11, 6, 7)$ and $(9, 4, 7)$. ∎

We can build a chart like the primitive triangle chart of Section 2.3. Here a and b can be any numbers such that $a > b$ and $b/a < \sqrt{2}/2$ and r and s represent the x- and the y-coordinates of R. And, of course, $x^2 - 2y^2 = z^2$. We recognize three solutions to the Pell equation $x^2 - 2y^2 = 1$ here; $x = 3$, $y = 2$, $x = 17$, $y = 12$, and $x = 99$, $y = 70$.

Hyperbola Chart

a	b	r	s	x	y	z
2	1	3	2	3	2	1
3	1	11/7	6/7	7	6	7
3	2	17	12	17	12	1
4	1	9/7	4/7	9	4	7
5	1	27/23	10/23	27	10	23
5	2	33/17	20/17	33	20	17
5	3	43/7	30/7	43	30	7
6	1	19/17	6/17	19	6	17
7	1	51/47	14/47	51	14	47
7	2	57/41	28/41	57	28	41
7	3	67/31	42/31	67	42	31
7	4	81/17	56/17	81	56	17
7	5	99	70	99	70	1

Let's finish with the example at the beginning: the Diophantine equation $2x^2 + 4y^2 = 3z^2$. It has infinitely many solutions and three of them were offered: $(2, 1, 2)$, $(2, 5, 6)$, and $(22, 29, 38)$. We can do this without a picture. We can simply find one point, form a line with rational slope through the point, and see where it intersects the curve.

Example 3.4.16

We search for an easy solution for the equation $2x^2 + 4y^2 = 3$. Without drawing this —it is an ellipse— we note that $(1, 1/2)$ is a solution. This, by the way, yields the solution $(2, 1, 2)$ to the equation $2x^2 + 4y^2 = 3z^2$. We then construct a line through $(1, 1/2)$ with a rational slope. Without help of an x- or y-intercept, we cannot be sure that we will get a second rational point. If we draw a line of slope 1 through $(1, 1/2)$, we get the line $y = x - 1/2$. Substituting this into the equation for the ellipse, we get $3x^2 - 2x - 1 = 0$. So $x = -1/3$, $y = -5/6$. These rationals yield the solution we know: $(-2, -5, 6)$ to the equation $2x^2 + 4y^2 = 3z^2$. Let us find another solution using a line of slope 2. The line through $(1, 1/2)$ is $y = 2x - 3/2$. Substituting this into the equation for the ellipse, we get $3x^2 - 4x + 1 = 0$. So $x = 1/3$, $y = -5/6$. This solution is $(2, -5, 6)$, which is essentially the same as the one before; it does not yield new information. Let's try again with a line passing through $(1, 1/2)$ of slope 3. That line is $y = 3x - 5/2$. Substituting this into the equation for the ellipse, we get $19x^2 - 30x + 11 = 0$. So $x = 11/19$, $y = -29/38$. This yields the solution $(22, -29, 38)$ to $2x^2 + 4y^2 = 3z^2$, which is the third solution we were told about.   ■

A picture, of course, helps so we draw Figure 3.6.

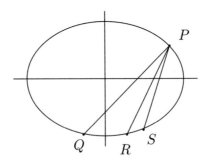

Figure 3.6

Figure 3.6 is an ellipse with the points $P = (1, 1/2)$, $Q = (-1/3, -5/6)$, $R = (1/3, -5/6)$, and $S = (11/19, -29/38)$. As you can see, it will be easy to generate lots of solutions by choosing rational points on the x-axis, joining them with P, and finding the intersection of the line with the ellipse.

This geometric method of solving Diophantine equations is called the **method of sweeping lines**. It seems too good to be true; simply find a single point on the curve and construct a line through the point with a rational slope so that it intersects the curve in a second point. That will automatically be a rational point also, and then we can go ahead and find integral solutions as we have outlined previously. But hold it. It does not always work. This works for quadratic equations because we always end up solving a quadratic in x with one rational root. It is automatic that the other root is rational as well. But it doesn't always work for cubics, as we shall now see.

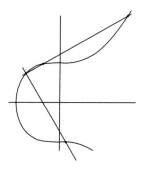

Figure 3.7

Example 3.4.17

Figure 3.7 depicts the curve that satisfies the equation $y^2 = x^3 + 17$. By trial and error we may find one rational point on the curve, $(-2, 3)$. Drawing a line of slope 1 through this point, as shown in Figure 3.7 gives us the line $y = x + 5$. Substituting this into the equation of the curve, we get $x^3 - x^2 - 10x - 8 = 0$. We do not know how to solve cubics yet (that will come in Chapter 4) but we know that $x + 2$ is a factor. Factoring, we get

$$x^3 - x^2 - 10x - 8 = (x + 2)(x^2 - 3x - 4).$$

This gives two more roots: $x = -1$ and $x = 4$. So we have two more points on the cubic $(-1, 4)$, $(4, 9)$. This looks promising. Let's try a another line, say of slope -3, as shown in the figure. The line through $(-2, 3)$ of slope -3 is $y = -3x - 3$. Substituting in and factoring yields

$$(x + 2)(x^2 - 11x + 4).$$

The quadratric formula gives

$$x = (11 \pm \sqrt{105})/2.$$

These are not rational numbers. Generally speaking, we will get a cubic equation for different slopes of lines through $(-2, 3)$, and the cubic is not guaranteed to factor into three rational linear terms. In order to ensure that you will find a new rational point on the curve, you must form a line passing through two known rational points on the curve. ■

We bring up the subject of finding rationals on a cubic for a reason. Finding the number of rational roots of Diophantine equations of degree 3 is a deep and significant area within the field of algebraic geometry. Such curves are called elliptic curves, and their study provided the ideas that are behind the proof of Fermat's last theorem.

EXERCISES

1. Prove this statement: If one root of a quadratic equation with rational coefficients is known to be a rational number, then the other root is also rational.

2. On each of the following arcs R, R' of the unit circle, find at least three rational points. Also find the rational point(s) on the arc that has (have) the smallest denominator.

    (a) $R = (4/5, 3/5)$, $R' = (3/5, 4/5)$

    (b) $R = (12/13, 5/13)$, $R' = (4/5, 3/5)$

    (c) $R = (15/17, 8/17)$, $R' = (21/29, 20/29)$

    (d) $R = (33/65, 56/65)$, $R' = (65/97, 72/97)$

    (e) $R = (48/73, 55/73)$, $R' = (28/53, 45/53)$

3. Prove Theorem 3.4.8: If R: (r, s) is a rational point on the unit circle and $r = b/a$, $s = c/a$, then the line PR, where P: $(-1, 0)$, has a y-intercept of $c/(a+b)$; that is, $Q = (0, c/(a+b))$.

4. In Example 3.4.11 we looked for points on the unit circle that solved the equation $x^2 + y^2 = 1$ and were nearly on the line $y = x$. Recall that $R_4 = (119/169, 120/169)$. Find R_n, where $n = 5$, 6, 7, and 8. How close are the x- and y-coordinates for R_8?

5. Prove or disprove:

   (a) $x^2 + y^2 = kz^2$ has an integral solution if and only if k can be represented as the sum of two squares.

   (b) $x^2 - ky^2 = -1$ has an integral solution if and only if k can be represented as the sum of two squares.

6. Formulate and prove a theorem that reads like this:

   If R: (r, s) is a rational point on the circle $x^2 + y^2 = 2z^2$ and $r = b/a$, $s = c/a$ then the line PR where P: $(-1, -1)$ has a y intercept of . . . .

7. Prove Theorem 3.4.14: If P, Q, R are described as in Figure 3.5, which depicts $x^2 - 2y^2 = z^2$, and if $q = b/a$, then R is the point (x, y), where coordinates are: $x = (a^2 + 2b^2)/(a^2 - 2b^2)$, $y = 2ab/((a^2 - 2b^2)$.

8. Draw pictures of the Diophantine equations that follow. Letting $P = (0, -1)$ and $Q = (0, 1/2)$, find R. Find several other solutions to the equations. Find general solutions to the given related equations.

   (a) $x^2 + 2y^2 = 1$, $x^2 + 2y^2 = z^2$
   (b) $x^2 + 3y^2 = 1$, $x^2 + 3y^2 = z^2$
   (c) $x^2 - 3y^2 = 1$, $x^2 - 3y^2 = z^2$

9. For each of the three equations in Exercise 8:

   (a) Make up theorems corresponding to Theorem 3.4.14.
   (b) Make up charts similar to the hyperbola chart.

10. Find five more solutions to the Diophantine equation $2x^2 + 4y^2 = 3z^2$.

11. Find five solutions to each the following Diophantine equations:

(a) $x^2 + 2y^2 = 3z^2$

(b) $2x^2 + 8y^2 = z^2$

(c) $x^2 + y^2 = 5z^2$

(d) $x^2 - 7y^2 = 2z^2$

12. For the elliptic curve $y^2 = x^3 + 17$,

(a) Use other rational slopes and find more solutions.

(b) Using the several rational points you have found on the curve, generate lots of new solutions.

13. Consider the following elliptic curves. Find integer points on each curve. Some points can be found by inspection, others by drawing lines joining your found points and intersecting the curve. Find points with large integer coordinates this way.

(a) $y^2 = x^3 - 4x^2 + 16$

(b) $y^2 = x^3 - 2x + 5$

(c) $y^2 = x^3 - 2x + 5$

14. Another way of studying the rational points on the unit circle is by examining the slopes of the lines through the origin that pass through the rational points. A line $y = mx$ has a **Pythagorean slope** if it intersects the unit circle in two rational points.

(a) Prove the following theorem: If a and b are positive fractions such that $a - b = 2$, then the fraction $(a + b)/ab$ is a Pythagorean slope and $(ab, a+b, ab+2)$ is a Pythagorean triple.

(b) Complete the last column of the following chart for Pythagorean slopes, which asks for a point in the first quadrant of the circle.

a	b	$a+b$	ab	$ab+2$	Point
3	1	4	3	5	
4	2	6	8	10	
5	3	8	15	17	
6	4	10	24	26	
5/2	1/2	3	5/4	13/4	
7/3	1/3	8/3	7/9	25/9	
13	5/4	9/2	65/16	97/16	

(c) Relate the a and b columns of this chart with those of Pythagorean chart in Section 2.3.

15. (a) Find an a and b, where $a - b = 2$ and $(a + b)/ab$ is the Pythagorean slope:

   (i) 33/56

   (ii) 48/575

   (iii) 612/35

   (b) If a and b generate the Pythagorean slope $(a+b)/ab$, then what two numbers c and d, two units apart, generate the reciprocal slope $ab/(a + b)$?

16. Here is one more concept concerning the unit circle and Pythagorean triples.

   Angle α is a **Pythagorean angle** if it may be formed out of the lines joining AO and OB meeting at the origin O, where A and B are rational points on the unit circle (see Figure 3.8). For example, $\arctan(3/4)$ (which is about 36.9 deg) is a Pythagorean angle because $A = (4/5, 3/5)$ and $B = (1, 0)$.

   Let $B = (1, 0)$, $A = (7/25, 24/25)$, $C = (8/17, 15/17)$. Let $\alpha = \angle AOB$, $\beta = \angle COB$.

   (a) Find a point D such that $\alpha + \beta = \angle DOB$.

   (b) Find E such that $\alpha - \beta = \angle EOB$.

   (c) Find F such that $2\alpha = \angle FOB$.

17. Prove that if α and β are Pythagorean angles, then

   (a) $\alpha + \beta$ is a Pythagorean angle.

   (b) $\alpha - \beta$ is a Pythagorean angle.

   (c) 2α is a Pythagorean angle.

   (d) What can you say about $\frac{1}{2}\alpha$?

18. Find $\angle AOB$ in degrees where A and B are

   (a) $(5/13, 12/13)$, $(8/17, 15/17)$

   (b) $(7/25, 24/25)$, $(12/37, 35/37)$

   (c) $(20/29, 21/29)$, $(21/29, 20/29)$

   (d) $(476/485, 93/485)$, $(93/485, 476/485)$.

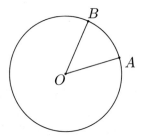

Figure 3.8

Chapter 4

The Real Numbers

In the three previous chapters we examined familiar numbers: the natural numbers, integers, and rational numbers. But the real line consists of other numbers, too; numbers that are not rational. Such numbers are called **irrational**. The irrational numbers are the most difficult to study. We know that all real numbers can be written as decimal expansions. Since the decimal expansion of a rational number must be either terminating or eventually repeating, the decimal expansion of a nonrational number must be infinite and non-periodic. Thus the irrational numbers are difficult to exhibit precisely because we have no way of describing their decimal behavior, in general. Of course, we can make up a decimal expansion for a particular irrational number; for example, let $N = 0.10110111011110\ldots$, where it is understood that one more 1 is put between successive 0s. Clearly it fits the description of a nonrational; but, generally speaking, the range of numbers that fall in the category of nonrepeating, nonterminating decimals is more than you could ever imagine. In fact, we shall learn that, in a technical sense, there are more irrational numbers than there are rational numbers—and we know that there are infinitely many rationals.

Philosophically the irrationals, as a family of numbers, cause ontological problems. Since it is impossible to describe the exact decimal expansion of such numbers, their very reality can be called into question. In fact, it was not until the mid-1800s that mathematicians confronted this problem directly. The upshot of this confrontation was that the irrational numbers were postulated to exist. In other words, the mathematicians axiomatized their way out of the dilemma by saying that all decimal expansions exist as numbers even if we don't know precisely what they are. This axiom is called the completeness axiom. We introduced it in Section 3.1.

While it is virtually impossible to distinguish any characteristics that an irrational might possess by examining its decimal expansion, irrationals do arise in particular contexts. We classify those contexts here as algebraic, geometric, and analytic. A number may have roots in more than one world. Perhaps the most famous irrational number is π; it has a geometric definition but it surfaces, sometimes in incredible ways, in many other branches of mathematics as well. While it is not completely known in its decimal form, it can be beautifully represented by various infinite series. The number e, the base of the natural logarithms, is also a famous irrational number. Its natural home is the world of analysis. The number, $\sqrt{2}$, is an algebraic entity; it is a solution of the polynomial equation $x^2 = 2$, but it also has roots in geometry. Over 2000 years ago, Greek mathematicians knew that it was impossible to measure the length of the diagonal of a square and the length of the side of a square with a common unit. This translates into the fact that $\sqrt{2}$ is irrational.

Our intention is to examine irrational numbers in their natural habitats. And there are subhabitats too; we shall define constructible numbers, arithmetic numbers, polygon numbers, algebraic numbers, and transcendental numbers. Two of these categories, arithmetic numbers and algebraic numbers, have general definitions within the field of complex numbers. Since we have restricted ourselves to a study of real numbers our definitions of these two types of numbers will naturally be restricted to the domain of real numbers. This observation should be kept in mind and we will repeat it in this Chapter at appropriate times.

4.1   Algebraic Representations

A number is represented arithmetically if it can be expressed through equations made up of arithmetic operations. The arithmetic operations are addition, subtraction, multiplication, division, exponentiation, and the taking of roots. In fact, such numbers have a name.

Definition 4.1.1 *A number x is called an* **arithmetic number** *if x can be expressed as a finite sequence of addition, subtraction, multiplication, division, exponentiation, and root taking of natural numbers.*

Notation: Let $\mathbb{A}$ represent the class of real arithmetic numbers.

Notice that the word "arithmetic" is used as an adjective here. The spelling does not distinguish it from its usual form as a noun. Only the context and the pronunciation can distinguish the two. As an adjective "arithmetic" is pronounced like "algebraic"; that is, with the accent on the third syllable.

Clearly the numbers of Chapters 1, 2, and 3 are arithmetic. The natural numbers are arithmetic; the integers can be derived by the subtraction of natural numbers, and the rationals can be derived by the division of natural numbers and integers. Of course, this is just the tip of the iceberg.

Example 4.1.2

All of the following numbers are arithmetic:

$$\sqrt{2}, \ \sqrt{3}, \ \sqrt{2}+\sqrt{3}, \ \sqrt[3]{2}, \ \sqrt[3]{2+\sqrt{2}}, \ \sqrt{-1},$$

$$\sqrt[5]{2 + \sqrt[3]{\frac{7}{16}} - 5},$$

$$\frac{\sqrt[11]{9} + \sqrt[31]{3 + \sqrt{17}}}{7 - 52\sqrt[4]{23}}.$$

■

One number in the example is not a real number; it is the imaginary number i. Since we are dealing with real numbers in this book, we will not examine complex numbers even though some of them are arithmetic.

Proving which arithmetic numbers are irrational is not a trivial task but it is one we shall investigate here. For example, you probably know $\sqrt{2}$ is irrational. So are $\sqrt{3}$ and $\sqrt[3]{2}$ and, for that matter, lots of other numbers like these. But roots of numbers that may look irrational can be rational (for example $\sqrt{2}/\sqrt{8}$ is $1/2$).

Theorem 4.1.3 *Let m be a natural number and suppose $m \neq k^n$ for any natural number k; then $\sqrt[n]{m}$ is irrational.*

Proof Suppose that $\sqrt[n]{m} = r/s$. Therefore $m = (r/s)^n$ and so $ms^n = r^n$. Suppose also that the fraction r/s is reduced to lowest terms. Since $ms^n = r^n$, it follows that $m \mid r^n$. Since r and s are relatively prime, it follows that $r^n \mid m$. Thus $m = r^n$ which contradicts the hypothesis of the theorem. So $ms^n \neq r^n$ and hence $\sqrt[n]{m}$ is irrational.   □

This theorem provides lots of examples of irrational numbers. Let us see what we can say about the the sums, products, and powers of such irrational numbers. In general, this is a difficult challenge, although some answers are easy to come by.

Theorem 4.1.4 *Let $N = r + s\sqrt[n]{m}$. If $\sqrt[n]{m}$ is irrational and r and s are rational, then N is irrational.*

Proof Suppose that N is rational. Then $\sqrt[n]{m} = (N - r)/s$. But this is impossible because $(N - r)/s$ is a rational number, contradicting the fact that $\sqrt[n]{m}$ is irrational.   □

Example 4.1.5

(a) The product of two irrational arithmetic numbers need not be irrational: $\sqrt{2} \times (1/\sqrt{2}) = 1$. That $1/\sqrt{2}$ is irrational follows because, if $1/\sqrt{2} = r$, a rational number, then $\sqrt{2} = 1/r$, also a rational number. This is a contradiction.

(b) The sum of two irrational numbers need not be irrational: $\sqrt{2} + (1 - \sqrt{2}) = 1$; $1 - \sqrt{2}$ is irrational by Theorem 4.1.4.

(c) An irrational raised to an irrational power need not be irrational. Consider $\sqrt{2}^{\sqrt{2}}$. If this is rational, we have our example. It turns out, however, it is irrational by Theorem 4.4.18. So raising this number to the $\sqrt{2}$ power, we get $(\sqrt{2}^{\sqrt{2}})^{\sqrt{2}} = 2$ and we have our example.   ∎

Of course, the preceding examples are a bit contrived. Suppose we are faced with multiplying $\sqrt{3} \times \sqrt[3]{2}$. It is not clear what we have.

Example 4.1.6

The product $\sqrt{3} \times \sqrt[3]{2}$ is irrational. Suppose it were rational; that is, $\sqrt{3} \times \sqrt[3]{2} = r/s$. So $s((\sqrt{3})(\sqrt[3]{2})) = r$ and therefore $s^6 3^3 2^2 = r^6$. We may suppose that r/s is in lowest terms and so $\gcd(r, s) = 1$. It follows that $r^6 \mid 3^3 2^2$, which is, of course, impossible. So $\sqrt{3} \times \sqrt[3]{2}$ is irrational. ∎

A tougher challenge is determining the nature of sums of irrational numbers. The following example takes on the task of determining irrationality of two relatively simple sums.

Example 4.1.7

(a) Let $N = \sqrt{2} + \sqrt{3}$ and suppose N is rational. We derive a contradiction from this assumption as follows: $N - \sqrt{2} = \sqrt{3}$. Squaring both sides we get, $(N - \sqrt{2})^2 = N^2 - 2\sqrt{2}N + 2 = 3$. So $N^2 - 1 = 2\sqrt{2}N$ and thus $(N^2 - 1)/2N = \sqrt{2}$. Since N is rational, the lefthand side of the equation is rational. Thus $\sqrt{2}$ is rational, a contradiction.

(b) Let $N = \sqrt[3]{2} + \sqrt[3]{3}$ and suppose that N is rational. Now $N^3 = 2 + 3\sqrt[3]{4}\sqrt[3]{3} + 3\sqrt[3]{2}\sqrt[3]{9} + 3$. So

$$N^3 - 5 = 3(\sqrt[3]{12} + \sqrt[3]{18}) = 3(\sqrt[3]{6}\sqrt[3]{2} + \sqrt[3]{6}\sqrt[3]{3}) = 3\sqrt[3]{6}N.$$

Thus $(N^3 - 5)/(3N) = \sqrt[3]{6}$. But the lefthand side is rational so the righthand side must also be rational, a contradiction. ∎

As we can see, the family, $\mathbb{A}$, of arithmetic numbers is somewhat formless; numbers can look like almost anything that has an addition, subtraction, multiplication, division, exponent, or root symbol appearing any number of times. Let us look at a different set of numbers that are very similar to the arithmetic numbers. We call these the algebraic numbers, and they are the solutions to polynomial equations with integer coefficients.

Definition 4.1.8 *A number, x, is an* **algebraic number** *if x is a solution to the equation*

$$a_n x^n + a_{n-1} x^{n-1} + a_{n-2} x^{n-2} + \cdots + a_2 x^2 + a_1 x + a_0 = 0,$$

where the a_i are integers.

Notation: We shall denote the class of real algebraic numbers by $\mathbb{B}$.

Remark Notice that the restriction of integers on the coefficients, a_i, can be changed to read "rationals" and there would be no difference in our set of algebraic numbers. This is because any polynomial equation with rational coefficients can be changed to one with integral coefficients by simply multiplying through both sides by a common multiple of the denominators of the coefficients. The righthand side remains 0 while the lefthand side is a polynomial with integer coefficients.

With a bit of thought we can see that the natural numbers, the integers, the rationals, and the roots of these numbers are all algebraic.

Theorem 4.1.9 *All natural numbers, integers, rational numbers, and roots of these are algebraic numbers.*

Proof The solution to the polynomial equation $x + -a_0 = 0$ shows that the natural number or the integer, a_0, is algebraic. The solution to $a_1 x + -a_0 = 0$ shows that the rational a_0/a_1 is algebraic. The solution to $a_n x^n + -a_0 = 0$ shows that an nth root of a_0/a_n is algebraic. $\square$

As with arithmetic numbers, we notice that there are algebraic numbers that are not real numbers. The polynomial equation $x^2 = -1$ does not have a solution that is a real number; its solutions are $\pm i$. Complex numbers satisfying Definition 4.1.8 are also algebraic numbers. They are not included in our set $\mathbb{B}$ simply because this book restricts itself to real numbers. Let us see if the arithmetic numbers we examined are algebraic numbers as well.

Example 4.1.10

(a) Let $N = \sqrt{2} + \sqrt{3}$. From Example 4.1.7 (a), $N^2 - 1 = 2\sqrt{2}N$. Squaring, we get $N^4 - 2N^2 + 1 = 8N^2$, which gives $N^4 - 10N^2 + 1 = 0$. So N is algebraic.

(b) Let $N = \sqrt[3]{2} + \sqrt[3]{3}$. From Example 4.1.7 (b), $(N^3 - 5) = 3\sqrt[3]{6}N$. Cubing both sides, we get $N^9 - 15N^6 + 75N^3 - 125 = 162N^3$. So N satisfies the equation $N^9 - 15N^6 - 87N^3 - 125 = 0$. $\blacksquare$

It is natural to wonder what the other real roots of these polynomial equations might be. It turns out that $\sqrt{2} + \sqrt{3}$ is one of four real solutions to $x^4 - 10x^2 + 1 = 0$, the other three are $-\sqrt{2} + \sqrt{3}$, $\sqrt{2} - \sqrt{3}$, and $-\sqrt{2} - \sqrt{3}$. It looks as though they are sort of like conjugates of each other. You can test this by multiplying these four linear polynomials and reclaiming the quartic, $x^4 - 10x^2 + 1$:

$$(x - (\sqrt{2} + \sqrt{3}))(x - (\sqrt{2} - \sqrt{3}))(x + (\sqrt{2} + \sqrt{3}))(x + (\sqrt{2} - \sqrt{3})).$$

It turns out that $\sqrt[3]{2} + \sqrt[3]{3}$ is the only real solution to $N^9 - 15N^6 - 87N^3 - 125 = 0$.

Sometimes it is possible to look at the polynomial equation that yields an algebraic number and tell whether it is rational or irrational. The rational root theorem is helpful here.

Theorem 4.1.11 Rational Root Theorem *If $a_n x^n + a_{n-1} x^{n-1} + \cdots + a_2 x^2 + a_1 x + a_0 = 0$, where a_i are integers and if r/s, a fraction in lowest terms, is a solution, then $s \mid a_n$ and $r \mid a_0$.*

Proof Suppose that r/s is a root and it is reduced to lowest terms. Then

$$a_n(r/s)^n + a_{n-1}(r/s)^{n-1} + \cdots + a_2(r/s)^2 + a_1(r/s) + a_0 = 0.$$

Multiplying through this equation by s^n, we get

$$a_n r^n + a_{n-1} r^{n-1} s + \cdots + a_2 r^2 s^{n-2} + a_1 r s^{n-1} + a_0 s^n = 0.$$

Since $s \mid a_i r^{n-i} s^i$ for $i > 0$, we see that $s \mid a_n r^n$, and since r and s are relatively prime, $s \mid a_n$. By an analogous argument, $r \mid a_0 s^n$ and again, since r and s are relatively prime $r \mid a_0$.   □

Example 4.1.12

Let $N = \sqrt{2} + \sqrt[3]{2}$. Then $\sqrt[3]{2} = N - \sqrt{2}$, so

$$2 = (N - \sqrt{2})^3 = N^3 - 3N^2\sqrt{2} + 3N(2) - 2\sqrt{2}.$$

Rewriting this we get $N^3 + 6N - 2 = (3N^2 - 2)\sqrt{2}$. Squaring both sides yields

$$N^6 + 2N^3(6N - 2) + 36N^2 - 24N + 4 = (9N^4 - 12N^2 + 4)(2).$$

This yields the polynomial equation

$$N^6 - 6N^4 - 4N^3 + 60N^2 - 24N - 4 = 0.$$

If $N = r/s$, a rational number, then $s = \pm 1$ and $r = \pm 4, \pm 2$, or ± 1.

Plugging these integers into the polynomial equation, we see that none works so, by the rational root theorem, N is irrational.   ■

Let us now organize our examination of the solutions to polynomial equations with integer coefficients; that is equations of the form

$$f(x) = a_n x^n + a_{n-1} x^{n-1} + \cdots + a_2 x^2 + a_1 x + a_0 = 0,$$

where a_i are integers. Our study begins with the familiar quadratic equations. The numbers that are solutions to quadratic equations have been known in some fashion by mathematicians for more than 2000 years. We learn them in high school algebra; the formula for finding the solutions is the quadratic formula.

The Quadratic Formula *Solutions to the equation $ax^2 + bx + c = 0$, where $a \neq 0$, are r and s, where*

$$r = \frac{1}{2a}(-b + \sqrt{b^2 - 4ac}), \quad s = \frac{1}{2a}(-b - \sqrt{b^2 - 4ac}).$$

In order to verify that the quadratic formula indeed gives the roots of a quadratic, we can simply plug the roots into $ax^2 + bx + c$ and see if we get 0. We do. We also know from high school algebra how the roots were obtained; they were obtained by completing the square. The quadratic formula works without restriction on the coefficients; however, the algebraic numbers and the arithmetic numbers are the ones that arise from quadratics with rational coefficients. We'll make a note of that with the following theorem.

Theorem 4.1.13 *Solutions to quadratic polynomial equations with rational coefficients are algebraic numbers and arithmetic numbers.*

Next in line is the class of cubic equations; then come the quartics, the quintics, and so on. The search for general solutions to these classes of polynomial equations forms one of the most interesting chapters in mathematics history. We shall examine this search briefly here and urge you to read a book on the history of mathematics for a full account.

The Italian mathematicians of the Renaissance period are credited with the discovery of explicit formulas for finding the roots of a general cubic and a general quartic. At least five mathematicians were involved in these discoveries: Scipione del Ferro, Antonio Fior,

Tartaglia, Cardano, and Ferrari. Scipione (1465–1526), a professor at the University of Bologna, discovered a formula that could express the roots of a cubic of the form: $x^3 + px + q = 0$. This was about 1515. Unfortunately, he did not publish his results; however he did share them with his student, Antonio Fior. Thirty years later Tartaglia (1500–1557) found a general solution to the cubic $x^3 + ax^2 + b = 0$ and rediscovered Scipione's solution. Not believing that Tartaglia had found a cubic formula, Fior challenged him to a public cubic-solving contest in 1545. Tartaglia won the contest; he solved all of Fior's problems and Fior solved none of his. While Tartaglia was not eager to share his secret cubic-solving abilities with anyone, he did confide them to Cardano after securing a promise that they would remain confidential. Cardano proceeded to publish the cubic formula as his own in his famous work *Ars Magna* (1545). To Cardano's credit, he didn't merely copy the earlier work of Scipione and Tartaglia but expanded on it. He seemed to recognize that there were indeed three roots to a cubic and two might be complex. At this time in history imaginary numbers were still very strange objects, as this translation of a quote from *Ars Magna* indicates:

> To divide 10 into two parts such that the product is 40 is manifestly impossible; nevertheless we will operate with these sophisticated quantities. But further work is as subtle as it would be useless.

Tartaglia, incensed over Cardano's breaking of his promise of secrecy, challenged Cardano's prize student, Ferrari, to a public problem-solving contest in 1548. In 1540 Ferrari (1522–1565) had shown how to reduce the problem of solving a general quartic to a problem of solving a cubic. This time, Tartaglia lost the contest, and, unfortunately, in history he lost the authorship of his solution to the cubic; Cardano is credited with the discovery. Here are formulas for general cubics and quartics.

A Cubic Formula *The solutions to $x^3 + px + q = 0$, where $p \neq 0$, are r_1, r_2, and r_3, where*

$$r_1 = \alpha - p/3\alpha$$
$$r_2 = \omega\alpha - p/3\omega\alpha$$
$$r_3 = \omega^2\alpha - p/3\omega^2\alpha$$

and where $\omega = \frac{1}{2}(-1 + \sqrt{3}\,(i)$ and $\alpha^3 = -\frac{1}{2}q + \sqrt{(\frac{q}{2})^2 + (\frac{p}{3})^3}$.

We should notice that the cubic formula does not address the solution of the general cubic; $ax^3 + bx^2 + cx + d = 0$. But it is not difficult to change any cubic to be of the form $x^3 + px + q = 0$. Simply replace the variable, x, with $x - b/3a$. Then

$$a(x - b/3a)^3 + b(x - b/3a)^2 + c(x - b/3a) + d =$$
$$ax^3 + (-b + (b)x^2 + (-b^2/3a + (c)x + 2b^3/27a^2 - bc/3a + d.$$

Dividing through by a, which doesn't affect the roots, yields $x^3 + px + q$. Once the latter equation is solved, simply translate the solutions by $b/3a$. This trick of translating can be used on quadratic equations, too. Given the general form $ax^2 + bx + c$, replacing the variable, x, by $x - b/2a$ rids us of the linear term.

In applying the cubic formula, there are a couple of things to observe. It doesn't matter which of the three cube roots of α^3 you pick to call α in the above calculation. Also, you should recall some facts about the cube roots of 1. Since both ω and ω^2 are cube roots of 1, there are some easily derived equalities: $\omega^3 = 1$, $\omega^{-1} = \omega^2$, $\omega^{-2} = \omega$, $1 + \omega + \omega^2 = 0$, and $1 + 1/\omega + 1/\omega^2 = 0$.

A Quartic Formula *The solutions to $x^4 + px^2 + qx + r = 0$ are the quadratic solutions to*

$$x^2 - \tfrac{1}{2}(-p + \sqrt{p^2 - 4r}) \text{ and } x^2 - \tfrac{1}{2}(-p - \sqrt{p^2 - 4r}), \text{ if } q = 0$$

$$x^2 + mx + n = 0 \text{ and } x^2 + ux + v = 0 \text{ if } q \neq 0, \text{ where}$$

$$m^2 = t, t \text{ is a solution to the equation: } y^3 + 2py^2 + (p^2 - 4r)y - q^2 = 0,$$

$$n = \tfrac{1}{2m}(m^3 + pm - q), u = -m, \text{ and } v = m^2 + p - n.$$

This formula, like the cubic formula, is not in the most general form. But, as with the cubic, the general form $ax^4 + bx^3 + cx^2 + dx + e$ becomes a quartic of the form $x^4 + px^2 + qx + r$ with the replacement of the variable, x, by $x - b/4a$.

As with the quadratic formula so, too, the cubic and the quartic formulas are true without regard to limitations on the coefficients. As with the quadratic formula, we can verify that these formulas give the desired result by simply plugging the respective solutions into the cubic and into the quartic and obtaining 0. This will take some doing, however. Finally, we note that the roots of cubics and quartics are arithmetic and algebraic numbers if the coefficients of the cubic and of the quartic are rational numbers.

Theorem 4.1.14 *Solutions to cubic polynomial equations and quartic polynomial equations with rational coefficients are arithmetic numbers and algebraic numbers.*

Example 4.1.15

(a) Consider the equation $x^3 + x - 1 = 0$. Using the cubic formula, we find $p = 1$, and $q = -1$. So $\alpha^3 = \frac{1}{2}\left(1 + \sqrt{1 + \frac{4}{27}}\right)$. Let α be the real cube root of this, so $\alpha = \sqrt[3]{\frac{1}{2}\left(1 + \sqrt{1 + \frac{4}{27}}\right)}$. Our real solution looks like this as an arithmetic number:

$$\alpha - p/(3\alpha) = \sqrt[3]{\frac{1}{2}\left(1 + \sqrt{1 + \frac{4}{27}}\right)} - 1 \bigg/ \left(3\sqrt[3]{\frac{1}{2}\left(1 + \sqrt{1 + \frac{4}{27}}\right)}\right).$$

As a decimal this number is $0.682327803828\ldots$.

The complex solutions look like this:

$$\omega\alpha - p/(3\omega\alpha) \text{ and } \omega^2\alpha - p/(3\omega^2\alpha).$$

(b) Consider the equation $x^3 - 3x + 1 = 0$. Using the cubic formula, we have $p = -3$, $q = 1$, and $\alpha^3 = \frac{1}{2}(-1 + \sqrt{3}(i))$. Some calculation yields

$$r_1 = \sqrt[3]{\frac{1}{2}(-1 + \sqrt{3}(i))} + 1\bigg/\left(\sqrt[3]{\frac{1}{2}(-1 + \sqrt{3}(i))}\right)$$
$$= \alpha + 1/\alpha = 1.532088886\ldots$$
$$r_2 = \alpha^4 + 1/\alpha^4 = -1.87938524\ldots$$
$$r_3 = \alpha^7 + 1/\alpha^7 = 0.347296355\ldots.$$

Since these are arithmetic numbers, we should be able to write them in arithmetic form as well as decimal form. This is not a trivial task, as you will find if you do Exercise 13 following this section. Notice that all three roots are real. We would never suspect that from looking at their arithmetic forms, but we can see this on a graphing calculator.

(c) Consider the quartic equation $x^4 - 2x^3 - 5x^2 + 10x - 3 = 0$. First, we translate the quartic to get rid of the cube term. Exchanging x for $x + 1/2$, we obtain $x^4 - (13/2)x^2 + 4x + 9/16$. Now m^2 is the solution to $y^3 - 13y^2 + 40y - 16 = 0$, and luckily the rational root test, or mere inspection, reveals that 4 is a solution to this cubic equation. Choosing $m = 2$ (if $m = -2$ the results will be the same), we find that $n = -9/4$, $v = -1/4$, and $u = -2$. The theorem tells us that the solutions to the quartic equation are just the solutions to the quadratic equations: $x^2 + 2x - 9/4 = 0$ and $x^2 - 2x - 1/4 = 0$. The quadratic formula gives us four solutions and translating them $+1/2$ units we finally get the four solutions to the original quartic equation: They are, in arithmetic form

$$\frac{1}{2}(-1 - \sqrt{13}), \; \frac{1}{2}(3 - \sqrt{5}), \; \frac{1}{2}(-1 + \sqrt{13}), \; \frac{1}{2}(3 + \sqrt{5}).$$

In decimal form they are, respectively

$-2.3027756\ldots, \; .3819660\ldots, \; 1.3027756\ldots, \; 2.6180340\ldots.$ ∎

It's not nice to pass judgment on these formulas, but they are nearly impossible to work with. Example 4.1.15 (c) was rigged so we could get through it. And Example 4.1.15 (b) was also tailor made. If it is upsetting to learn how difficult these formulas can be to work through, we can at least be thankful that there are formulas. Listen to this next piece of news. There is no general formula for quintics. This is not to say that such a formula has yet to be discovered; it is to say that there is no formula, none. Sure, we may solve certain quintics, like $x^5 - 7 = 0$: $x = \sqrt[5]{7}$, but we can't solve them in general with a formula built out of roots. Let us finish our story.

When we left off, the Italian Renaissance mathematicians were struggling with solving cubic and quartic polynomials. More than two centuries passed as mathematicians searched for a solution to the general quintic. In 1799, Ruffini, an Italian mathematician-physicist, proved that there would be no more need to search; there would be no general formula for quintics. His proof lacked rigor, and he is not credited with the discovery. This discovery is credited to N. H. Abel, (1802–1829), a Norwegian mathematician, who provided a rigorous proof. This was an amazing achievement. Heretofore, mathematicians showed their genius by solving complicated problems and creating very specific formulas. This is what Tartaglia and Cardano had done with the cubic formula. The idea that something was not just difficult to solve but, by its very nature, impossible to solve shed a new light on mathematics. What Abel proved was that there is no constructible solution to a general quintic equation. By this we mean that there is no formula constructed out of combinations of square roots, cube roots, fourth roots, or, for that matter, nth roots of rational numbers that will represent the solution to $ax^5 + bx^4 + cx^3 + dx^2 + ex + f = 0$. The French mathematician Evariste Galois (1811−1832) completed the saga by producing a theory that ties the possibility of solving cubics and quartics with the impossibility of solving quintics. Today Galois theory is still studied and admired as one of the most beautiful theories of mathematics.

Until the nineteenth century mathematicians tried hard to make their numbers arithmetic. They felt that numbers did not exist unless they could be found through a series of mechanical operations of addition, subtraction, multiplication, division, exponentiation, and root taking. We see that the real solutions of any polynomial equations with integer coefficients of degree 1, 2, 3, or 4 are algebraic numbers that are arithmetic.

Since the general quintic cannot be solved, there must be quintic polynomial equations whose solutions are not arithmetic. Indeed there are. While we are not in position to prove it, we can state a surprising theorem by Chowla and Bhalotra, proved in 1942, that allows us to create lots of examples.

Theorem 4.1.16 *Suppose that* $x^5 + ax^3 + bx^2 + cx + d$ *has no rational roots and cannot be factored into a cubic and a quadratic with integral coefficients. If* a *and* b *are even and* c *and* d *are odd, then the solutions to* $x^5 + ax^3 + bx^2 + cx + d = 0$ *are not arithmetic.*

Example 4.1.17

(a) Let $f(x) = x^5 + x + 1 = 0$. There is one real solution to this equation and it is $-.754877666247\dots$. Look at this number carefully. It is an algebraic number that is not arithmetic. We leave it as an exercise to show that $f(x)$ is not the product of a cubic and a quadratic.

(b) Let $f(x) = x^5 - 4x^3 + 3x - 1 = 0$. There are five real solutions to this equation (your graphing calculator will show this) and $f(x)$ looks to be of the right form but, in fact, it factors into a cubic and a quadratic: $x^5 - 4x^3 + 3x - 1 = (x^3 - x^2 - 2x + 1)(x^2 + x - 1)$. Thus these five solutions are arithmetic. ∎

We see that there are algebraic numbers that are not arithmetic. It is true, however, that all arithmetic numbers are algebraic. In set notation we may write $\mathbb{A} \subset \mathbb{B}$. The proof of this is not that easy, and we have not studied algebraic structures enough to pursue it here.

Before moving along we should cite a most famous result in algebra called the fundamental theorem of algebra. This theorem is to

algebra what the fundamental theorem of arithmetic is to arithmetic.
They both deal with factorization.

The Fundamental Theorem of Algebra *There are exactly n, not
necessarily distinct, complex solutions to the polynomial equation:*

$$a_n x^n + a_{n-1} x^{n-1} + \cdots + a_2 x^2 + a_1 x + a_0 = 0$$

where the coefficients, a_i, are complex numbers.

While we are not studying solutions of polynomial equations with
complex coefficients, this is a theoretical result that, for us, unleashes
a myriad of possible numbers that might qualify as algebraic numbers; after all, there are as many as n real number solutions to every
polynomial equation of degree n with integer coefficients. This mas-
terpiece theorem tells us that when it comes to solving polynomials,
there are no numbers beyond the complex numbers that can exist.
The insight that the complex numbers are the set that comprised the
total solutions to polynomial equations was proved in 1799 by Carl
Friedrich Gauss in his Ph.D. dissertation.

We now recognize that there are lots of real algebraic numbers.
But they will be difficult to access because there is no general formula
for finding solutions of polynomial equations of degree higher than
four. Thus there is no directive as to how to express such numbers in
terms of successive uses of arithmetic operations. But there are whole
sets of arithmetic and algebraic numbers that do have a recognizable
form. Let us discuss this now. This is appropriate because the study
of algebra involves the study of abstract structure, and algebraic
numbers provide an ideal setting for such a study. First we need
some definitions.

Definition 4.1.18 *An* **integral domain** *is made up of a set of ob-
jects, S, along with two operations $+$ and $\times$ that satisfy the following:*
Closure
    (i)   For all a, b in S, $a + b$ is in S.
    (ii)   For all a, b in S, $a \times b$ is in S.
Commutativity
    (iii)   For all a, b in S, $a + b = b + a$.
    (iv)   For all a, b in S, $a \times b = b \times a$.
Associativity
    (v)   For all a, b, c in S, $(a + b) + c = a + (b + c)$.

(vi)   For all a, b, c in S, $(a \times (b)) \times c = a \times (b \times (c))$.

Existence of an Identity

(vii)   There exists 0 in S such that for all a in S, $a + 0 = a$.

(viii)   There exists $1 \neq 0$ in S such that for all a in S, $a \times 1 = a$.

Existence of an Additive Inverse

(ix)   For all a in S, there is a b in S such that $a + b = 0$.

Cancellation

(x)   For all a, b, c in S, if $a \neq 0$ and $a \times b = a \times c$, then $b = c$.

Distributivity

(xi)   For all a, b, c in S, $a \times (b + c) = a \times b + a \times c$.

Definition 4.1.19 *A **field** is made up of a set of objects, S, along with two operations $+$ and $\times$ that satisfy the axioms of an integral domain along with the axiom of existence of multiplicative inverses.*

Existence of a Multiplicative Inverse

(xii) For all $a \neq 0$, a in S, there is b in S such that $a \times b = b \times a = 1$.

Notation: The integral domain and the field described above is often denoted by $(S, +, \times)$.

We should note that technically, axiom (x) can be replaced by axiom (xii) for the axioms for a field since (xii) implies (x). Notice also that integral domains provide the natural setting for the concepts of prime number, factoring, remainders, and congruence arithmetic. In a field, there are no remainders after division; after all, division of a by b is exact; it is the number expressed by a/b. A field, on the other hand, is a nice complete package; it is a mathematical system that has enough elements that all the basic arithmetic operations, except root finding, can be carried out successfully.

The integers are the best-known example of an integral domain. The real numbers and the rational numbers are the best-known examples of fields. We shall assign notation to these systems.

Notation: Let $\mathbb{R}$ stand for the field of real numbers, $\mathbb{Q}$ stand for the field of rational numbers, and $\mathbb{Z}$ stand for the integral domain of integers. These letters are standard designations.

Although we are not in a position to prove it here, the algebraic numbers and the arithmetic numbers also form fields. Using set notation, we have the following inclusions: $\mathbb{Q} \subset \mathbb{A} \subset \mathbb{B} \subset \mathbb{R}$.

Now we shall gather together like types of algebraic numbers and construct more examples of fields.

Example 4.1.20

Consider numbers of the form $r + s\sqrt{2}$, where r and s are rational. We can add them, subtract them, multiply them, and divide them and the results are numbers of the same form. Besides closure, the other laws of a field hold because these are real numbers and they form a field.

(a) Consider $(2 + 3\sqrt{2}) + (1 - 5\sqrt{2})$. This is $3 - 2\sqrt{2}$.

(b) Consider $(2 + 3\sqrt{2}) \times (1 - 5\sqrt{2})$. Using elementary algebra techniques, we get the product $-28 - 7\sqrt{2}$.

(c) Consider $(2 + 3\sqrt{2})/(1 - 5\sqrt{2})$. Multiplying the numerator and denominator by $(1 + 5\sqrt{2})$, the conjugate of the denominator, we get

$$(2 + 3\sqrt{2})(1 + 5\sqrt{2})/(1 - 5\sqrt{2})(1 + 5\sqrt{2})$$

$$= (32 + 13\sqrt{2})/(-49) = (-32/49) - (13/49)\sqrt{2}. \quad \blacksquare$$

This is a particular example of a general situation.

Definition 4.1.21 *The* **quadratic field**, $\mathbb{Q}(\sqrt{k})$, *is the set of numbers of the form* $r + s\sqrt{k}$, *where k is a natural number that is not a perfect square and r and s are rational. Addition and multiplication are carried out in the familiar setting of the real numbers; that is*

$$(r + s\sqrt{k}) + (t + u\sqrt{k}) = (r + t) + (s + u)\sqrt{k},$$

$$(r + s\sqrt{k}) \times (t + u\sqrt{k}) = (rt + ksu) + (ru + st)\sqrt{k}.$$

Theorem 4.1.22 $\mathbb{Q}(\sqrt{k})$ *is a field.*

Proof The laws of commutativity, associativity, and distributivity all hold because they hold in the parent field of the real numbers and are thus inherited by $\mathbb{Q}(\sqrt{k})$. Obviously, 0 and 1 are in $\mathbb{Q}(\sqrt{k})$, $0 = 0 + 0\sqrt{k}$, $1 = 1 + 0\sqrt{k}$. The numbers are closed under addition and multiplication and finally, every number $r + s\sqrt{k}$ has an additive inverse, $-r + -s\sqrt{k}$, and every nonzero number has a multiplicative inverse; the inverse of $r + s\sqrt{k}$, where $r, s \neq 0$ is

$$(1/(r^2 - ks^2))(r - s\sqrt{k}). \qquad \square$$

As a sidebar, notice that the set of numbers of the form $n + m\sqrt{k}$, where n and m are integers, is an integral domain. So we can theoretically speak of concepts like primes and factoring in this domain. We leave it as a project (5.10) to analyze divisibility properties of a certain integral domain of this type.

Quadratic fields give us a glimpse into the types of numbers that make up a multitude of irrational numbers. We are seeing that subsets of the irrationals combine with the rationals to form fields. These fields are algebraic universes that lead parallel existences. The fields $\mathbb{Q}(\sqrt{k})$ and $\mathbb{Q}(\sqrt{j})$ are very much alike, yet they contain entirely different irrationals; their entire overlap is the rationals.

Theorem 4.1.23 *If j and k are not perfect squares and $\gcd(j, k) = 1$, then $\sqrt{j}$ is not in the field $\mathbb{Q}(\sqrt{k})$.*

The proof is left as an exercise.

We have barely scratched the surface of the subject of the fields that contain irrationals. We will examine it a bit more before giving a general formulation. The following example shows that we can use the idea of supplementing an existing field with an alien element again and again to form new fields.

Example 4.1.24

Consider the field $\mathbb{Q}(\sqrt{2})$ and $\sqrt{3}$. Since $\sqrt{3}$ is not in $\mathbb{Q}(\sqrt{2})$, we may build the set of numbers $a + b\sqrt{3}$, where a and b live in the field $\mathbb{Q}(\sqrt{2})$. So $a = r + s\sqrt{2}$, $b = t + u\sqrt{2}$ and a typical element appears as

$$(r + s\sqrt{2}) + (t + u\sqrt{2})\sqrt{3} = r + s\sqrt{2} + t\sqrt{3} + u\sqrt{6}.$$

We may denote this field by $[\mathbb{Q}(\sqrt{2})](\sqrt{3})$. Addition is clear, but multiplication and division can pose a trial. Consider the numbers $x = \sqrt{2} + \sqrt{3} + \sqrt{6}$ and $y = \sqrt{2} - 2\sqrt{3} - \sqrt{6}$. Their sum is easy enough: $x + y = 2\sqrt{2} - \sqrt{3}$. Their product $x \times y$ is more difficult:

$$(\sqrt{2} + \sqrt{3} + \sqrt{6})(\sqrt{2} - 2\sqrt{3} - \sqrt{6}) =$$

$$2 - 2\sqrt{6} - \sqrt{12} + \sqrt{6} - 6 - \sqrt{18} + \sqrt{12} - 2\sqrt{18} - 6 = -10 - 9\sqrt{2} - \sqrt{6}.$$

Finally, we calculate x/y. One way to do this is to find the multiplicative inverse of y. Let it be $a\sqrt{2} + b\sqrt{3} + c\sqrt{6} + d$. So

$$(\sqrt{2} - 2\sqrt{3} - \sqrt{6})(a\sqrt{2} + b\sqrt{3} + c\sqrt{6} + (d)) = 1.$$

Multiplying this out, the lefthand side becomes

$$2a + b\sqrt{6} + 2c\sqrt{3} + d\sqrt{2} - 2a\sqrt{6} - 6b - 6c\sqrt{2}.$$
$$-2\sqrt{3}d - 2a\sqrt{3} - 3b\sqrt{2} - 6c - d\sqrt{6}$$

We may set up the following equalities:

$$
\begin{aligned}
2a - 6b - 6c &= 1 \\
(-6c - 3b + (d)\sqrt{2} &= 0 \\
(2c - 2a - 2d)\sqrt{3} &= 0 \\
(b - 2a - (d)\sqrt{6} &= 0.
\end{aligned}
$$

Solving simultaneously, we get $a = c = -1/16$, and $b = -1/8$. So the multiplicative inverse of $\sqrt{2} - 2\sqrt{3} - \sqrt{6}$ is $\frac{1}{2}\sqrt{2} + \frac{1}{4}\sqrt{3} - \frac{1}{4}\sqrt{6} - \frac{3}{4}$. Now back to our division problem:

$$(\sqrt{2} + \sqrt{3} + \sqrt{6})/(\sqrt{2} - 2\sqrt{3} - \sqrt{6})$$
$$= (\sqrt{2} + \sqrt{3} + \sqrt{6}) \times \frac{1}{2}\sqrt{2} + \frac{1}{4}\sqrt{3} - \frac{1}{4}\sqrt{6} - \frac{3}{4}$$
$$= \frac{1}{4} - \frac{3}{4}\sqrt{2} - \frac{1}{4}\sqrt{3}. \qquad\blacksquare$$

There is a general way of describing the process of building fields. The polynomials over a field form an integral domain. Some polynomials are like prime numbers; they do not factor; $p(x) = x^2 - 2$ is an example of such a polynomial over the rationals. These polynomials are called irreducible. When an irreducible polynomial introduces a root to a field, like the introduction of $\sqrt{2}$ to the rationals, a foreign element is thrown into the mix. In order to absorb it a new expanded field is formed. The expanded field includes all the new members that result from using the alien element in arithmetic operations. The following is a description of what happens.

(i) Start with a field, F.

(ii) Find a polynomial, call it $p(x)$, of degree $n > 1$ that cannot be factored over F.

(iii) Give a name, α, to a root of the irreducible polynomial. So $p(\alpha) = 0$.

(iv) Build polynomials in α of degree less than n. These are the numbers of the new field, call it $F(\alpha)$.

(v) Addition for the numbers in $F(\alpha)$ is the familiar polynomial addition. Multiplication of two numbers, $f(\alpha)$ and $g(\alpha)$, is simply polynomial multiplication if the degree of the product is $< n$.

If the degree of $f(\alpha) \times g(\alpha)$ is $\geq n$, then the condition that $p(\alpha) = 0$ must be brought in. This can be done efficiently by performing the division $f(x)g(x) \div p(x)$, finding the remainder, and evaluating it at α.

Example 4.1.25

(a) Recall Example 4.1.20: $F = \mathbb{Q}$, $p(x) = x^2 - 2$, $\alpha = \sqrt{2}$; numbers in $\mathbb{Q}(\sqrt{2})$ are of the form $r + s\sqrt{2}$ and $(2 + 3\sqrt{2})(1 - 5\sqrt{2}) = -28 - 7\sqrt{2}$. From the general point of view outlined above, we perform multiplication this way:

$$(2 + 3x)(1 - 5x) \div (x^2 - 2) = (-15x^2 - 7x + 2) \div (x^2 - 2) = -15 + R$$

where R stands for the remainder: $R = -7x - 28$.

Substituting in $\sqrt{2}$ for x, we get the wanted answer: $-7\sqrt{2} - 28$.

(b) Consider $\mathbb{Q}(\sqrt[3]{2})$. Following the procedure above, we have

(i) $F = \mathbb{Q}$

(ii) $p(x) = x^3 - 2$

(iii) Let $\alpha = \sqrt[3]{2}$

(iv) The numbers in $\mathbb{Q}(\sqrt[3]{2})$ are of the form $r + s\sqrt[3]{2} + t\sqrt[3]{2^2}$.

Let us find $a \times b$, where $a = 2 + \sqrt[3]{2} + 3\sqrt[3]{4}$ and $b = -1 + 2\sqrt[3]{2} - \sqrt[3]{4}$.

(v) $(2 + \sqrt[3]{2} + 3\sqrt[3]{4}) \times (-1 + 2\sqrt[3]{2} - \sqrt[3]{4}) =$
$-2 + 4\sqrt[3]{2} - 2\sqrt[3]{4} - \sqrt[3]{2} + 2\sqrt[3]{4} - \sqrt[3]{8} - 3\sqrt[3]{4} + 6\sqrt[3]{8} - 3\sqrt[3]{16}$
Using the facts that $\sqrt[3]{8} = 2$ and $\sqrt[3]{16} = 2\sqrt[3]{2}$, we obtain $8 - 3\sqrt[3]{2} - 3\sqrt[3]{4}$.

Using the technique of polynomial divison, we obtain

$$(2 + x + 3x^2)(-1 + 2x - x^2) = -3x^4 + 5x^3 - 3x^2 + 3x - 2.$$
$(-3x^4 + 5x^3 - 3x^2 + 3x - 2) \div (x^3 - 2) = -3x + 5 + R$ and $R = -3x^2 - 3x + 8$.

Substituting in $\sqrt[3]{2}$ for x we get $-3\sqrt[3]{4} - 3\sqrt[3]{2} + 8$ which corresponds to our answer above.

(c) Consider $\mathbb{Q}(\alpha)$¡ where α is a root of $x^2 - 2x - 1$. Numbers are of the form $r + s\alpha$. Let us find the product $(1 + \alpha)(2 + \alpha)$. The answer is, of course, $2 + 3\alpha + \alpha^2$. Noting that $\alpha^2 = 1 + 2\alpha$, we make the substitution and obtain $3 + 5\alpha$. We get the same result if we use the technique of polynomial division.

$$(1 + x)(2 + x) = 2 + 3x + x^2; (x^2 + 3x + 2) \div (x^2 - 2x - 1) = 1 + R$$

$R = 5x + 3$. So our answer is $5\alpha + 3$.

We can also solve for α and plug the result into $2 + 3\alpha + \alpha^2$. This is inconvenient; there are two roots to the equation $x^2 - 2x - 1$; $1 \pm \sqrt{2}$. Choosing one of the roots, say $1 + \sqrt{2}$, we obtain $8 + 5\sqrt{2}$. Using the other root, $1 - \sqrt{2}$, we get $8 - 5\sqrt{2}$.   ■

Note that example (c) contains numbers of the form $r + s\sqrt{2}$, so we can ask if this field is the same as $\mathbb{Q}(\sqrt{2})$. The answer is yes; the naming of the elements is different but the arithmetic operations are the same on the same set of elements. We leave this as an exercise. Here is another example of a field whose elements have different representations.

Example 4.1.26

Consider the field of Example 4.1.24, $[\mathbb{Q}(\sqrt{2})](\sqrt{3})$. Its set of numbers is $a + b\sqrt{3}$, where a and b live in the field $\mathbb{Q}(\sqrt{2})$. So $a = r + s\sqrt{2}$, $b = t + u\sqrt{2}$ and a typical element appears as

$$(r + s\sqrt{2}) + (t + u\sqrt{2})\sqrt{3} = r + s\sqrt{2} + t\sqrt{3} + u\sqrt{6}.$$

Contrast this with the field $\mathbb{Q}(\sqrt{2} + \sqrt{3})$. We know that this is the field $\mathbb{Q}(\alpha)$, where α is a root of $x^4 - 10x^2 + 1$. Numbers are of the form $a + b\alpha + c\alpha^2 + d\alpha^3$, where

$$\alpha = \sqrt{2} + \sqrt{3}, \ \alpha^2 = 5 + 2\sqrt{6}, \ \alpha^3 = 11\sqrt{2} + 9\sqrt{3}.$$

So we may relate the two different notations for the same element like this Let $r + s\sqrt{2} + t\sqrt{3} + u\sqrt{6}$ be in $[\mathbb{Q}(\sqrt{2})](\sqrt{3})$ and $a + b\alpha + c\alpha^2 + d\alpha^3$ be in $\mathbb{Q}(\sqrt{2} + \sqrt{3})$. Then, going from $[\mathbb{Q}(\sqrt{2})](\sqrt{3})$ to $\mathbb{Q}(\sqrt{2} + \sqrt{3})$, we have

$$a = r + 5t, \ b = s + 11u, \ c = s + 9u, \ d = 2t.$$

Translating from $\mathbb{Q}(\sqrt{2} + \sqrt{3})$ to $[\mathbb{Q}(\sqrt{2})](\sqrt{3})$, we have

$$r = (1/2)(-5d + 2a), \ s = (1/2)(11c - 9b), \ t = (1/2)d, \ u = (1/2)(b - (c)).$$

These two fields are the same, algebraically speaking. You can prove that in the exercises.   ■

We have barely scratched the surface of this subject of building fields. It is true that an irreducible polynomial over a field F does always generate a new field. We leave you with this in the hope that you will want to pursue this in an advanced algebra course. To whet your appetite, notice that the use of remainders after division is a valuable tool for studying the integral domain of polynomials. The definition of congruence modulo a fixed polynomial is what you would expect: $f(x) \equiv g(x) \pmod{p(x)}$ if and only if $p(x) \mid f(x) - g(x)$. In fact, the world of congruence arithmetic is as relevant and useful to the world of polynomials as it is to the world of integers and the world of natural numbers that we studied in Chapters 1 and 2.

EXERCISES

1. Prove or disprove: If a is an integer and $\sqrt[n]{a}$ is rational, then $\sqrt[n]{a}$ is an integer.

2. Suppose that a and b are natural numbers such that a is not a perfect mth power and b is not a perfect nth power. Then $\sqrt[m]{a} \times \sqrt[n]{b}$ is irrational.

   (a) If this is true, prove it.

   (b) If this is false, prove it and make conditions on the hypothesis so that it becomes true.

3. (a) Prove that $\sqrt{3} + \sqrt{5}$ is irrational.

   (b) Prove that $\sqrt{3} + \sqrt{5}$ is algebraic by finding a polynomial equation it must satisfy. Find the other roots to this polynomial.

4. (a) Prove or disprove: If r and s are natural numbers that are not a perfect square then $\sqrt{r} + \sqrt{s}$ is irrational.

   (b) Prove that $\sqrt{r} + \sqrt{s}$ is algebraic by finding a polynomial equation it must satisfy. Find the other roots to this polynomial.

5. Prove that the following numbers are algebraic by finding a polynomial that each satisfies and determine whether it is rational or irrational.

   (a) $\sqrt[3]{2} + 3\sqrt[3]{4}$

(b) $\sqrt{2} + \sqrt[4]{2}$

(c) $\sqrt{2} + \sqrt{3} + \sqrt{5}$

6. Consider the equation $x^3 - x = 0$. We don't need the cubic formula to know that the three solutions are 0, 1, -1, but use the cubic formula anyway. The solutions will look complicated, so make a guess which of your three solutions in the formula is 0, is 1, and is -1. Then prove it.

7. Using elementary algebra and the rational root theorem, find all the rational and irrational solutions to the following equations:

(a) $x^3 - 3x + 2 = 0$

(b) $x^3 + x^2 - 3x - 2 = 0$

(c) $2x^3 + x^2 - 6x - 3 = 0$

(d) $3x^3 + x^2 + x - 2 = 0$

(e) $x^4 - 3x^3 + 6x - 4 = 0$

8. Use the cubic formula to find the real solutions to the following equations. Write your answers both as arithmetic numbers and in decimal form.

(a) $x^3 - 6x - 6 = 0$

(b) $x^3 + 6x + 2 = 0$

(c) $x^3 + 12x + 12 = 0$

(d) $x^3 - 12x + 20 = 0$

9. Use the quartic formula to find the real solutions to the following equations. Put your answers both in arithmetic form and decimal form.

(a) $x^4 + x^2 + 1 = 0$

(b) $x^4 + 2x + 1 = 0$

10. Show that $f(x) = x^5 + x + 1$ of Example 4.1.17 (a) cannot be factored into a cubic and a quadratic.

11. Write the five solutions to the quintic $f(x) = x^5 - 4x^3 + 3x - 1 = 0$ in 4.1.15 (b) and put them in arithmetic form.

12. Find a quintic that has

    (a) three real roots, all arithmetic
    (b) one real arithmetic root
    (c) three real roots, all nonarithmetic
    (d) five real roots, all nonarithmetic

13. The real roots of the cubic $x^3 - 3x + 1$ are all arithmetic numbers: Write $r_1 = \sqrt[3]{\frac{1}{2}(-1 + \sqrt{3}(i))} + 1 \left/ \left(\sqrt[3]{\frac{1}{2}(-1 + \sqrt{3}(i))} \right) \right.$ as an arithmetic number.

14. Find a division formula for general members of the set

    (a) $\mathbb{Q}(\sqrt{3})$
    (b) $\mathbb{Q}(\sqrt{5})$
    (c) $\mathbb{Q}(\sqrt{6})$

15. Prove Theorem 4.1.23: If j and k are not perfect squares and $\gcd(j, k) = 1$, then $\sqrt{j}$ is not in the field $\mathbb{Q}(\sqrt{k})$. If $\gcd(j, k) \neq 1$, is the theorem true?

16. Prove: $\sqrt[3]{2}$ is not in the field $\mathbb{Q}(\sqrt{k})$ for any natural number k.

17. Show that the field of Example 4.25 (c) is the same as the field $\mathbb{Q}(\sqrt{2})$.

18. Consider the field $F = \mathbb{Q}(\sqrt[3]{2})$.

    (a) Express a general member of F.
    (b) Multiply in F : $(2 + \sqrt[3]{9}) \times (1 + 2\sqrt[3]{3} + \sqrt[3]{9})$.
    (c) Find the multiplicative inverse of $1 + \sqrt[3]{3} + \sqrt[3]{9}$.

19. Consider the field $F = [\mathbb{Q}(\sqrt{2})](\sqrt{5})$.

    (a) Express a general member of F.
    (b) Find the multiplicative inverse of $1 + \sqrt{2} + \sqrt{5}$.

20. Consider the field $F = [[\mathbb{Q}(\sqrt{2})](\sqrt{3})](\sqrt{5})$.

    (a) Express a general member of F.

(b) Multiply in $F : (2 + \sqrt{6}) \times (\sqrt{10} + \sqrt{15})$

(c) Find the multiplicative inverse of $\sqrt{2} + \sqrt{3} + \sqrt{5}$.

21. Consider the field $\mathbb{Q}(\alpha)$, where α is a solution to $x^2 - 3x + 1 = 0$.

   (a) Multiply: $(1 + \alpha) \times (1 - \alpha)$.

   (b) Divide: $(1 + \alpha) \div (1 - \alpha)$.

   (c) Compare $\mathbb{Q}(\alpha)$ with $\mathbb{Q}(\sqrt{5})$ as follows: If x is a general element of $\mathbb{Q}(\alpha)$, write what it looks like in $\mathbb{Q}(\sqrt{5})$.

22. Two fields $(F, +, \times)$ and $(F', \oplus, \otimes)$ are **isomorphic** if there is a one-to-one mapping f from F onto F' such that $f(x + y) = f(x) \oplus f(y)$ and $f(x \times y) = f(x) \otimes f(y)$. Let $F = \mathbb{Q}(\sqrt{2})$ and $F' = \mathbb{Q}(\alpha)$, where α is a root of $x^2 - 2x - 1$, say $1 + \sqrt{2}$. Find an isomorphism $f : r + s\sqrt{2} \mapsto r' + s'\alpha$.

23. Let $F = [\mathbb{Q}(\sqrt{2})](\sqrt{3})$ and $F' = \mathbb{Q}(\alpha)$, where α is a root of $x^4 - 10x^2 + 1$. Let us say $\alpha = \sqrt{2} + \sqrt{3}$. Find an isomorphism:

$$f : r + s\sqrt{2} + t\sqrt{3} + u\sqrt{6} \mapsto r' + s\alpha + t'\alpha^2 + u'\alpha^3.$$

4.2   Geometric Representations

In the previous section we examined irrational numbers from the algebraic point of view. In this section we look at numbers that can be expressed in geometric terms. The discovery of irrational quantities was probably first recognized in the context of geometry. More than 2000 years ago Greek mathematicians were aware of the fact that it was impossible to measure the length of the diagonal of a square and the length of the side of a square with a common unit. That is, if d were the length of the diagonal and s the length of the side of the square, then there is no unit length, t such that $d = at$ and $s = bt$ for natural numbers a and b. If there were it would follow from the Pythagorean theorem that $s^2 + s^2 = 2s^2 = d^2$ so $2b^2 = a^2$. But there are no natural numbers for which this is true. Furthermore, mathematicians believed that the ratio of the circumference to the diameter of a circle is not expressible in a ratio of whole numbers either. Because Greek philosophers believed in the harmony of the ideal world, the fact that geometric ratios arising in such simple and beautiful objects as the square and the circle cannot be expressed in terms of whole numbers was devastating. Nowadays we simply say that the ratio of the diagonal to the side of a square is $\sqrt{2}$, and the ratio of the circumference to the diameter of a circle is π, but this would offer no solace to the Greek mathematician. By whatever name you call these ratios, they are alien to ratios of natural numbers.

As we saw in the previous section, the number, $\sqrt{2}$, being a solution to the polynomial equation $x^2 = 2$, has a natural home in the field of algebra as well as in the field of geometry. It turns out that the world of algebra cannot provide a home for π. That ratio is not a solution to any standard polynomial equation and there is no neat algebraic niche to fit it in. The letter π was adopted by Euler for this ratio and comes from the Greek word "perimetros." As with the square root notation, having a letter representing this ratio does not, in any way, make easier the understanding of what the number really is. Happily π is a Greek letter, perhaps preserving its mystery and keeping it at a distance from those who think that "the name is the thing."

The numbers that arise naturally in elementary geometry are not as complicated as those that we studied in algebra. For us, as for the Greeks, elementary geometry will consist of straight lines and circles. Our measurements will be taken along straight lines, around polygons, and around and across circles. There are also areas and volumes to be measured. We will be especially concerned with comparisons, or ratios, between measurements. And it will be our constant mission to measure our objects with whole numbers. As we have already noted, this cannot always be done. Pairs of lengths of line segments that cannot be measured with a single unit are called incommensurable. Incommensurability, in measurement, is the geometric version of irrationality in numbers.

Definition 4.2.1 *Two lengths r and s are* **commensurable** *if there is a length t such that $r = kt$ and $s = lt$, where k and l are natural numbers.*

Here is a geometric argument that shows that the world of geometry has incommensurable lengths.

Theorem 4.2.2 *Given the isosceles right triangle $\triangle ACB$, the lengths of the hypotenuse AB and the side AC are incommensurable.*

Proof Suppose that the hypotenuse and the side of an isosceles right triangle are commensurable. Figure 4.1 shows such a triangle, $\triangle ACB$. Suppose that s units measures the hypotenuse and r units measures the side. Suppose also that the unit length is the longest possible, thus making r and s relatively prime numbers. Call this unit U. Since all isosceles right triangles are similar, we can assume that r and s are the smallest numbers that can be used to measure both lengths. The following argument shows that the hypotenuse measurement s is an odd number. If s were even, then we could form the triangle $\triangle ADC$. This triangle is also an isosceles right triangle. With the very same U as the one used for $\triangle ACB$, we can measure the hypotenuse and the side; the numbers are $s/2$ and r, respectively. But this is a contradiction because we said that the numbers r and s are the smallest measuring numbers.

Now triangle $\triangle EAB$ is similar to $\triangle ACB$. The same unit, U, that measures $\triangle ACB$ also measures $\triangle EAB$. The numbers are s for the side and $2r$ for the hypotenuse. Since the hypotenuse measurement is even, we can assume that a unit that is twice the length of U will measure both the side and the hypotenuse of $\triangle EAB$. That would imply that s is even. This is a contradiction. Thus we conclude that s and r are not commensurable.   □

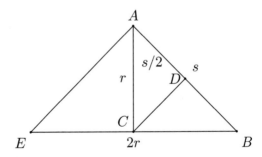

Figure 4.1

This theorem says that there are not natural numbers r and s such that $s/r = 2r/s$. That is, there are not numbers r and s such that $s^2 = 2r^2$. Putting it still another way, there is no fraction s/r such that $(s/r)^2 = 2$. And so, finally, $\sqrt{2} \neq s/r$ for any natural numbers, r and s.

Let us continue this geometric approach and carry it out as the Greeks would by constructing lengths with straightedge and compass. Saying it another way, let us see what lengths we can make from a finite sequence of straight lines and circles. Such lengths will be called constructible. Note again that "length" means straight-line length.

Definition 4.2.3 *A* **length** *is* **constructible** *if it can be constructed from a finite number of uses of a straightedge and compass.*

Notation: The set of constructible lengths is denoted by $\mathbb{S}^+$.

The "$+$" notation indicates that the constructible lengths are positive.

Let us turn to creating an arithmetic for the constructible lengths. Given two lengths, we see that comparing sizes and adding is natural. And so too is subtracting lengths as long as we take a smaller length away from a larger. Let us look into multiplication and division.

Theorem 4.2.4 *If r and s are constructible lengths, then so are $r \times s$ and r/s.*

Proof Suppose that we have lengths r and s and a unit length, call it "1," given to us.

For multiplication, consult Figure 4.2a. On two straightedges, meeting at point A, measure off lengths of 1 and s along one and measure off along a length of r on the other. Placing points B, C, and D at the ends of respective segments, 1, r, and s, connect B to C and construct, using a compass, a line parallel to BC passing through D. It meets line AC at E. Now AE is of length rs. This follows because $\triangle ACB$ is similar to $\triangle AED$, which yields the following equal ratios ($|XY|$ indicates the length of segment XY):

$$r : 1 = (|AC| : |AB|) = (|AE| : |AD|) = (|AE| : s).$$

Similarly for division, use the same construction with A, B, C, and D, only this time connect C and D and draw a line parallel to CD passing through line AC at F, as shown in Figure 4.2b. Then we have similar triangles $\triangle ADC$ and $\triangle ABF$, so $r : s = (|AF| : 1)$. Thus $AF = r/s$. $\square$

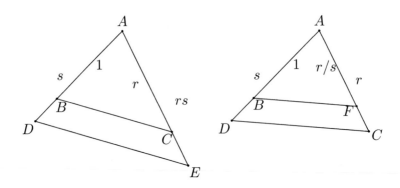

Figure 4.2a   **Figure 4.2b**

Remark Notice the importance of the concept of "unit" in the description of multiplication and division. It would be nice to dispense with this concept by using an alternate definition for the multiplication of two segments; after all, we didn't use a unit when adding or subtracting segments. We could do the following Place the segments so they join at right angles, build a rectangle, and call the product the area of the rectangle generated by the segments. But then we have a problem: How do we compare the size of the product with the length of a side? In fact, we can't. To make sense we must make the product of two lengths another length. In order to do this the concept of similarity of triangles is used. This leads to ratios of lengths and this, in turn, leads to the use of a unit.

We can also find the square root of a constructible length.

Theorem 4.2.5 *If r is a constructible length, then so is $\sqrt{r}$.*

Proof Suppose r is given. Figure 4.3 shows how the construction of $\sqrt{r}$ is accomplished. From point A mark off a length of 1, label it B, and then continue from B along the same line, mark off the length r and label it with C. So $|AB| = 1$ and $|BC| = r$. Next construct a circle with diameter AC. At B construct a perpendicular and let D mark the intersection of that perpendicular with the circle. The triangle $\triangle ADC$ is a right triangle. The length of BD is $\sqrt{r}$. This is because $\triangle ABD$ is similar to $\triangle DBC$ so $(|BD| : 1) = (r : |BD|)$. $\square$

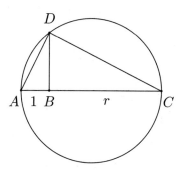

Figure 4.3

Example 4.2.6

Here are examples of constructible lengths.

$$\sqrt{2 + \sqrt{5}}$$

$$\sqrt{5 + \sqrt{7 + \sqrt{9 + \sqrt{11}}}}$$

$$\sqrt[8]{1 + \sqrt[4]{1/13 + \sqrt{102} - \sqrt[16]{43}}}$$

$$\frac{\sqrt{1/7 + 14\sqrt[32]{123} + \sqrt[4]{7}}}{\sqrt[8]{-11 + \sqrt{401} - \sqrt{79/19}}}$$

∎

As with arithmetic numbers, it is not easy to classify the kinds of numbers that we can construct. As the preceding example demonstrates, the numbers in $\mathbb{S}^+$ can look pretty bizarre. We know this Constructible lengths are made up of combinations of sums, differences that are positive, products, quotients, and square roots of natural numbers. Are we missing any? To help us with this we shall employ a technique that has been used before; that is, we will look at a larger set that contains the constructible lengths as a subset but that has useful properties that the constructible lengths do not have.

In particular, we will have our eye on finding a field that contains $\mathbb{S}^+$. Keeping the spirit of geometric construction we proceed as follows: Our tools are the straightedge and the compass and our canvas is the Cartesian plane. We shall begin with points in the plane (r, s), where r and s are natural numbers. We shall draw lines joining such points and draw circles with centers at such points and radii that are natural numbers. Letting L denote the set of such lines and C, the set of such circles, we shall proceed by building new points (t, u) from the following intersections: pairs of distinct lines in L, a line in L with a circle in C, and pairs of distinct circles in C. Adding these newly generated points to the mix, we generate new sets of lines, L', a new set of circles, C', and a new set of points. Continuing the process of intersecting, as new points are formed the set of numbers that is finally generated is closed under subtraction as well as addition, multiplication, division, and the taking of square roots. Thus this set includes $\mathbb{S}^+$ and more. We shall call these numbers constructible numbers. Basically the constructible numbers include all the constructible lengths and their negatives.

Definition 4.2.7 *A number is **constructible** if it can be constructed from the intersections of the lines and circles that can be formed from natural number lattice points (r, s) and natural number radii.*

Notation: Let $\mathbb{S}$ denote the set of constructible numbers.

Theorem 4.2.8

  1. *If t and u are in $\mathbb{S}$, then $t + u$ is also in $\mathbb{S}$.*

  2. *If t and u are in $\mathbb{S}$, then $t - u$ is also in $\mathbb{S}$.*

  3. *If t and u are in $\mathbb{S}$, then $t \times u$ is also in $\mathbb{S}$.*

  4. *If t and u are in $\mathbb{S}$ and $u \neq 0$, then t/u is also in $\mathbb{S}$.*

  5. *If t is in $\mathbb{S}$, then $\sqrt{t}$ is also in $\mathbb{S}$.*

Proof For (1) we note that the line joining $(0, t)$ to $(1, t)$ intersects the line joining $(u, 0)$ to $(2u, u)$ at the point $(t + u, t)$.

For (3) we note that the line joining $(0, t)$ to $(1, t)$ and the line joining $(0, 0)$ to $(u, 1)$ intersect at the point (tu, t).

We leave the remainder of the proof as an exercise. $\square$

Corollary 4.2.9 *The set of constructible numbers form a field under the usual addition and multiplication of real numbers.*

This follows directly from the fact that $\mathbb{S}$ is a subset of the field of real numbers and is closed under addition, subtraction, multiplication, and division.

Let us now examine what the constructible numbers look like. We use our high school analytic geometry for this. Recall that lines in the Cartesian plane can be written in the form $y = ax + b$ and $x = c$ and circles in the form $x^2 + y^2 + cx + dy = e$.

Theorem 4.2.10

1. If (r, s) and (t, u) are two points in the coordinate plane, where r, s, t, u are in $\mathbb{S}$, and if $ax + by = c$ is the line joining the points, then a, b, c are in $\mathbb{S}$.

2. If (r, s) is the center and R is the radius of a circle, where r, s, R are in $\mathbb{S}$, then $x^2 + y^2 + ax + by = c$ describes the circle and a, b, c are in $\mathbb{S}$.

The proof is left as an exercise.

Using this theorem, we may examine what kinds of numbers inhabit $\mathbb{S}$ by intersecting lines with lines, lines with circles, and circles with circles.

Theorem 4.2.11

1. If (x, y) is the intersection of two lines $y = ax+b$ and $y = cx+d$ in the plane and a, b, c, $d \in \mathbb{S}$, then x and y are in $\mathbb{S}$.

2. If (x, y) is an intersection of $y = ax + b$ with the circle $x^2 + y^2 + cx + dy = e$, where a, b, c, d, $e \in \mathbb{S}$, then x and y are in $\mathbb{S}$.

3. If (x, y) is an intersection of two circles $x^2 + y^2 + ax + by = c$ and $x^2 + y^2 + dx + ey = f$ and a, b, c, d, e, $f \in \mathbb{S}$, then x and y are in $\mathbb{S}$.

Proof We leave the proof of (1) and (3) as an exercise. For (2), substituting $y = ax + b$ into the equation of the circle, we get

$$x^2 + (ax)^2 + 2abx + b^2 + cx + dax + db =$$

$$(1 + a^2)x^2 + (2ab + ad + (c)x + (b^2 + db) = e.$$

So we have $x^2 + rx + s = 0$, where $r = (2ab + ad + (c)/(1 + a^2)$ and $s = (b^2 + db - (e)/(1 + a^2)$. So r, s, $\in \mathbb{S}$. Thus $x = \frac{1}{2}(-r \pm \sqrt{r^2 - 4s})$ so $x \in \mathbb{S}$. The fact that $y \in \mathbb{S}$ is proved the same way. $\qquad\square$

So what Theorem 4.2.11 tells us is that the various intersections of lines and circles create numbers that are built from addition, subtraction, multiplication, division, and taking square roots of natural numbers. Since numbers of this type are formed by the same processes as those already formed (that is, by adding, subtracting, multiplying, dividing, and taking square roots), we can say that $\mathbb{S}$ contains no numbers other than the constructible numbers and their negatives. Since only square roots are involved in building numbers, we can say that a constructible number is made of of a string of addition, subtraction, multipication, division, and nth root symbols where $n = 2^k$ for some natural number k.

We have studied mathematical structures in the previous section and found that the integers are an integral domain, and the rational, the arithmetic, and the algebraic numbers formed fields. And here we have another field, $\mathbb{S}$. The following inclusion relations hold:

$$\mathbb{Z} \subset \mathbb{Q} \subset \mathbb{S} \subseteq \mathbb{A} \subset \mathbb{B}.$$

There remains a question whether there is an arithmetic number that is not in $\mathbb{S}$. It appears the answer would be yes; in fact, $\sqrt[n]{m}$, where $n \neq 2^k$ for some k and m is a natural number, should be such an example. In particular, it appears that $\sqrt[3]{2}$ is not constructible. This is not crystal clear because, perhaps, a finite string of natural numbers mixed with 2^kth roots might somehow equal $\sqrt[3]{2}$. In fact, this cannot happen; $\sqrt[3]{2}$ is not constructible. We prove this fact using the technique of building fields by adjoining new elements as we described in Section 4.1; in particular, in Examples 4.1.24, 4.1.25, and 4.1.26.

Example 4.2.12

(a) Consider the arithmetic number $\sqrt{(1/2) + 2\sqrt[4]{11} + \sqrt{7}}$. We shall build up a field that contains this number. Beginning with the field of rational numbers, let $F_1 = \mathbb{Q}$. Let $F_2 = F_1(\sqrt{11})$. So F_2 contains numbers of the form $a + b\sqrt{11}$, where a and b are in F_1. Let $F_3 = F_2(\sqrt{\sqrt{11}})$. So members of F_3 look like $a + b\sqrt[4]{11}$, where a and b are in F_2. Let $F_4 = F_3(\sqrt{7})$. Finally, let

$$F_5 = F_4(\sqrt{((1/2) + 2\sqrt[4]{11} + \sqrt{7})}).$$

Notice that, at the $j+1$ stage, $F_{j+1} = F_j(\sqrt{k_j})$, $k_j \in F_j$ but $\sqrt{k_j} \notin F_j$.

(b) Consider a constructible number from Example 4.2.6:

$$\frac{\sqrt{1/7 + 14\sqrt[32]{123} + \sqrt[4]{7}}}{\sqrt[8]{-11 + \sqrt{401} - \sqrt{76/19}}}$$

This number is in F_{14}, where F_j, for $1 \leq j \leq 14$, is defined as follows:

$$F_1 = \mathbb{Q}, \quad F_2 = F_1(\sqrt{123}), \quad F_3 = F_2(\sqrt[4]{123}), \quad F_4 = F_3(\sqrt[8]{123}),$$

$$F_5 = F_4(\sqrt[16]{123}), \quad F_6 = F_5(\sqrt[32]{123}), \quad F_7 = F_6(\sqrt{7}), \quad F_8 = F_7(\sqrt[4]{7}),$$

$$F_9 = F_8(\sqrt{1/7 + 14\sqrt[32]{123} + \sqrt[4]{7}}), \quad F_{10} = F_9(\sqrt{401}),$$

$$F_{11} = F_{10}(\sqrt{76/19}) \quad F_{12} = F_{11}(\sqrt{-11 + \sqrt{401}}),$$

$$F_{13} = F_{12}(\sqrt[4]{-11 + \sqrt{401}} - \sqrt{76/19}),$$

$$F_{14} = F_{13}(\sqrt[8]{-11 + \sqrt{401}} - \sqrt{76/19}).$$

Notice again that at the $j+1$ stage $F_{j+1} = F_j\sqrt{k_j}$, $k_j \in F_j$ but $\sqrt{k_j} \notin F_j$. ∎

Theorem 4.2.13 *If x is a solution to the equation $x^3 - 2 = 0$, then $x \notin \mathbb{S}$.*

Proof Suppose that $\sqrt[3]{2} \in \mathbb{S}$. Then there exists a sequence of fields $F_1, \ldots, F_{j+1}$, where $F_i = \mathbb{Q}$, and for $1 \leq i \leq j$ we have $F_{i+1} = F_i(\sqrt{k_i})$, $k_i \in F_i$, $\sqrt{k_i} \notin F_i$, $\sqrt[3]{2} \in F_{j+1}$, and $\sqrt[3]{2} \notin F_j$. Let x be a solution to $x^3 - 2 = 0$ and let $x = p + q\sqrt{k_j}$, where $p, q \in F_j$. We now show that $y = p - q\sqrt{k_j}$ is also a solution to $x^3 - 2 = 0$ and that real roots come in pairs. Since there is only one real cube root of 2, this cannot be. Thus $\sqrt[3]{2}$ is not constructible.

Let $x = p + q\sqrt{k_j}$. Since $x^3 - 2 = 0$, we have

$$x^3 - 2 = (p + q\sqrt{k_j})^3 - 2 = p^3 + 3p^2q\sqrt{k_j} + 3pq^2k_j + q^3k_j\sqrt{k_j} - 2$$

$$= (p^3 + 3pq^2 k_j - 2) + (3p^2 q + q^3 k_j)\sqrt{k_j} = 0 + 0\sqrt{k_j}.$$

So

$$(p^3 + 3pq^2 k_j - 2) = 0 \text{ and } (3p^2 q + q^3 k_j) = 0.$$

Now

$$y^3 - 2 = (p - q\sqrt{k_j})^3 - 2 = p^3 - 3p^2 q\sqrt{k_j} + 3pq^2 k_j - q^3 k_j \sqrt{k_j} - 2$$
$$= (p^3 + 3pq^2 k_j - 2) - (3p^2 q + q^3 k_j)\sqrt{k_j} = 0 + 0\sqrt{k_j} = 0. \quad \square$$

The Greeks posed three challenges that remained unsolved until the mid-1800s. Specifically, using a straightedge and compass, they wondered whether it is possible to

1. Duplicate a cube; that is, construct the side of a cube whose volume is twice that of a given cube

2. Trisect an angle

3. Square a circle; that is, construct a square equal in area to that of a given circle

The search for solutions to these challenges, much like the search for the proof of Fermat's last theorem, spawned new fields of mathematics. For us, the solutions to these challenges involve irrational numbers: The first number is a solution to the equation $x^3 - 2 = 0$; the second is a solution to $8x^3 - 6x - 1 = 0$; the third involves π.

Theorem 4.2.14 *It is impossible, using straightedge and compass, to*

   1. Duplicate a cube

   2. Trisect any given angle

Proof (1) In order to build a cube whose volume is twice the volume of a cube with side of length x, we must build a cube with a volume of $2x^3$. The side of that cube would be $\sqrt[3]{2}x$. Theorem 4.2.13 shows that this cannot be done with straightedge and compass because $\sqrt[3]{2}$ is not constructible.

For (2) note first that we can construct a $60°$ angle using a compass and straightedge. Simply mark off successive lengths of the radius of a given circle around its circumference. It will cut the circle into six equal parts, each of which marks off $60°$. We shall show it is impossible to trisect a $60°$ angle by showing that the length of

$\cos 20°$ is not constructible. The triple angle formula from trigonometry states: $\cos 3\theta = 4\cos^3\theta - 3\cos\theta$. Using radian notation, we let $\theta = \pi/9$. So $3\theta = \pi/3$ and $\cos 3\theta = 1/2$. So we have $1/2 = 4\cos^3(\pi/9) - 3\cos(\pi/9)$; that is, $8\cos^3(\pi/9) - 6\cos(\pi/9) - 1 = 0$. Thus, $\cos(\pi/9)$ is a solution to the polynomial equation $8x^3 - 6x - 1 = 0$. We show it has no constructible solutions in much the same way as the proof of Theorem 4.2.13.

Suppose, as in the proof of Theorem 4.2.13, that x is a solution to a cubic and x is of the form $x = p + q\sqrt{k_j}$. Let $y = p - q\sqrt{k_j}$. Since $8x^3 - 6x - 1 = 0$ we have

$$8(p + q\sqrt{k_j})^3 - 6(p + q\sqrt{k_j}) - 1$$
$$= 8(p^3 + 24pq^2k_j - 6p - 1) + (24p^2q + q^3k_j - q)\sqrt{k_j} = 0.$$

So

$$(8p^3 + 24pq^2k_j - 6p - 1) = 0 \text{ and } (24p^2q + q^3k_j - q) = 0.$$

Consider the expression $8y^3 - 6y - 1 = 0$. Since

$$8(p - q\sqrt{k_j})^3 - 6(p - q\sqrt{k_j}) - 1$$
$$= 8(p^3 + 24pq^2k_j - 6p - 1) - (24p^2q + q^3k_j - q)\sqrt{k_j}$$

this, too $= 0$ so

$$8x^3 - 6x - 1 = 8(x - (p + q\sqrt{k_j})(x - p + q\sqrt{k_j})(x - r),$$

where r is the third real root. Equating the constant coefficients, yields $8(p^2 - q^2k_j)r = -1$. Thus $r = -1/8((p^2 - q^2k_j)$, which tells us that r is rational. But the rational root theorem shows us that this cubic equation has no rational solutions. So the assumption that it has a constructible root is wrong.

We conclude that the angle $\theta = 60°$ cannot be trisected. □

The third classical problem, squaring the circle, involves constructing a square whose area is the same as that of a given circle. This problem predates the Greek mathematicians by at least a 1000 years. In the *Rhind Papyrus*, around 1500 B.C., the Egyptian scribe Ahmes stated that the area of a circle is equal to that of a square with side of length $(8/9)$th the diameter of the circle. If this were true π would be constructible. But it is not true. We discuss the nature of π in Section 4.4.

One more note about these challenges. The notion of employing tools such as the straightedge and compass to construct mathematical entities is curious. The straightedge and compass are physical

objects, and this can be misleading. Circles, squares, and angles
are theoretical concepts. If we are really concerned about physically
meeting any of these challenges, there are instruments that can do
the job. And even theoretically we can conjure up "instruments" to
accomplish this. In fact, Archimedes proved that with a ruler (that
is, a straightedge with markings), one could trisect an angle. And
with instruments that can trace out conics, the circle can be squared.
And if we are allowed to move into three dimensions, a cube can be
duplicated. The Greek mathematicians recognized this, but that was
not their point. Mathematics is theoretical and the rules of the game
are not whimsical. Lines and circles are the fundamental concepts of
geometry and constructible lengths were generated from successive
drawings of lines and circles.

The trisection problem leads to another fascinating question that
the Greeks left to future mathematicians: For what n is it possible,
with straightedge and compass, to inscribe a regular n-gon in a circle?
The Greeks knew how to build such polygons for $n = 3$, 4, 5, 6 and
any multiple of these by 2^n. But they could not inscribe a regular
7-gon.

We shall coin a definition for numbers that measure the length
of sides of regular polygons.

Definition 4.2.15 *The nth* **polygon number** *measures the length
of the side of an n-gon inscribed in a unit circle. We assume $n > 2$.*

Notation: The nth polygon number is denoted by P_n. The set of
nth polygon numbers for $n > 2$ is denoted by $\mathbb{P}$.

Example 4.2.16 Consider a square inscribed in the unit circle. The
length, s, of the side of such a square is a polygon number. Since the
radius of the circle is 1, we have the diagonal of the square measuring
2. So the Pythagorean Theorem yields $s^2 + s^2 = 2^2$. Solving for s, we
get $s = \sqrt{2}$. So $\sqrt{2}$ is the polygon number P_4. It is a constructible
number as well.   ∎

The natural setting for the study of regular n-gons inscribed in
a circle is trigonometry. Trigonometry relates straight line lengths
to circular measure. Recall the trigonometric functions of $\cos \theta$ and
$\sin \theta$. Looking at the unit circle in Figure 4.4, P is the point (x, y)
and θ is the angle made by the line OP with the x-axis; we say that
$x = \cos \theta$, and $y = \sin \theta$.

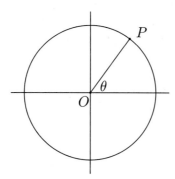

Figure 4.4

As in Section 4.1, we will find it useful to carry out our examination using complex numbers. Recall from trigonometry that if we take the x-y-plane to be the complex plane, then the point (x, y) becomes the single complex number $z = x + iy = \cos\theta + i\sin\theta$. The key to our examination is DeMoivre's theorem.

DeMoivre's Theorem If $z = \cos\theta + i\sin\theta$ and $z' = \cos\phi + i\sin\phi$, then $z \times z' = \cos(\theta + \phi) + i\sin(\theta + \phi)$ and $z^n = \cos(n\theta) + i\sin(n\theta)$.

The proof follows from the trigonometric formulas for $\cos(\theta \pm \phi)$ and $\sin(\theta \pm \phi)$.

Inscribing a regular n-gon amounts to dividing a circle into n equal parts. For us this translates into dividing the unit circle with n complex numbers that are equally spaced about the circumference. This, in turn, translates into solving the polynomial equation $z^n = 1$ in complex numbers. You may recall that these n solutions are called the nth roots of unity.

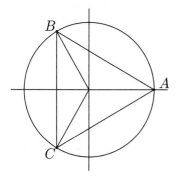

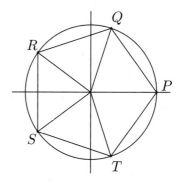

Figure 4.5a   **Figure 4.5b**

Theorem 4.2.17 *The n complex solutions to the equation $z^n = 1$ are of the form $x + iy$, where $x = \cos(2k\pi/n)$, $y = \sin(2k\pi/n)$, and $k = 0, 1, 2, \ldots n - 1$.*

Proof Plugging in $x + iy$ for z and using DeMoivre's theorem, we get $z^n = \cos(n2k\pi/n) + i\sin(n2k\pi/n) = \cos(2k\pi) + i\sin(2k\pi)$. And this expression $= 1$.   □

Example 4.2.18

(a) Figure 4.5a shows an equilateral triangle inscribed in a unit circle. The three complex numbers A, B, and C are arranged $120° = 2\pi/3$ apart and around the circle. So the numbers are $\cos 2k\pi/3 + i\sin 2k\pi/3$, where $k = 0, 1, 2$. So $A = 1 + i0$, $B = -1/2 + i\sqrt{3}/2$, and $C = -1/2 - i\sqrt{3}/2$. From what we have learned, these three numbers should be solutions to the equation $z^3 = 1$. We can check this out algebraically: $z^3 - 1 = (z - 1)(z^2 + z + 1)$. The quadratic formula solves $z^2 + z + 1 = 0$ with the two other solutions: B and C.

(b) Figure 4.2.5b shows a regular pentagon inscribed in a unit circle. The five complex numbers P, Q, R, S, and T are arranged $72°$ apart and around the circle. So $Q = \cos(2\pi/5) + i\sin(2\pi/5)$, $R = \cos(4\pi/5) + i\sin(4\pi/5)$, $S = \cos(6\pi/5) + i\sin(6\pi/5)$, and $T = \cos(8\pi/5) + i\sin(8\pi/5)$. These numbers are solutions to the equation $z^5 = 1$. Now we try to factor $z^5 - 1 = 0$. We know that $z^5 - 1 = (z - 1)(z^4 + z^3 + z^2 + z + 1)$ and we have the quartic formula we could use on $z^4 + z^3 + z^2 + z + 1$. But there is a simpler procedure to find the solutions z. Since the four complex numbers are arranged in conjugate pairs $a + ib$, $a - ib$ and $c + id$, $c - id$, we have factors

$$(z - (a + i(b)))(z - (a - i(b)))(z - (c + i(d)))(z - (c - i(d))) =$$
$$(z^2 - 2az + 1)(z^2 - 2cz + 1).$$

Here we let

$$a = \cos(2\pi/5), \; b = \sin(2\pi/5), \; c = \cos(4\pi/5), \; d = \sin(4\pi/5).$$

Multiplying $(z^2 - 2az + 1)(z^2 - 2cz + 1)$ we get

$$z^4 - (2a + 2c)z^3 + (4ac + 2)z^2 - (2a + 2c)z + 1.$$

So $2a + 2c = -1$. Since c represents the cos of an angle twice that of a, we may use the formula that $\cos 2\theta = 2\cos^2 \theta - 1$. This tells us that $a + (2a^2 - 1) = -1/2$. Solving for a using the quadratic formula, we get $a = (-1 \pm \sqrt{5})/4$. It turns out that

$$a = (-1 + \sqrt{5})/4 \text{ and } c = (-1 - \sqrt{5})/4.$$

So $\cos 72° = (-1 + \sqrt{5})/4$ and $\cos 144° = (-1 - \sqrt{5})/4$. ∎

Finding $\cos x$ and $\sin x$ allows us to find the length of the side of a regular n-gon. Here is how.

Theorem 4.2.19 *The length of the side, s, of a regular n-gon inscribed in a unit circle is $s = \sqrt{2 - 2\cos(2\pi/n)}$.*

Proof Consider the right triangle PRQ in Figure 4.6. We label the length of PR with r, the length of QR with t, and the length of PQ with s. We see that $r = \sin\theta$ and $t = 1 - \cos\theta$. The Pythagorean theorem gives us $t^2 + r^2 = s^2$. Given that $\theta = 2\pi/n$ and $\sin^2\theta + \cos^2\theta = 1$, we have our result. □

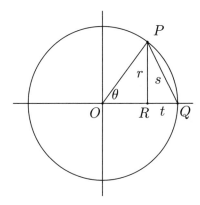

Figure 4.6

Corollary 4.2.20 *If $\cos 2\pi/n$ is constructible, then the nth polygon number is constructible.*

Corollary 4.2.21 *The following polygon numbers are $P_3 = \sqrt{3}$, $P_4 = \sqrt{2}$, $P_5 = \sqrt{(5 - \sqrt{5})/2}$, $P_6 = 1$.*

These corollaries tell us that an equilateral triangle, a square, a regular pentagon, and a regular hexagon can be inscribed in a circle using a straightedge and compass. We leave it as an exercise to create these constructions. Also, we see that the constructibility of a regular n-gon comes down to the nature of $\cos(2\pi/n)$. If $\cos(2\pi/n)$ is a constructible number, then it is possible to construct a regular n-gon. Let us examine a bit more the nature of numbers of the form $\cos(2\pi/n)$.

Theorem 4.2.22 *The number* $\cos(2\pi/n)$ *is algebraic.*

Proof We know that $\cos(2\pi/n) + i\sin(2\pi/n)$ and $\cos(2\pi/n) - i\sin(2\pi/n)$ are solutions to the polynomial equation $z^n - 1 = 0$. So they are algebraic (complex) numbers and so their sum is algebraic. Their sum is $2\cos(2\pi/n)$, and hence $\cos(2\pi/n)$ is also algebraic. □

Corollary 4.2.23 *The polygon numbers are algebraic numbers; that is,* $\mathbb{P} \subset \mathbb{B}$.

It turns out that some numbers of the form $\cos(2\pi/n)$ are rational, but not many. In the first quadrant the only such angles are $0°$, $60°$, and $90°$; $\cos 0° = 1$, $\cos 60° = 1/2$, $and \cos 90° = 0$. This follows from a trigonometric identity that can be proved by induction. The proof is left as an exercise (with hints given).

Lemma 4.2.24 *The following equality is true where the coefficients* a_j *are integers:* $2\cos(n\theta) = (2\cos\theta)^n + a_{n-1}(2\cos\theta)^{n-1} + \cdots + a_1(2\cos\theta) + a_0$.

Theorem 4.2.25 *If* $\theta = 2\pi/n$ *and* $\cos\theta$ *is a rational number, then* $\cos = 0, \pm1, or \pm 1/2$.

Proof Letting $\theta = 2\pi/n$, and $x = 2\cos\theta$, the lemma tells us that

$$2\cos n\theta = 2\cos 2\pi = 2 = x^n + a_{n-1}x^{n-1} + \cdots + a_1 x + a_0.$$

So, by the rational root theorem, x is an integer. But $\cos\theta$ is limited between 1 and -1, so x must be either 0, ±1 or ±2. So $\cos\theta$ is either $0, \pm1$, or $\pm1/2$.   □

Example 4.2.26
Let us search for an algebraic representation for $\cos(2\pi/7)$. This comes down to solving the equation $z^7 - 1 = 0$. This factors into

$$z^7 - 1 = (z-1)(z^6 + z^5 + z^4 + z^3 + z^2 + z + 1).$$

Since solutions come in conjugate pairs $(z - (a + ib))(z + (a + ib))$, we know that the product, $z^2 + 2az + 1$, is a factor. So

$$z^6 + z^5 + z^4 + z^3 + z^2 + z + 1 = (z^2 - 2a_1 z + 1)(z^2 - 2a_2 z + 1)(z^2 - 2a_3 z + 1).$$

But there is a problem: how do we find a_1, a_2, and a_3? My calculator can find roots and it tells me that $a_1 = 0.623489801859\ldots$. I don't really know much else about this number. Is it rational? It doesn't look like it. Is it constructible? I don't think so. Is it arithmetic? Project 5.19 will provide an answer. ∎

Are there polygon numbers that we know are not constructible? The answer is yes.

Example 4.2.27

Recall Theorem 4.2.14 which said that it was impossible to trisect a general angle with a straightedge and compass and, in particular, $60°$ could not be trisected. It follows that $\cos 20°$ or $\cos \pi/9$ is not constructible. Thus P_{18} is not constructible. It follows that P_9 is also not constructible because if you can construct P_9 you can surely construct P_{18}. ∎

So now that we have analyzed the polygon numbers, we ask the question that the Greek mathematicians posed: For which n is P_n constructible? We know P_3, P_4, P_5, and P_6 are. We don't know about P_7. It was Gauss who provided a startling solution. At the age of 17 he became absorbed with this problem of inscribing regular p-gons in a circle where p is a prime number. He proved that if p is a prime that is also a Fermat number, then such a p-gon can be constructed. For him this was the discovery he cherished most; in fact, he asked that a regular heptadecagon (17-sided polygon) be inscribed on his tombstone. For some reason the request was not granted, but the polygon is engraved on the monument to him in Braunschweig, Germany, the place of his birth. Recall Fermat numbers from Section 1.3; a Fermat number is a number of the form $2^{2^n} + 1$. Also recall that only five such numbers so far have been found to be prime: 3, 5, 17, 257, and 65537. Armed with this information, we may sum up our findings as follows.

Theorem 4.2.28 *The polygon number P_n is constructible if*

1. $n = p$, where p is a Fermat prime

2. $n = r - s$, where P_r and P_s are constructible

3. $n = m/2$, where P_m is constructible

4. $n = r \times s$, where $\gcd(r, s) = 1$ and P_r and P_s are constructible

Proof Gauss took care of condition (1).

Condition (2) is clear geometrically. If you wish algebraic confirmation, recall that $\cos(\alpha - \beta) = \cos\alpha\cos\beta + \sin\alpha\sin\beta$ and observe that $\sin\alpha$ and $\sin\beta$ are constructible numbers, since $\cos\alpha$ and $\cos\beta$ are constructible, and also that constructible numbers are closed under addition and multiplication.

Condition (3) is easy to understand geometrically; it is simply bisecting an angle with straightedge and compass (as opposed to trisecting an angle). Algebraically, it follows from the formula $\cos\frac{\theta}{2} = \sqrt{(\cos\theta - 1)/2}$.

Condition (4) uses Theorem 1.1.11. Since $\gcd(r, s) = 1$, there are integers a and b such that $ar - bs = 1$. So $(1/rs)(ar - bs) = a/s - b/r = 1/n$. So letting $\alpha = 2\pi a/s$ and $\beta = 2\pi b/r$, we note that $\theta = \alpha - \beta = 2\pi/n$. Thus $\cos\theta$ is constructible because $\cos\alpha$ and $\cos\beta$ are constructible. $\qquad\square$

Let us summarize the interrelationship of our classifications of numbers. In this section we have introduced three new types of numbers: the constructible numbers, denoted by $\mathbb{S}^+$; the field containing the constructible numbers, $\mathbb{S}$; and polygon numbers, $\mathbb{P}$. So we now have these three along with the two types from Section 4.1, the arithmetic numbers, $\mathbb{A}$, and algebraic numbers, $\mathbb{B}$; and the rationals, $\mathbb{Q}$, and the real numbers, $\mathbb{R}$, from Chapter 3. The following set theoretic inclusions hold:

$$\mathbb{Q} \subset \mathbb{S} \subset \mathbb{A} \subset \mathbb{B} \subset \mathbb{R}.$$

We have found examples of numbers showing that the inclusions are proper. Consider the following numbers a, b, c, and d, where

$$a = \sqrt{2},\ b = \sqrt[3]{2},\ c = \text{the real solution to } x^5 + x + 1 = 0,\ d = \pi$$

$$a \in \mathbb{S},\ a \notin \mathbb{Q};\quad b \in \mathbb{A},\ b \notin \mathbb{S};\quad c \in \mathbb{B},\ c \notin \mathbb{A};\quad d \in \mathbb{R},\ d \notin \mathbb{B}.$$

The polygon numbers don't fit into this inclusion scheme so easily. There are almost no rationals in $\mathbb{P}$. We know $\mathbb{P} \subset \mathbb{B}$, but $\mathbb{P} \not\subset \mathbb{S}$ and $\mathbb{P} \not\subset \mathbb{A}$. For example, $\cos(2\pi/5)$ is irrational and constructible,

$\cos(\pi/9)$ is arithmetic but not constructible, and $\cos(2\pi/7)$ is algebraic but perhaps not arithmetic and surely not constructible. Project 5.19 allows you to discover this.

EXERCISES

1. Explain how to construct the following with a straightedge and compass:

   (a) a square
   (b) an equilateral triangle
   (c) a regular octagon
   (d) a regular hexagon

2. Consider Figure 4.7: $ABCD$ is a square, one unit on a side, and E is the intersection of the extended side AB with the circle centered at A with radius AC. Find the following lengths:

   (a) AE
   (b) DE
   (c) CE

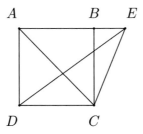

Figure 4.7

3. Consider Figure 4.8: AB is length r, BC is of length s, DB is perpendicular to AC, and BE is perpendicular to DC. Find the following lengths:

   (a) DB
   (b) DC
   (c) DA
   (d) EB

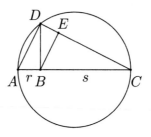

Figure 4.8

4. The rectangle of Figure 4.9 is called the **golden rectangle**. It is constructed as follows: $ABCD$ is a square, the point E is the midpoint of side AB, F is the intersection of the extended side AB with the circle centered at E with radius EC, and G is the fourth point on the rectangle formed by D, A, and F.

(a) What is the ratio $AF : FG$?

(b) What is the ratio $FG : BF$?

(c) How are the two ratios in (a) and (b) related?

(d) What can you say about $ADGF$ and $BFGC$?

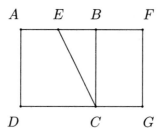

Figure 4.9

5. Using Figure 4.9, draw a circle with center at A and radius AF. Then draw a chord on that circle of length EF.

(a) What kind of regular polygon can you construct with chords of this length?

(b) Explain how you could construct a regular pentagon in just a few steps.

6. Construct, in as few steps as you can, the following lengths:

   (a) $\sqrt{21}$

   (b) $\sqrt[4]{2}$

   (c) $2/(\sqrt{2}+1)$

   (d) $\sqrt{5+\sqrt{5}}$

   (e) $(\sqrt{5}-1)/4$

   (f) $\sqrt{2-\sqrt{2}}$

7. (a) Prove Theorem 4.2.10 (1). If (r,s) and (t,u) are two points in the coordinate plane where r, s, t, and u are in the field $\mathbb{S}$, and if $ax+by=c$ is the line joining the points, then a, b, c are in $\mathbb{S}$.

   (b) Prove Theorem 4.2.10 (2). If (r,s) is the center and R is the radius of a circle where r, s, and R are in $\mathbb{S}$, then $x^2+y^2+ax+by=c$ describes the circle with center (r,s) and radius R and a, b, c are in $\mathbb{S}$.

8. (a) Prove Theorem 4.2.11 (1). If (x,y) is the intersection of two lines $y=ax+b$ and $y=cx+d$ in the plane and a, b, c, and d are constructible, then x and y are constructible.

   (b) Prove Theorem 4.2.11 (3). If (x,y) is an intersection of two circles $x^2+y^2+ax+by=c$ and $x^2+y^2+dx+ey=f$ and a, b, c, d, e, $f \in \mathbb{S}$, then x and y are in $\mathbb{S}$.

9. For each number below find the smallest field that contains that number by adjoining elements to $\mathbb{Q}$. Also find polynomial equations for which each of the constructible numbers is a solution.

   (a) $\sqrt{1+\sqrt{2}}$

   (b) $2+\sqrt{2+\sqrt{2}}$

   (c) $\sqrt{2}+\sqrt{3}+\sqrt{5}$

   (d) $\sqrt{5-3\sqrt{1+\sqrt{3}}}$

   (e) $\sqrt[4]{2+\sqrt[4]{2}}$

10. Prove or disprove: A cubic equation with rational coefficients that has no rational solution has no constructible solution.

11. Next to each number write $\mathbb{Q}$ (rational), $\mathbb{S}$ (constructible), $\mathbb{A}$ (arithmetic), and $\mathbb{B}$ (algebraic) where appropriate. If none apply write N.

    (a) $\cos(60°)$

    (b) $\cos(40°)$

    (c) $\cos(12°)$

    (d) $\cos((360/7)°)$

    (e) $\cos(5°)$

    (f) $\cos(3°)$

    (g) $\cos(1°)$

    (h) $\cos(7° \, 30 \, \text{minutes})$

12. Find a polynomial equation whose solutions are the following:

    (a) $\cos(\pi/12)$

    (b) $\cos(\pi/5)$

    (c) $\cos(4\pi/5)$

    (d) $\cos(4\pi/9)$

13. A **Pythagorean angle** is an acute angle of a Pythagorean triangle. Prove that a Pythagorean angle is not of the form $2\pi/n$.

14. Prove Lemma 4.2.24

    (a) First show that $\cos(n+1)\theta = 2\cos(n\theta) - 2\cos(n-1)\theta$.

    (b) Prove: $2\cos(n\theta) = (2\cos\theta)^n + a_{n-1}(2\cos\theta)^{n-1} + \cdots + a_1(2\cos\theta) + a_0$, where the coefficients a_j are integers.

15. List all n, where $3 \leq n \leq 100$, such that a regular n-gon can be constructed.

16. Prove Corollary 4.2.21: The following polygon numbers are $P_3 = \sqrt{3}$, $P_4 = \sqrt{2}$, $P_5 = \sqrt{(5 - \sqrt{5})/2}$, $P_6 = 1$.

17. For the following regular polygons, find P_n. Express your answer as a constructible number.

    (a) octagon, P_8

    (b) decagon, P_{10}

    (c) dodecagon, P_{12}

18. Find the polygon numbers P_n for the following n. Express your answer as a decimal. Multiply P_n by n and see how close to 2π you get.

    (a) $n = 7$

    (b) $n = 9$

    (c) $n = 11$

    (d) $n = 13$

19. Prove by induction that the constructible numbers are algebraic and are solutions to polynomial equations of degree 2^n.

4.3 Analytic Representations

An analytic representation is one that expresses a number as the limit of a sequence. We shall find that this representation will present a whole new set of challenges, quite different from the algebraic and the geometric approaches. It is nearly futile to extract algebraic, geometric, or polygonal information from analytic representations. Recall from Section 3.1 that the real numbers are expressible in terms of decimal expansions. The real number x written in decimal form looks like this:

$$x = a_n a_{n-1} \ldots a_1 a_0 . a_{-1} a_{-2} \ldots a_{-k} \ldots .$$

This is shorthand for

$$a_n 10^n + a_{n-1} 10^{n-1} + \cdots + a_1 10 + a_0 + a_{-1} 10^{-1} + cldots + a_{-k} 10^{-k} + \cdots .$$

This is a sequence of rational numbers that continues forever. Well, not quite; there are decimal expansions that terminate. Such decimal expansions represent fractions. We also learned that those decimal expansions that eventually repeat also represent fractions. This means that all the numbers whose decimal representations neither

terminate nor eventually repeat are not rational. What we are saying is that the decimal representation of irrational numbers is going
to be tough to deal with. We can concoct examples of decimal expansions that represent irrational numbers like the one at the beginning
of this Chapter but, generally speaking, it is virtually impossible
to exhibit numbers that follow no pattern. And it is impossible to
discover anything in the decimal expansion that would herald its significance. To drive this point home, here is a ridiculous example. It
lists the decimal beginnings of some familiar numbers.

Example 4.3.1

(a) $1.41421356237\ldots$ (constructible, but not rational)

(b) $1.25992104989\ldots$ (arithmetic, but not constructible)

(c) $0.309016994375\ldots$ (polygon and constructible)

(d) $0.939692620786\ldots$ (polygon, but not constructible)

(e) $0.754877666247\ldots$ (algebraic, but not arithmetic)

(f) $3.141592653589\ldots$ (real, but not algebraic)

(g) $36.4621596072\ldots$ (real, but otherwise unknown)

You probably recognize one number, or at most two, from the
above decimal expansions. Here is a key to their identity:

(a) $\sqrt{2}$

(b) $\sqrt[3]{2}$

(c) $\cos(2\pi/5)$

(d) $\cos(\pi/9)$

(e) the real solution to $x^5 + x + 1 = 0$

(f) π

(g) π^π ∎

The strength of the decimal expansion is that, with this representation, it is possible to locate the placement of the real number on
the number line with precision. There is a fundamental existential
problem with infinite strings of numbers that don't follow a pattern.
Without a way of telling what number goes where in the expansion,

there is no telling exactly where the number is situated. But we can get close, and that is what this section is all about—getting close. There are infinite processes that do follow a pattern, and they allow us to get arbitrarily close to irrational numbers. Let us begin with these.

Example 4.3.2

(a) 0.3, 0.33, 0.333, 0.3333, 0.33333, ...

(b) 1, $(1 + 1/2)^2$, $(1 + 1/3)^3$, $(1 + 1/4)^4$, $(1 + 1/5)^5$, ...

(c) $1/1$, $2/1$, $3/2$, $5/3$, $8/5$, $13/8$, ...

(d) 1, $2\sqrt{2}$, $2^2\sqrt{2 - \sqrt{2}}$, $2^3\sqrt{2 - \sqrt{2 + \sqrt{2}}}$,

$$2^4\sqrt{2 - \sqrt{2 + \sqrt{2 + \sqrt{2}}}}, \ldots$$

(e) $1 - 1/2 + 1/3 - 1/4 + 1/5 - \cdots$

(f) $1/1! + 1/2! + 1/3! + 1/4! + \cdots$

(g)
$$\frac{2 \times 2 \times 4 \times 4 \times 6 \times 6 \times \cdots}{1 \times 3 \times 3 \times 5 \times 5 \times 7 \times \cdots}$$

(h) $[1;\ 2,\ 2,\ 2,\ 2,\ 2,\ \ldots]$

(i)
$$1 + \cfrac{1^2}{2 + \cfrac{3^2}{2 + \cfrac{5^2}{2 + \cfrac{7^2}{2 + \ddots}}}}$$

∎

The first example is simply a sequential look at the decimal expansion of $1/3$. Examples (b), (c), and (d) are also sequences of numbers; (e) and (f) are infinite series; (g) is an infinite product; (h) represents an infinite continued fraction; and (i) is an infinite fraction.

The idea behind these patterned processes is to approach, arbitrarily close to some limit, some exact number. We do this through the idea of sequences. In the first four examples, the sequences are already displayed; in the next five examples we must turn the arithmetic process into a sequence.

Example 4.3.3

This example refers to Example 4.3.2. In (a), (b), (c), and (d) the numbers, as given, are arranged in an infinite sequence. In each case we write down the first five terms of the sequence — a_1, a_2, a_3, a_4, a_5 — in decimal form to see if we can trace some movement.

(a) 0.3, 0.33, 0.333, 0.3333, 0.33333

(b) 1, 2.25, 2.3704, 2.4414, 2.4883

(c) 1, 2, 1.5, 1.6667, 1.6

(d) 2.8284, 3.0614, 3.1214, 3.1365, 3.1403

In (e) and (f) the series of numbers is transformed into a sequence of partial sums. We list the decimal equivalents below.

(e) 1, $1 - 1/2$, $1 - 1/2 + 1/3$, $1 - 1/2 + 1/3 - 1/4$,
    $1 - 1/2 + 1/3 - 1/4 + 1/5$

   1, 0.5, 0.8333, 0.5833, 0.7833

(f) $1/1!$, $1/1! + 1/2!$, $1/1! + 1/2! + 1/3!$, $1/1! + 1/2! + 1/3! + 1/4!$,
    $1/1! + 1/2! + 1/3! + 1/4! + 1/5!$

   1, 1.5, 1.6667, 1.7083 1.7167

In (g) the infinite product is transformed into a sequence of partial products.

(g)

$$\frac{2}{1}, \quad \frac{2 \times 2}{1 \times 3}, \quad \frac{2 \times 2 \times 4}{1 \times 3 \times 3}, \quad \frac{2 \times 2 \times 4 \times 4}{1 \times 3 \times 3 \times 5}, \quad \frac{2 \times 2 \times 4 \times 4 \times 6}{1 \times 3 \times 3 \times 5 \times 5}$$

   2, 1.3333, 1.7778, 1.4222, 1.7067

In (h) the continued fraction is transformed into a sequence of convergents.

(h) [1], [1; 2], [1; 2, 2], [1; 2, 2, 2], [1; 2, 2, 2, 2]

   1, 1.5, 1.4, 1.4167, 1.4138

In (i) the infinite quotient is transformed into a sequence of partial quotients.

(i)

$$1^2, \quad 1+\frac{1^2}{2}, \quad 1+\frac{1^2}{2+\dfrac{3^2}{2}}, \quad 1+\frac{1^2}{2+\dfrac{3^2}{2+\dfrac{5^2}{2}}}, \quad 1+\frac{1^2}{2+\dfrac{3^2}{2+\dfrac{5^2}{2+\dfrac{7^2}{2}}}}$$

$$1, \quad 1.5, \quad 1.1538, \quad 1.3816, \quad 1.1977 \qquad \blacksquare$$

The sequences we see here are all tending to a number, although, with the exception of (a) and (h), it is not at all clear what that number is. The sequence (a) converges to $1/3$, and the sequence (h) converges to $\sqrt{2}$ as we learned in Section 3.3. But the underlying question is this: When can we be assured that a sequence of numbers will converge to a limit? What follows may seem like a crash course in analysis but much of it is familiar from your study of calculus. To feel completely comfortable with the concepts, a course in advanced analysis is required, but you can continue this section, with understanding, in relative comfort.

Definition 4.3.4 *A sequence of numbers, $a_1, a_2, \ldots, a_n, \ldots$* **converges to a limit** *L if, for a given number $\epsilon > 0$ there exists a natural number N such that if $n > N$, then $|a_n - L| < \epsilon$.*

Notation: The sequence of numbers, $a_1, a_2, \ldots, a_n, \ldots$ is denoted by $\{a_n\}$.

Definition 4.3.5 *A* **Cauchy sequence** *is a sequence $\{a_n\}$ such that, given $\epsilon > 0$, there is an N such that $|a_m - a_p| < \epsilon$ for all $m, p > N$.*

The Completeness Axiom Every Cauchy sequence of rational numbers has a limit.

This is the traditional form of the completeness axiom. The following theorem ties it in with the completeness axiom of Section 3.1.

Theorem 4.3.6 *The decimal representation of a number is a Cauchy sequence of rational numbers.*

Proof   Recall, once more, that we can write a real number x as follows: Notice that it is in slightly different notational form from Definition 3.1.1:

$$x = A.a_1a_2\ldots a_n a_{n+1}\ldots.$$

Here A represents the whole number part of x. Let $\{b_n\}$ be the sequence where

$$b_n = A.a_1a_2\ldots a_n.$$

Now b_n is clearly a rational number because it is represented by a decimal that is terminating.

Let $\epsilon > 0$ and N be such that $10^{-N} < \epsilon$. Now if m and $p > N$, letting $m < p$, we have

$$|b_m - b_p| = 0.0\,0\,0\ldots a_{m+1}\ldots a_p\,0\,0\ldots < 10^{-N} < \epsilon.$$

Thus $\{b_n\}$ is a Cauchy sequence.   □

So, technically speaking, real numbers exist. This is because every number that is defined by a Cauchy sequence exists, and since reals numbers are represented by decimal expansions and decimal expansions are Cauchy sequences, real numbers exist. Finding the limit of a Cauchy sequence, however, is another matter. There may be those who feel queasy about numbers whose decimal expansions are not fully known. They may feel suckered by the completeness axiom; after all, it simply handles the dilemma of existence by mandating it. Others may think there is absolutely no problem whatsoever; so let's get on with our study.

Let us return to Example 4.3.3. If we could only prove that these were all Cauchy sequences, we would be assured that they are approaching an honest-to-goodness number. An analysis of these sequences shows they are of three types: increasing, decreasing, and alternating between increasing and decreasing.

Definition 4.3.7 *A sequence* $\{a_n\}$ *is*

(a) **increasing** *if* $a_{n+1} \geq a_n$

(b) **decreasing** *if* $a_{n+1} \leq a_n$

(c) **alternating** *if the sequence of even terms,* $\{a_{2n}\}$, *is increasing and the sequence of odd terms,* $\{a_{2n+1}\}$, *is decreasing or vice versa*

(d) **bounded above** *by* M *if* $a_n < M$ *for all* n

(e) **bounded below** *by* m *if* $a_n > m$ *for all* n

We adopt these definitions for simplicity of exposition. Our " increasing" sequence is often called "nondecreasing" and our "decreasing" sequence is often referred to as "nonincreasing." Sequences that we have referred to as "alternating" have no standard name as far as we know.

Theorem 4.3.8

1. *An increasing sequence $\{a_n\}$ that is bounded above converges.*

2. *A decreasing sequence $\{a_n\}$ that is bounded below converges.*

3. *An alternating sequence converges if the decreasing terms are all larger than the increasing terms and if, given $\epsilon > 0$, there is an N such that $|a_n - a_{n+1}| < \epsilon$ for $n > N$.*

For the interested student who is seeing these concepts for the first time we recommend looking in a basic analysis book for proofs of these statements.

Example 4.3.9

(a) Consider the sequence of Example 4.3.3(e):

$$1, \; 1-1/2, \; 1-1/2+1/3, \; 1-1/2+1/3-1/4, \; 1-1/2+1/3-1/4+1/5 \cdots.$$

Clearly it is alternating because each successive member of the sequence is alternately adding on and subtracting off a number of the form $1/n$. To be precise, the even terms, a_{2n+2}, come from subtracting off $1/(2n+2)$ from the previous term a_{2n+1} and the odd terms, a_{2n+1}, come from adding $1/(2n+1)$ to the previous term a_{2n}. Finally, $|a_n - a_{n+1}| = 1/(n+1)$. So given $\epsilon > 0$, there exists an N such that $|a_n - a_{n+1}| = 1/(n+1) < \epsilon$ for $n > N$. The N in question can be a natural number $> 1/\epsilon$. This is because if $n > N$, then we have the following inequalities:

$$1/(n+1) < 1/n < 1/N < \epsilon.$$

So this sequence converges by Theorem 4.3.8 (3). While it may be a mystery what that number is, it is surely between 0.5833 and 0.7833. The answer is that the number is $\ln(2)$; that is, the natural logarithm of 2. We will discuss this number further in Section 4.4. Here it is to 11 place decimal accuracy: 0.69314718056.

(b) Consider the sequence from Example 4.3.3 (f):

$$1/1!, \; 1/1! + 1/2!, \; 1/1! + 1/2! + 1/3!, \; 1/1! + 1/2! + 1/3! + 1/4!, \ldots.$$

It is monotonically increasing; every successive term is larger than the preceding one because we are adding on a positive term each time. Notice that this upwardly mobile sequence never exceeds 2 because it is less than the sequence

$$1, \; 1 + 1/2, \; 1 + 1/2 + 1/4, \; 1 + 1/2 + 1/4 + 1/8 + \cdots.$$

You can check it out; the terms being added onto the given sequence are $1/n!$; the terms being added onto the comparison sequence are $1/2^{n-1}$ and $n! \geq 2^{n-1}$. But the sequence

$$1, \; 1 + 1/2, 1 + 1/2 + 1/4, \ldots, 1 + 1/2 + 1/4 + 1/8 + \cdots + 1/2^{n-1}, \ldots$$

is $1, 3/2, 7/4, \ldots, (2^n - 1)/2^{n-1}, \ldots$ and $(2^n - 1)/2^{n-1} = 2 - 1/2^{n-1} < 2$. So the original sequence converges by Theorem 4.3.8. As in (a), the actual number here is new to us; it is $e - 1$. We will discuss this number further in Section 4.4 also. To 11 place accuracy it is 1.71828182846. ∎

Notice that sequences in Example 4.3.2 (a), (b), (d), and (f) are monotically increasing; sequences (c), (e), (g), (h), and (i) are alternating. The conditions of Theorem 4.3.8 (1) are met in every case, so these sequences all converge to unique numbers. Here are those numbers both to 11 place decimal accuracy and exactly:

(a) $0.33333333333 \approx 1/3$

(b) $2.71828182846 \approx e$

(c) $1.61803398875 \approx (1 + \sqrt{5})/2$

(d) $3.14159265359 \approx \pi$

(e) $0.69314718056 \approx \ln(2)$

(f) $1.71828182846 \approx e - 1$

(g) $1.57079632679 \approx \pi/2$

(h) $1.41421356237 \approx \sqrt{2}$

(i) $1.27323954474 \approx 4/\pi$

That these numbers are the limits is the result of several different fields of study; notably analysis and number theory. We have discussed (a) and (h) and those who remember their calculus will recognize the limits for (b), (e), and (f). Even if we can't find an exact limit, it would be nice to know how close we are getting. This, of course, depends upon how many terms of the sequence we can handle and what our threshold of accuracy is. This is an important and substantial area of study. We touch on it here and return to the subject with the examination of continued 'fractions.

Example 4.3.10

(a) The sequence of Example 4.3.3 (e) converges rather slowly to its limit $L = \ln(2)$. By this we mean that given a particular $\epsilon > 0$, the N in the definition of convergence must be relatively large. As we look at the first five terms of the sequence, we can see that we are not close to $\ln(2)$. In fact, the sequence jumps over and back with steps of size $1/n$, leaving L somewhere in the middle. If we let $\epsilon = 1/1000$, then we may let $N = 1000$, and clearly if $n > N, |a_n - \ln(2)| < \epsilon$. Actually, we can do considerably better than $N = 1000$, but still the N must be large. We leave it as an exercise to find a smaller value for N that will still work.

(b) The sequence of Example 4.3.3 (f) converges relatively quickly to its limit, $e - 1$. Let $\epsilon = 1/1000$. Then we may let $N = 5$ because

$$|1/1! + 1/2! + 1/3! + 1/4! + 1/5! + 1/6! - (e - 1)| < \epsilon$$

and the sequence is increasing. ∎

A rich source of Cauchy sequences are infinite continued fraction expansions. Let us return to our study that we began in Section 3.3. Recall that an infinite continued fraction is a number of the form

$$a_0 + \cfrac{1}{a_1 + \cfrac{1}{a_2 + \cfrac{1}{a_3 + \cfrac{1}{\ddots}}}}$$

$$a_{n-1} + \frac{1}{a_n}$$

$\therefore$

where $a_i > 0$ for all $i > 0$, $a_0 \geq 0$.

We denote this infinite continued fraction by:

$$[a_0; a_1, a_2, a_3, \ldots, a_n, \ldots].$$

Its ith convergent, p_i/q_i, is the fraction $[a_0; a_1, a_2, a_3, \ldots, a_i]$.

Our first order of business to is prove that infinite continued fraction expansions tend toward a limit.

Theorem 4.3.11 *The convergents of an infinite continued fraction*

$$[a_0; a_1, a_2, a_3, \ldots, a_n, \ldots]$$

form a Cauchy sequence.

Proof It follows from Corollary 3.3.14 that, for even n

$$p_0/q_0 < p_2/q_2 < p_4/q_4 < \ldots < p_n/q_n$$

and, for odd n

$$p_n/q_n < \ldots < p_5/q_5 < p_3/q_3 < p_1/q_1.$$

By Theorem 3.3.13, $p_{n+1}/q_{n+1} - p_n/q_n = (-1)^n/q_{n+1}q_n$. So the conditions of Theorem 4.3.8 (3) are met. $\square$

If the continued fraction is infinite the limit must be irrational. This is true because finite continued fractions are rational.

Theorem 4.3.12 *The infinite continued fraction*

$$[a_0; a_1, a_2, a_3, \ldots, a_n, \ldots]$$

converges to an irrational number.

We now begin a study of the irrational numbers that are limits of certain continued fractions. We will make it easy on ourselves by examining those infinite continued fractions that have a pattern; in particular, we will study infinite continued fractions that are eventually periodic.

Notation: If the infinite continued fraction

$$[a_0; a_1, a_2, a_3, \ldots, a_n, a_1, a_2, a_3, \ldots, a_n, a_1, \ldots]$$

repeats we will write it as $[a_0; \overline{a_1, a_2, a_3, \ldots, a_n}]$. We assume that n is the smallest number for which there is repetition.

Example 4.3.13

(a) Consider the infinite continued fraction

$$1 + \cfrac{1}{1 + \cfrac{1}{1 + \cfrac{1}{1 + \ddots}}}$$

In shorthand, this is the continued fraction $[1; \overline{1}]$. We may find its convergents just as we did in Section 3.2 using quotient tables.

		1	1	1	1	1	1	1	...
p_k	0	1	1	2	3	5	8	13	21 ...
q_k	1	0	1	1	2	3	5	8	13 ...

So its convergents are

$$1/1, \ 2/1, \ 3/2, \ 5/3, \ 8/5, \ 13/8, \ 21/13, \ldots.$$

Notice that this is the sequence of Example 4.3.2 (c). Notice that the numerators are $1, 2, 3, 5, 8, 13, 21, \ldots$, while the denominators lag behind and are $1, 1, 2, 3, 5, 8, 13, \ldots$. These numbers are the famous Fibonacci numbers we saw in Section 1.4. Finally, notice that this is an alternating sequence: The odd terms are increasing; the even terms are decreasing. We may write it as

$$1/1 < 3/2 < 8/5 < 21/13 < \ldots < x < \ldots < 13/8 < 5/3 < 2/1$$

with the limit, x, trapped in the middle.

(b) Let us consider the continued fraction $[1; \overline{1, 2, 3}]$.

		1	1	2	3	1	2	3	...
p_k	0	1	1	2	5	17	22	61	205 ...
q_k	1	0	1	1	3	10	13	36	121 ...

Its convergents are

$$1/1, \ 2/1, \ 5/3, \ 17/10, \ 22/13, \ 61/36, \ 205/121, \ldots.$$

As in (a), the number, x, is caught in the middle.

$$5/3 < 22/13 < 205/121 < \ldots < x < \ldots < 61/36 < 17/10 < 2/1. \quad \blacksquare$$

This example displays sequences of approximations, but in each case there is a mystery as to exactly what number, x, these sequences are approximating. You remember that we have seen the answer to (a); it is $(1 + \sqrt{5})/2$. It turns out that we can solve for the limit x for all of these repeating continued fractions.

Example 4.3.14

(a) Let x be the continued fraction

$$1 + \cfrac{1}{1 + \cfrac{1}{1 + \cfrac{1}{1 + \ddots}}}$$

So $x - 1 = 1/x$, which yields $x^2 - x - 1 = 0$. From the quadratic formula we obtain $x = (1 \pm \sqrt{5})/2$. The negative solution does not apply here. So we have proven again what we knew before: The continued fraction $[1; \overline{1}]$ represents the irrational number $(1 + \sqrt{5})/2$. Notice that $(1 + \sqrt{5})/2 \approx 1.61803399$ and the convergent $21/13 \approx 1.61538462$. This is close. The number $(1 + \sqrt{5})/2$ is known as the famous golden mean, or golden ratio, the same one examined in the exercises of Section 4.2. It is a famous enough number to merit its own symbol, ϕ. There are more possibilities for exploring the golden mean in the projects.

(b) Consider the continued fraction $x = [1; \overline{1, 2, 3}]$ again.

$$x = 1 + \cfrac{1}{1 + \cfrac{1}{2 + \cfrac{1}{3 + \cfrac{1}{1 + \ddots}}}} \quad ; \quad x - 1 = \cfrac{1}{1 + \cfrac{1}{2 + \cfrac{1}{3 + (x - 1)}}}$$

So we have

$$x = 1 + \cfrac{1}{1 + \cfrac{1}{2 + \cfrac{1}{x + 2}}}.$$

We can unwind this continued fraction and find an equation with x. It is easier to unwind if we use quotient tables.

		1	1	2		$x + 2$
p_k	0	1	1	2	5	$5(x+2) + 2$
q_k	1	0	1	1	3	$3(x+2) + 1$

So $p_k/q_k = (5x+12)/(3x+7) = x$. This yields the quadratic equation $3x^2 + 2x - 12 = 0$, and therefore $x = -1/3 + \sqrt{37}/3 \approx 1.69425422$. The convergent $205/121 \approx 1.6942149$ is pretty close. ∎

These examples indicate that continued fractions that are periodic converge to irrational numbers that are solutions to a quadratic equation $ax^2 + bx + c = 0$ with integer coefficients. We'll call these types of numbers quadratic numbers.

Definition 4.3.15 *A **quadratic number** is an irrational number that is a solution to a quadratic equation with integer coefficients.*

It follows from the quadratic formula that a quadratic number is of the form $r + s\sqrt{n}$, where r and s are rational numbers and $s \neq 0$ and n is a natural number that is not a perfect square.

If the continued fraction is immediately periodic, then we can find the convergents from the following table of quotients.

Theorem 4.3.16 *If $[a_0; \overline{a_1, a_2, a_3, \ldots, a_k}]$ is an infinite continued fraction, then the following table of quotients, when filled out, yields the solution*

		a_0	a_1	a_2	a_3	$\ldots$	a_{k-1}	$x - a_0 + a_k$
p_k		0	1					
q_k		1	0					

The "proof" follows by inspecting the continued fraction:

$$a_0 + \cfrac{1}{a_1 + \cfrac{1}{\ddots \\ a_{k-1} + \cfrac{1}{a_k + (x - a_0)}}}.$$

Theorem 4.3.17 *The continued fraction $[a_0; \overline{a_1, a_2, a_3, \ldots, a_k}]$ converges to a quadratic number.*

The reason for this is that when the table of quotients in Theorem 4.3.16 is filled out we have $x = (rx + s)/(tx + u)$ for some integers r, s, t, and u, and this leads to a quadratic equation.

We find that if the continued fraction is eventually periodic, the situation is the same.

Theorem 4.3.18 *If the continued fraction*

$$[a_0; a_1, a_2, a_3, \ldots, a_k, \overline{a_{k+1}, \ldots, a_m}]$$

eventually repeats, then the table of quotients looks like this:

	y	a_{k+1}	a_{k+2}	$\cdots$	a_{n-1}	$x - y + a_m$
p_k		0	1			
q_k		1	0			

where y is derived by subtracting off the nonrepeating part of the continued fraction.

Let's see how this works in a simple example.

Example 4.3.19

Consider the continued fraction $[1 : 2, \overline{3}]$. So

$$x = 1 + \cfrac{1}{2 + \cfrac{1}{3 + \cfrac{1}{3 + \ddots}}}.$$

Now, subtracting off the nonrepeating part, step by step we get

$$1/(1/(x - 1) - 2)) = 3 + \cfrac{1}{3 + \cfrac{1}{3 + \cfrac{1}{3 + \ddots}}}.$$

Substituting y for $1/(1/(x-1)-2))$, we obtain $y = 3 + 1/y$. Thus y satisfies the quadratic $y^2 - 3y - 1 = 0$ and so $y = 3/2 + \sqrt{13}/2$. Substituting x for y and noting that $y = (x - 1)/(-2x + 3)$, we get

$$x = (1 + 3y)/(1 + 2y) = 5/6 + \sqrt{13}/6. \qquad \blacksquare$$

Theorem 4.3.20 *If* $[a_0; a_1, a_2, a_3, \ldots, a_k, \ldots]$ *is a continued fraction that eventually repeats, then it converges to a quadratic number.*

Here is an informal showing of this. Note that the number y in Theorem 4.3.18 is of the form $(rx + s)/(tx + u)$. Since y is also a solution to an immediately repeating continued fraction, Theorem 4.3.17 tells us it is of the form $a + \sqrt{b}$, where a and b are rationals. Setting these two representations of y equal, we find that y is of the form $a' + \sqrt{b'}$.

So we see that eventually repeating continued fractions converge to solutions of quadratic equations. The converse is also true; that is, if x is a real solution to a quadratic equation, then it can represented by a periodic continued fraction. We will not prove this.

Theorem 4.3.21 *A continued fraction converges to a quadratic number if and only if it is eventually periodic.*

Let us examine the simplest of these solutions, the number $\sqrt{n}$, where n is not a perfect square. We can do this with a hand calculator just as we did in Example 3.3.8. Recall how it works. You need only use the keys that perform division, inverting (or taking reciprocal), and subtraction. We will represent these keys by $\div$, x^{-1}, and $-$, respectively. Inside the angled brackets we show what the calculator screen displays.

Example 4.3.22

(a) Consider $\sqrt{5}$. Here is the procedure with the calculator.

$\sqrt{5} = \langle\ 2.236067977\ \rangle\ -2 = \langle\ 0.236067977\ \rangle\ x^{-1}$
$\langle\ 4.236067977\ \rangle\ -4 = \langle\ 0.236067977\ \rangle.$

$$\sqrt{5} \ = \ 2 + \cfrac{1}{4 + \cfrac{1}{4 + \ \ddots}}$$

That is, $\sqrt{5} = [2; \overline{4}]$.

(b) Consider $\sqrt{3}$.

$\sqrt{3} = \langle\ 1.732050808\ \rangle\ -1 = \langle\ 0.732050808\ \rangle\ x^{-1}\ \langle\ 1.366025404\ \rangle$
$-1 = \langle\ 0.366025404\ \rangle\ x^{-1}\ \langle\ 2.732050808\ \rangle\ -2 = \langle\ 0.732050808\ \rangle.$

$$\sqrt{3} = 1 + \cfrac{1}{1 + \cfrac{1}{2 + \cfrac{1}{1 + \cfrac{1}{2 + \ddots}}}}$$

So $\sqrt{3} = [1; \overline{1, 2}]$.

(c) Let us try a longer example. Consider $\sqrt{31}$.

$\sqrt{31} = \langle\, 5.567764363\, \rangle\ -5 = x^{-1}\ \langle\, 1.76129406\, \rangle\ -1 = x^{-1}$
$\langle\, 1.313552873\, \rangle\ -1 = x^{-1}\ \langle\, 3.189254788\, \rangle\ -3 = x^{-1}$
$\langle\, 5.283882181\, \rangle\ -5 = x^{-1}\ \langle\, 3.522588121\, \rangle\ -3 = x^{-1}$
$\langle\, 1.913552872\, \rangle\ -1 = x^{-1}\ \langle\, 1.094627394\, \rangle\ -1 = x^{-1}$
$\langle\, 10.56776435\, \rangle\ -10 = x^{-1}\ \langle\, 1.76129411\, \rangle$

The repetition has begun, sort of. While the first appearance of the number 1.76129406 is not the same as the second appearance, 1.76129411, they are close. They would be the same if the calculator could carry procedure to more than 12 digits. The rounding error has introduced this discrepancy, and it will get larger if we continue the process. We may conclude:

$$\sqrt{31} = [5; \overline{1, 1, 3, 5, 3, 1, 1, 10}]. \qquad \blacksquare$$

Recall again Pell's equations from Section 2.3; we were looking for integer solutions to $x^2 - ky^2 = \pm 1$. In Section 3.2 we pointed out that the solutions r, s offer us fractions that closely approximate $\sqrt{k}$. Let us carry on this examination.

Definition 4.3.23 *The fraction r/s is called a* **Pell convergent** *of k if $r^2 - ks^2 = \pm 1$.*

Theorem 4.3.24 *If r/s is a Pell convergent, then $|r/s - \sqrt{k}| < 1/(2s^2)$.*

Proof Since r/s is a Pell convergent, we have $|r^2 - ks^2| = 1$. Therefore,

$$|r^2/s^2 - k| = |r/s - \sqrt{k}||r/s + \sqrt{k}| = 1/s^2.$$

It follows that

$$|r/s - \sqrt{k}| \times 2 < 1/s^2$$

because $r > s$ and $k > 1$ and so $|r/s + \sqrt{k}| > 2$. The result then follows. □

Recall also that it was difficult to find solutions to Pell's equations. Since we have an algorithm for finding square roots, we can try it out on solutions to Pell's equations. We can check it against the Pell chart in Section 2.3.

Example 4.3.25

(a) Consider the convergents of the continued fraction expansion of $\sqrt{5}$; that is, $[2; \overline{4}]$. We can get them with our table.

		2	4	4	4	4	4	
p_k	0	1	2	9	38	161	682	2889	...
q_k	1	0	1	4	17	72	305	1292	...

| $p_k^2 - 5q_k^2$ | | -1 | 1 | -1 | 1 | -1 | 1 | ... |

Notice that a fourth row has been added to the table evaluating the Diophantine equation $p_k^2 - 5q_k^2$ at each of the convergents. The convergents, p_k/q_k, beginning with $k = 0$ are, respectively,

$$1, \quad 9/4, \quad 38/17, \quad 161/72, \quad 682/305, \quad 2889/1292.$$

For example, for $k = 4$, $p_k/q_k = 682/305$, and $p_k^2 - 5q_k^2 = -1$. What we find is that p_k and q_k provide us with solutions for Pell's equation for each pair of entries in the table.

(b) Consider the convergents of the continued fraction expansion of $\sqrt{3}$. That is, $[1; \overline{1, 2}]$.

		1	1	2	1	2	1	2	1	
p_k	0	1	1	2	5	7	19	26	71	97	...
q_k	1	0	1	1	3	4	11	15	41	56	...

| $p_k^2 - 3q_k^2$ | | -2 | 1 | -2 | 1 | -2 | 1 | -2 | 1 | ... |

Notice here that not every pair of p_k and q_k satisfies Pell's equation $r^2 - 3s^2 = \pm 1$. In fact, the pairs (p_k, q_k) for k odd satisfy a Pell equation. As we already know from Theorem 2.3.8, there is no solution to the Pell equation $r^2 - 3s^2 = -1$.

(c) Consider the convergents of $\sqrt{31}$. The continued fraction expansion is $[5; \overline{1, 1, 3, 5, 3, 1, 1, 10}]$. The Pell chart does not include solutions to $r^2 - 31s^2 = 1$.

	5	1	1	3	5	3	1	1	10	...
p_k	5	6	11	39	206	657	863	1520	16063	...
q_k	1	1	2	7	37	118	155	273	4435	...

$p_k^2 - 3q_k^2$	−6	5	−3	2	−3	5	−6	1	−6	...

Notice that only one of these pairs of entries, 1520, 273, offers a solution to Pell.   ∎

We know that continued fraction convergents home in on a limit. Theorem 4.3.24 gives an indication of how accurate the Pell convergents are, $|p_n/q_n - \sqrt{k}| < 1/2q_n^2$. Of course, not all continued fraction convergents are Pell convergents. Corollary 3.3.15 shows how accurate we can expect continued fraction convergents to be. If p_n/q_n is a convergent of an infinite continued fraction that converges to the real number R, then $|p_n/q_n - R| < 1/(q_{n+1}q_n)$. Example 3.3.18 displays the accuracy of the convergents of $\sqrt{2}$. But then, continued fraction expansions of square roots are special. Let us try our hand at finding a fraction approximation to a real number, $\sqrt[3]{2}$ that is not a solution to a quadratic equation. Of course, we should remember that we are constructing the continued fraction from the calculator's knowledge of $\sqrt[3]{2}$, which is 11 places, so the fraction accuracy is limited.

Example 4.3.26
˙Consider $\sqrt[3]{2}$. The continued fraction begins like this:

$$[1; 3, 1, 5, 1, 1, 4, 1, 1, 8, 1, 14, \ldots]$$

	1	3	1	5	1	1	4	1	1	8	1	14
p_k	1	4	5	29	34	63	286	329	635	5429	6064	90325
q_k	1	3	4	23	27	50	227	277	504	4309	4813	71691

Now $96389/76504 \approx 1.25992104988$. We expect from Theorem 4.3.28 that the accuracy would be within $1/(76504 \times 148195) \approx 8.8 \times 10^{-11}$; that's nine-place accuracy. By our calculator, $\sqrt[3]{2} \approx 1.25992104989$, so it looks as though we have done a bit better. We recall from Sections 3.2 and 3.3 that continued fraction expansions yield the best rational approximations to given fractions in the sense that the denominators of the convergents are the smallest possible. This holds true for irrational numbers as well. So if we restrict our denominator to be < 100, the best fraction approximation to $\sqrt[3]{2}$

would be $63/50 = 1.26$. That's closer than 8×10^{-5}. The fraction with denominator < 1000 that most closely approximates $\sqrt[3]{2}$ is $635/504 \approx 1.2599263492$, and this differs by less than 4.2×10^{-7}. So that gives us six-decimal accuracy. ∎

In summary, continued fractions offer us the optimal way to hone in on irrational numbers with rational approximations. If the continued fraction is eventually repeating, there is the neat result that the irrational limit is the solution to a quadratic equation. In the language of Sections 4.1 and 4.2, these continued fractions converge to quadratic numbers that are also constructible, arithmetic, and algebraic. If the continued fraction does not repeat, there are interesting questions that remain unanswered. For example, it is believed, but has not been proven, that if a number is algebraic the entries in the continued fraction expansion are bounded. We shall see in the next section that the continued fraction expansion of the number e is not bounded and, indeed, e is not algebraic.

EXERCISES

1. Consider the sequence $\sqrt{2}, \sqrt{1 + \sqrt{2}}, \sqrt{1 + \sqrt{1 + \sqrt{2}}}, \ldots$.

   (a) Show that this sequence is monotonically increasing and bounded above.

   (b) Find the limit of the sequence.

2. Consider the following continued fraction–like sequence:

$$\sqrt{2} + \cfrac{1}{\sqrt{2} + \cfrac{1}{\sqrt{2} + \ddots}}.$$

   (a) Show that this sequence is monotonically increasing and bounded above.

   (b) Find the limit of the sequence.

3. Consider the following continued fraction–like sequence:

$$2 + \cfrac{1}{1 + \cfrac{1}{2 + \cfrac{2}{3 + \cfrac{3}{4 + \cfrac{4}{5 + \ddots}}}}}.$$

(a) Show that this sequence is monotonically increasing and bounded above.

(b) Find the limit of the sequence.

4. Consider the following continued fraction–like sequence:

$$\cfrac{1}{1 + \cfrac{1}{2 + \cfrac{9}{2 + \cfrac{25}{2 + \cfrac{49}{2 + \ddots}}}}}$$

(a) Show that this sequence is monotonically increasing and bounded above.

(b) Find the limit of the sequence.

5. Check out the speed of convergence of the infinite sequences of Example 4.3.3. In each case let $\epsilon = 1/1000$ and find an N such that if $n > N$, then $|a_n - L| < \epsilon$. Using this, rank the speed of convergence of the different sequences.

6. Prove Theorem 4.3.8.

(a) An increasing sequence $\{a_n\}$ that is bounded above converges.

(b) A decreasing sequence $\{a_n\}$ that is bounded below converges.

(c) An alternating sequence converges if the decreasing terms are all larger than the increasing terms and if, given $\epsilon > 0$, there is an N such that $|a_n - a_{n+1}| < \epsilon$ for $n > N$.

7. For as many of the alternating sequences in Example 4.3.3 as you can, show that the hypotheses of Theorem 4.3.8 are true; that is, its monotonically decreasing terms are all larger than the monotonically increasing terms and given $\epsilon > 0, |a_n - a_{n+1}| < \epsilon$ for $n > N$ for some N.

8. Given the following continued fractions, find the following three items: (i) the quadratic equation for which the limit is a solution, (ii) the quadratic number that is the limit of the convergents, and (iii) the convergent with smallest denominator that agrees with the limit to six-place accuracy.

   (a) $[2; \overline{2, 3}]$
   (b) $[1; \overline{1, 3, 5}]$
   (c) $[1; \overline{4, 3, 2, 1}]$

9. Given the following continued fractions, find the following three items: (i) the quadratic equation for which the limit is a solution, (ii) the quadratic number that is the limit of the convergents, and (iii) the convergent with smallest denominator that agrees with the limit to six-place accuracy.

   (a) $[2; 1, \overline{2}]$
   (b) $[1; 3, \overline{1}]$
   (c) $[1; 2, 2, \overline{1, 2}]$

10. Find the continued fractions that represent the following square roots:

   (a) $\sqrt{6}$
   (b) $\sqrt{7}$
   (c) $\sqrt{13}$
   (d) $\sqrt{29}$
   (e) $\sqrt{73}$
   (f) $\sqrt{61}$

(g) $\sqrt{97}$

11. For the given square roots, which of its convergents is a Pell convergent?

    (a) $\sqrt{7}$
    (b) $\sqrt{10}$
    (c) $\sqrt{22}$
    (d) $\sqrt{43}$
    (e) $\sqrt{46}$

12. Find a solution for the following Pell equations $x^2 - ky^2 = 1$ and also, where possible, $x^2 - ky^2 = -1$, where k is

    (a) 29
    (b) 37
    (c) 41
    (d) 53
    (e) 61 (recall Section 2.3)
    (f) 92 (recall Section 2.3)

13. If you have a computer, find the smallest solution to the Pell equation $x^2 - 313y^2 = 1$.

14. Finish this thought: If $[a_0; \overline{a_1, a_2, a_3, \ldots, a_k}]$ is the continued fraction expansion of $\sqrt{n}$, then, among the continued fraction convergents, the Pell convergents occur ....

15. Find the quadratic numbers that are limits for the following continued fraction expansions:

    (a) $[1; \overline{2}]$
    (b) $[1; \overline{3}]$
    (c) $[1; \overline{4}]$
    (d) $[1; \overline{5}]$
    (e) $[1; \overline{6}]$
    (f) Generally, what can you say about $[1; \overline{n}]$, where $n > 1$?

16. Find the continued fraction expansions of

    (a) $\sqrt{2}/2$
    (b) $2\sqrt{2}$
    (c) $\sqrt{2}/10$
    (d) $10\sqrt{2}$

17. Find a fraction with smallest denominator that agrees with the following to seven-place accuracy

    (a) $\sqrt[3]{3}$
    (b) $\ln(2)$
    (c) $\cos \pi/9$
    (d) π
    (e) π^2

18. Find the continued fraction expansions of the following numbers composed of e:

    (a) e
    (b) $\sqrt{e}$
    (c) $(e-1)/(e+1)$
    (d) Make up other numbers that use e and try to find patterns.

19. Consider the continued fraction $[1; \overline{1}]$.

    (a) Find the first 10 convergents.
        What can you say about
    (b) the difference between convergents p_n/q_n and q_n/p_n?
    (c) the difference between p_{n+2}/q_{n+2} and p_n/q_n?
    (d) the sum of q_n^2/p_n^2 and q_n/p_n?

20. Recalling that $\phi = (1+\sqrt{5})/2$, notice that $\phi^2 = \phi+1$. Express higher powers of ϕ in terms of ϕ itself.

    (a) ϕ^3
    (b) ϕ^4

(c) ϕ^5

(d) ϕ^6

(e) What do you notice?

21. Find the twelfth decimal in the decimal expansion of $\sqrt{3}$.

22. The continued fraction expansion of $\sqrt{2}$ has period one, while the expansion of $\sqrt{3}$ has period two. Experimenting with continued fraction expansions of square roots, answer the following. Give reasons for your answers if you can.

   (a) For which n does $\sqrt{n}$ have a continued fraction expansion of period one?

   (b) For which n does $\sqrt{n}$ have a continued fraction expansion of period two?

4.4   Searching for Transcendental Numbers

We have studied irrational numbers from three different perspectives: algebra, geometry, and analysis. We have distinguished different types of irrationals: quadratic numbers, constructible numbers, polygon numbers, arithmetic numbers, and algebraic numbers. We have met just about all types of known irrational numbers. But we haven't yet found a home for the two most famous irrational numbers of all, π and e. We are in good company. They were not proved to be irrational until the mid-1700s; this was accomplished by John Lambert (1728−1777). And they were not proved to be nonalgebraic until the late 1800s; Charles Hermite (1822−1901) proved in 1872 that e was not algebraic, Ferdinand Lindemann (1852−1939) proved in 1882 that π was not algebraic. In fact, it was not until 1844 that a nonalgebraic irrational number was known at all. It was constructed by the French mathematician J. Liouville (1809−1882). The proof that a new kind of number existed is fascinating; we shall return to it later. In the twentieth century the search for these new numbers has not produced a great many of them. In 1900, the German mathematician David Hilbert (1862−1943) presented his 23 famous problems at the International Congress in Paris, and problem 7 posed the question of whether the number $2^{\sqrt{2}}$ was algebraic or not. It was shown to be not algebraic in 1934 by the Russian mathematician Aleksander Gelfond (1906−1968). But today, more than 60 years later, many numbers are only suspected of being nonalgebraic. Here we shall carry out our own search for these rare, yet plentiful numbers. They are called transcendental numbers.

Definition 4.4.1 *A number is a* **transcendental number** *if it is not algebraic.*

Notation: We denote the set of real transcendental numbers by $\mathbb{T}$.

That such numbers exist, and exist in great numbers, was proved by the German mathematician Georg Cantor (1845−1918). His proof involves counting members of infinite sets. This involves taking a brief detour into set theory. We shall examine the sizes of different infinite sets. We know, for example, that the set of natural numbers is infinite, basically because there is no largest natural number. We shall say that any set is countable if we can attach a unique natural number to every element; that is, if we can count every member. This makes sense. What may be difficult, however, is the reasoning behind some of the proofs. Since we do not intend to give a background in set theory, reading of the proofs will be slow going for those who have not had experience with functions and one-to-one correspondences, but the concepts are not that difficult. Let us recall a few basic definitions. A function is a **one-to-one correspondence** if it sends distinct elements to distinct images; that is, if $f(a) = f(b)$ then $a = b$. A function maps A **into** B if the set $\{f(x) : x \in A\} \subseteq B$. A function maps A **onto** B if the set $\{f(x) : x \in A\} = B$.

Definition 4.4.2 *A set S is* **countable** *if there is a one-to-one function from S into $\mathbb{N}$, the set of natural numbers.*

Example 4.4.3

(a) The even positive numbers are countable. Let f be the counting function that assigns even positive numbers to natural numbers as follows:

$$2 \to 1, \quad 4 \to 2, \quad 6 \to 3, \ldots, 2n \to n.$$

So $f(2n) = n$.

(b) The integers are countable. Let I be the counting function that assigns to each integer a natural number as follows:

$$0 \to 1, \quad 1 \to 2, \quad -1 \to 3, \quad 2 \to 4, \quad -2 \to 5, \ldots.$$

So $I(0) = 1$ and for $n > 0, I(n) = 2n$ and $I(-n) = 2n + 1$. We will refer to this particular counting function I again.

(c) Let $S = \{x : 0 < x < 1 \text{ and } x \text{ is rational}\}$. This set of rationals between 0 and 1 is countable. Let R be the counting function that assigns to each rational in S a natural number as follows:

$$1/2 \to 1,\ 1/3 \to 2,\ 2/3 \to 3,\ 1/4 \to 4,\ 3/4 \to 6, \dots.$$

In general, $R(a/b) = a + (b - 1)(b - 2)/2$ where a/b is reduced to lowest terms. We shall refer to this function R again. Notice that R maps the set of rationals into, rather than onto, the set of natural numbers. That is because R does not assign a value to a/b if the fraction is not in lowest terms. So, for example, the number 5 is not the image of any fraction. Under the formula its pre-image would be $2/4$, but this fraction is not in lowest terms. ■

The following theorem is useful to us. It employs some ingenious counting methods; in particular, it relies on the fundamental theorem of arithmetic and the fact that there are infinitely many primes.

Theorem 4.4.4 *If S_1, S_2, \dots S_n, \dots is a countable collection of countable sets, then $\bigcup S_i$ is countable.*

Proof Here is how we shall assign the counting numbers. We assume that each of the sets, S_i, has a counting function (call it f_i) that counts the specific set, S_i. Let g count the members of $\bigcup S_i$ as follows: If $x \in \bigcup S_i$, then let $j = \min\{i : x \in S_i\}$. Let $g(x) = p_j^{f_j(x)}$. Here p_j represents the jth prime. Now this function g is one-to-one because, by definition, x is picked from a particular set location so j is unique and f_j is one-to-one. □

Theorem 4.4.5 *The set of rational numbers is countable.*

Proof This follows from Theorem 4.4.4 because the rational numbers are a countable set of countable sets. Adopting the temporary notation $(n, n + 1)$ to indicate the set of rational numbers between n and $n + 1$ and recalling that $\mathbb{Z}$ represents the set of integers, we may write

$$\mathbb{Q} = \mathbb{Z} \cup (0, 1) \cup (-1, 0) \cup (1, 2) \cup (-2, -1) \cup \dots.$$

Example 4.4.3 (b) shows this is a countable collection of sets, Example 4.4.3 (c) shows each of the sets is countable. □

Example 4.4.6

(a) Let us see what a counting function will look like that can count the rationals. It will be made up of the functions I and R from Example 4.4.3. Let $x = -3/5$. And let the following sets be

$$\mathbb{Z} = S_1, \quad (0,1) = S_2, \quad (-1,0) = S_3, \ldots.$$

We find that $x \in (-1,0) = S_3$. In fact, $x = -1 + 2/5$. Now the counting function R assigns 8 to $2/5$ because $8 = 2 + (3)(4)/2$. So $g(x) = 5^8$. We use 5 because it is the third prime.

(b) Let us find what number is assigned to $x = 64/17$. Now $x = 3 + 13/17$ so $x \in (3,4) = S_8$. Also, $R(13/17) = 13 + (15)(16)/2 = 133$. And 19 is the eighth prime number. So $g(x) = 19^{133}$. ∎

We now set about counting the algebraic numbers. Recall that an algebraic number is a solution to a polynomial equation with integer coefficients.

Definition 4.4.7 *The number x is an algebraic number of* **degree** *n if x is a solution to a polynomial of degree n with integer coefficients and n is the smallest degree polynomial for which this is true.*

Theorem 4.4.8 *The set of algebraic numbers of degree n is countable.*

Proof Let S be the set of algebraic numbers of degree n. We describe a function, g, that counts the members of S. Let s be an algebraic number of degree n. So s is a solution to the equation $a_0 + a_1 x + \cdots + a_n x^n = 0$. Recalling the function I: $I(0) = 1$ and for $n > 0, I(n) = 2n$ and $I(-n) = 2n + 1$, let

$$g(s) = 2^{I(a_0)} \times 3^{I(a_1)} \times 5^{I(a_2)} \times \cdots \times p_{n+1}{}^{I(a_n)} \times p_{n+i+1}$$

where s is the ith of the n possible real solutions (recall the fundamental theorem of algebra) numbered from smallest to largest and p_n is the nth prime number. If s is a multiple root we can count it multiple times, as we did when counting fractions in Example 4.4.3 (c). Also recall that I is the function that counts the integers as defined in 4.4.3 (b). □

Example 4.4.9

Consider the middle, or second, real solution, call it s, to the polynomial equation $x^5 + 0x^4 + 0x^3 + 0x^2 - 5x + 1 = 0$. There are three real solutions to this quintic. Incidentally, they are all nonarithmetic numbers, as Theorem 4.1.16 tells us. Using the scheme used in the proof of Theorem 4.4.8, let's find out what number will be assigned to s. Note that $I(1) = 2$, $I(0) = 1$, and $I(-5) = 11$. Notice also that 2 is the first prime, 3 is the second, and 13 is the sixth. We shall call our solution s the second solution. Thus

$$g(s) = 2^2 \times 3^{11} \times 5^1 \times 7^1 \times 11^1 \times 13^2 \times 19.$$

If we chose the smallest, or first, real root, the counting number would be

$$2^2 \times 3^{11} \times 5^1 \times 7^1 \times 11^1 \times 13^2 \times 17. \qquad \blacksquare$$

Theorem 4.4.10 *There are a countable number of algebraic numbers.*

Proof Since there are a countable number of algebraic numbers of degree n and a countable number of degrees, this follows from Theorem 4.4.4. $\qquad \square$

Now comes the surprise: We cannot count the real numbers; there are too many. Since we can count the algebraic numbers, the rest of the real numbers, the transcendentals, must not be countable. That means there are lots more transcendental numbers than algebraic numbers.

Theorem 4.4.11 (Cantor) *The real numbers cannot be counted.*

Proof Consider S, the set $\{x: 0 < x \le 1$, where x is a real number$\}$. Suppose we have found a counting function f from S to N. We shall use the following notation: If $x \in S$, then $f(x) = n$, for some natural number n, and we shall represent x by the decimal expansion

$$x = 0.a_{n,1}a_{n,2} \dots a_{n,k} \dots.$$

We shall assume that a number ending in all 9s will be designated by its equivalent that ends in all 0s. We say this for uniqueness of representation. Now we construct a real number as follows

$$r = 0.b_1 b_2 \dots b_k \dots,$$

where $b_k = 0$, if $a_{k,k} \neq 0$, $b_k = 1$, if $a_{k,k} = 0$.

Notice that r cannot be identical with any of the real numbers we have counted because it differs at the kth place. So our assumption that the counting function counted all real numbers was wrong. Thus there is no such function; the reals are uncountable. $\square$

Corollary 4.4.12 *The transcendental numbers cannot be counted.*

Now that we know that there are oodles of transcendental numbers, the search is on to find them. We have examined solutions of polynomial equations, numbers that can be constructed from straightedge and compass, polygon numbers, and limits of continued fraction expansions. We have not found a single transcendental number among them and, had we not been clued in that π and e are transcendental, we could not point to a single example. Of course, we can generate a few transcendentals from the ones we already know as this theorem shows.

Theorem 4.4.13 *If x is a transcendental number and y is algebraic, then $x+y$, $x-y$, xy, y/x, x/y, x^n, and $\sqrt[n]{x}$ are all transcendental numbers.*

Proof We proceed by assuming the contrary and deriving a contradiction. Here is how it works for the sum of two numbers. Let x be a transcendental number and y an algebraic number. Then suppose that z is algebraic, where $z = x+y$. It follows that $x = z - y$ and this implies x is algebraic; a contradiction. So z must be transcendental. (Now this is an easy proof compared to what has come before.) $\square$

So we know that arithmetic combinations π with algebraic numbers and e with algebraic numbers will yield transcendental numbers. We should note that Theorem 4.4.13 implies that $\sqrt{\pi}$ is transcendental. This fact shows that it is impossible, with straightedge and compass, to square the circle. That is, it is impossible to build a square with the same area as the area of a circle. This was one of the famous unsolved problems that the Greek mathematicians posed.

Before we unleash a whole host of transcendental numbers on you, numbers that have been proved transcendental only recently (in this century), let us study the first transcendental that was constructed. A couple of the proofs are tough going and are included for the sake of completeness, but the ideas behind them are in the spirit of our examination of continued fractions, best possible approximants, and the rates of convergence of rationals to reals. The number we study was found by the French mathematician Joseph Liouville (1809–1882); it is

$$1/10^{1!} + 1/10^{2!} + 1/10^{3!} + \cdots + 1/10^{n!} + \cdots$$
$$= 0.110001000000000000000000100\ldots0001000\ldots.$$

The 5th 1 is positioned in the 120th place.

Corollary 3.3.15 tells us that the nth convergent p_n/q_n of the real number R is closer to R than $1/q_n^2$. Theorem 4.3.24 tells us that, for square roots, $\sqrt{k}$, we can improve on this and find convergents that are closer to $\sqrt{k}$ than $1/2q_n^2$. In fact, while the proof is a bit beyond us, the truth of the matter is that given any real number R, there is a rational number p/q closer to R than $1/\sqrt{5}q^2$. Furthermore, the constant $\sqrt{5}$ cannot be improved upon because of our old friend, the golden mean, ϕ. The following theorem says what we mean.

Theorem 4.4.14

1. *Given any real number R, there is a rational number p/q such that $|R - p/q| < 1/(\sqrt{5}q^2)$.*

2. *Let $k > \sqrt{5}$. Given any positive integer q, $|\phi - p/q| > 1/kq^2$ for all rational numbers p/q.*

Part (2) of the theorem tells us that the speed of convergence to ϕ by fractions is necessarily restrained. This type of restraint holds for all algebraic numbers. As an example, we show that for $\sqrt{2}$ we may state the constraint like this.

Theorem 4.4.15 *Given any positive integer q, $|\sqrt{2} - p/q| > 1/3q^2$, for all rational numbers p/q.*

Proof For $q = 1$ the theorem holds right away because p may be 1 or 2 and, in either case, $|\sqrt{2} - p/q| > 1/3$. Now if $q > 1$, then let us assume, for the sake of being perverse, that $|p/q - \sqrt{2}| \leq 1/3q^2$. Thus $p/q < \sqrt{2} + (1/3q^2)$ for some q. Because we know that $\sqrt{2} < 10/7$ and since $q > 1$, we have $p/q < 10/7 + 1/12$. So we have $p/q + \sqrt{2} < 10/7 + 10/7 + 1/12 < 3$. Now

$$|p^2/q^2 - 2| = (p/q - \sqrt{2})(p/q + \sqrt{2})| < 1/3q^2 \times 3 = 1/q^2.$$

Therefore, $|p^2 - 2q^2| < 1$. Since p and q are natural numbers, it follows that $p^2 - 2q^2 = 0$ and so $p/q = \sqrt{2}$. But this cannot be because $\sqrt{2}$ is irrational. So $|\sqrt{2} - p/q| > 1/3q^2$. $\qquad\square$

Here is how the rate of convergence may be governed for general algebraic numbers.

Theorem 4.4.16 (Liouville) *Let z be an algebraic number of degree $n > 1$ and let $r_m = p_m/q_m$ be a sequence of rational numbers converging to z. Then, for a sufficiently large M, $|z - p_m/q_m| > 1/q_m^{n+1}$ for all $q_m > M$.*

Proof Suppose that z is a solution to the polynomial equation $f(x) = a_n x^n + a_{n-1} x^{n-1} + \cdots + a_2 x^2 + a_1 x + a_0 = 0$. Then

$$f(r_m)/(r_m - z) = (f(r_m) - f(z))/(r_m - z) =$$
$$a_n(r_m^{n-1} + r_m^{n-2} z + \cdots + r_m z^{n-2} + z^{n-1}) +$$
$$a_{n-1}(r_m^{n-2} + r_m^{n-3} z + \cdots + r_m z^{n-3} + z^{n-2}) + \cdots +$$
$$a_3(r_m^2 + r_m z + z^2) + a_2(r_m + z) + a_1.$$

Letting m be such that $|z - r_m| < 1$, we may say that, for sufficiently large m,

$$f(r_m)/(r_m - z) < n|a_n|(|z| + 1|)^{n-1} + (n - 1)|a_{n-1}|(|z| + 1|)^{n-2} + \cdots$$
$$+ 3|a_3|(|z| + 1|)^2 + 2|a_2|(|z| + 1|) + |a_1| = M.$$

Let $q_m > M$. Then $|z - r_m| > |f(r_m)|/M > |f(r_m)|/q_m$.
Now

$$|f(r_m)| = |(a_n p_m^n + a_{n-1} p_m^{n-1} q_m + \cdots + a_1 p_m q_m^{n-1} + a_0 q_m^n)/q_m^n|.$$

Note that r_m cannot be a solution to $f(x) = 0$ because if it were we could factor out $(x - r_m)$ and so z would necessarily be of lesser degree. Hence $f(r_m) \neq 0$. Furthermore, the numerator of this fraction is an integer so it must be at least 1. We conclude that $|z - r_m| > (1/q_m)(1/q_m^n) = 1/q_m^{n+1}$. $\qquad\square$

Using his theorem, Loiuville constructed a transcendental number. Notice that its decimal expansion is characterized by rapidly increasing stretches of zeros of length $m!$.

Theorem 4.4.17 *The number $z = 1/10^{1!} + 1/10^{2!} + \cdots + 1/10^{n!} + \cdots$ is transcendental.*

Proof Let $r_m = p_m/q_m = 1/10^{1!} + 1/10^{2!} + \cdots + 1/10^{m!} = p_m/10^{m!}$. Then $|z - r_m| < (10)(1/10^{(m+1)!})$. Now if z is an algebraic number of degree n, then Liouville's theorem says that $|z - r_m| > 1/10^{(n+1)m!}$ for sufficiently large m. So

$$1/10^{(n+1)m!} < (10)(1/10^{(m+1)!}) = 1/10^{(m+1)!-1}.$$

But this is false for $m > n$, so z is transcendental.   □

Now that we have seen that the rate of convergence of the continued fraction to an algebraic number is restrained, it makes sense to explore the entries in the continued fraction expansion. For quadratic numbers, the entries are periodic. It has been theorized, but not proved, that the entries are bounded for all algebraic numbers. Certainly bounded entries are consistent with a restrained rate of convergence. We shall note that the continued fraction expansion for some known transcendental numbers, in particular those based on the number e, are unbounded. Further we recall from Section 3.3 the dramatic changes in the entries of the expansion of π; the number 292 is the 5th entry in the continued fraction expansion. We shall leave this notion of restrained versus erratic behavior of continued fraction expansions for a project (Project 5.21).

Let us now open the flood gates of the transcendental dam. As mentioned in the introduction to Section 4.4, Hilbert's seventh problem was solved by Gelfond. But not only did Gelfond show that the number $2^{\sqrt{2}}$ is transcendental, he proved that a whole class of numbers like $2^{\sqrt{2}}$ is transcendental.

Theorem 4.4.18 (Gelfond) *The number z^y is transcendental if z is algebraic (not 0 or 1) and y is irrational and algebraic (it may be complex).*

Example 4.4.19

(a) Consider $10^{1/2}$. This number is irrational because if $10^{1/2} = p/q$, then $q^2 = 10p^2$. This is an impossibility because $10p^2$ has, in its prime factorization, an odd number of 2s and 5s while q^2 cannot. Of course, we know this number to be root constructible; after all, it is $\sqrt{10}$.

(b) Consider $10^{\sqrt{2}}$. This number is transcendental by Theorem 4.4.18.

(c) Consider $10^{\log 2}$. This number is 2, by definition.   ∎

Theorem 4.4.20 *If x is a natural number that is not a power of 10, then $log_{10} x$ is transcendental.*

Proof Let $y = \log_{10} x$. Suppose that y is not transcendental. Since x is not a power of 10, y cannot be rational. We leave the proof of this as an exercise. Thus y is irrational and algebraic. It follows from Theorem 4.4.18 that 10^y is transcendental. But $x = 10^y$ and x is a natural number. This contradiction tells us that $y = \log_{10} x$ is transcendental.   □

So we have uncovered a whole new line of transcendental numbers; $\log_{10} x$ for natural numbers x that are not powers of 10. So, for example, $\log 2 = 0.301029995664\ldots$, and $\log 3 = 0.47712125472\ldots$ are transcendental.

But the theorem that opens up the treasure chest of transcendentals is as follows.

Theorem 4.4.21 *If $z \neq 0$ is an algebraic (it may be complex) number, then e^z is transcendental.*

As in the case of Gelfond's theorem, the proof is well beyond this book; it can be found in advanced books on number theory.

Let us recall some facts about the functions e^z, $\cos x$ and $\sin x$ that we have picked up in a calculus course. We let z represent a complex number and x represent a real number.

(i) $e^z = 1 + z + z^2/2! + z^3/3! + \cdots + z^n/n! + \cdots$

(ii) $\cos x = 1 - x^2/2! + x^4/4! - x^6/6! + \cdots + (-1)^n x^{2n}/(2n)! + \cdots$

(iii) $\sin x = x - x^3/3! + x^5/5! - x7/7! + \cdots + (-1)^n x^{2n+1}/(2n+1)! + \cdots$

(iv) $e^{ix} = \cos x + i \sin x$

Theorem 4.4.22

1. Let x be an algebraic number. If $x \neq 0$, then $\cos x$ is a transcendental number. If $x \neq 1$, then $\cos^{-1}(x)$ is transcendental.

2. If $x \neq 0$, then e^x transcendental. If $x \neq 1$, then $\ln(x)$ is transcendental.

Proof If $\cos x$ were algebraic so, too, would $i \sin x$ be algebraic. Then their sum would be algebraic. But we know that $e^{ix} = \cos x + i \sin x$ and since x is an algebraic number and $x \neq 0$, Theorem 4.4.21 tells us that e^{ix} is transcendental. This contradiction establishes that $\cos x$ is transcendental. If $\cos^{-1}(x) = y$ were algebraic and $x \neq 1$ then $y \neq 0$ and $\cos y = x$ is transcendental. But x was assumed to be algebraic, so this contradiction proves that $\cos^{-1}(x)$ is transcendental if $x \neq 1$. We leave the rest of the proof as an exercise. $\square$

Theorem 4.4.23 e^π *is transcendental.*

Proof We know that $e^{i\pi} = \cos \pi + i \sin \pi = -1$, so $e^{-\pi} = (-1)^i = i^{2i}$ and $e^\pi = i^{-2i}$. Since i is algebraic it follows from Theorem 4.4.18 that i^{-2i} is transcendental. $\square$

Example 4.4.24
Here are examples of transcendental numbers. We can display most of these with a hand calculator.

(a) $\cos(\sqrt{2}) = 0.155943694765\ldots$

(b) $\cos^{-1}(4/5) = 0.643501108793\ldots$

(c) $e^\pi = 23.1406926328\ldots$

(d) $\ln(\sqrt[3]{2}) = 0.231049060187\ldots$

(e) $i^i = 0.20787957635076190854695\ldots$ ■

We have been flirting with complex numbers in the past two theorems and part (e) of the last example. In elementary algebra we have learned about adding, subtracting, multiplying, and dividing complex numbers, but raising complex numbers to complex powers belongs in an advanced course. This would be a good time to make plans to take such a course. Making sense of numbers such as i^i is actually a complex task—no pun intended.

It looks as though we can prove that most anything is transcendental. But there are many elementary numbers that, unbelievably, are not understood at all. Not only have the following numbers not been proved to be transcendental, it is not even known whether they are irrational. If you can believe this, they might be rational numbers. Here are some examples.

Example 4.4.25

It is not known whether the following numbers are rational, arithmetic, algebraic, or transcendental.

$$\pi + e, \quad \pi \times e, \quad \pi^e, \quad 2^\pi, \quad 2^e, \quad \pi^\pi, \quad e^e \qquad \blacksquare$$

Since e^z and the trigonometric functions of sine and cosine offer a power series representation, we can home in on a transcendental number with as great an accuracy as we wish. We need not be limited to the decimal expansion on our caculator.

Example 4.4.26

We know that $\cos(1)$ is transcendental. Now

$$\cos x = 1 - x^2/2! + x^4/4! + \cdots + (-1)^n x^{2n}/(2n)! + \cdots.$$

The calculator tells us that $\cos(1) = 0.540302305868\ldots$. The series tells us that

$$\cos(1) = 1 - 1/2! + 1/4! + \cdots + (-1)^n/(2n)! + \cdots.$$

If we want the series for $\cos(1)$ to 20-place accuracy, we need only go out 12 places because $1/22! = 8.9 \times 10^{-22}$. This is feasible. $\qquad \blacksquare$

While we have listed lots of exotic transcendental numbers, none of them has a decimal pattern that can be remembered. Here is one that does. The decimal built from the counting numbers is transcendental.

$$0.12345678910111213141516171819202122232425\ldots$$

We finish up our brief look at transcendentals by revisiting π and e one last time. The number e is the less well known of the two. It is not an everyday number like π is. It is known as the base of the natural logarithms, and it was born less than 300 years ago. As we have seen in Theorem 4.4.21, the series e^z is invaluable to us in our search for transcendentals. And the exponential function e^x is known to all calculus students and all students of science who study exponential growth. We have mentioned that e has an unbounded continued fraction expansion. Incredibly, its expansion follows a pattern.

$$e = [2; 1, 2, 1, 1, 4, 1, 1, 6, 1, 1, 8, \ldots]$$

So we may approximate e with fractions to as close as we like. Also, there are patterned continued fraction expansions based on terms using e. Here are two examples; we leave it as an exercise to find more.

$$(e - 1)/(e + 1) = [0; 2, 6, 10, 14, 18, \ldots]$$

$$(e^2 - 1)/(e^2 + 1) = [0; 1, 3, 5, 7, 9, \ldots]$$

We may also approximate e to as great an accuracy as we like with its series expansion:

$$e = 1 + 1 + 1/2! + 1/3! + \cdots + 1/n! + \cdots.$$

Here is e to 21 places:

$$e = 2.718281828459045235360\ldots.$$

Unquestionably, π is the most famous number in all of mathematics. It is a most natural of numbers to consider—the ratio of the circumference to the diameter of a circle. Not surprisingly, it occurs in formulas for circular objects in geometry. We all know them.

$C = 2\pi r$; C stands for the circumference of a circle with radius r.

$A = \pi r^2$; A stands for the area of a circle with radius r.

$V = (4/3)\pi r^3$; V stands for the volume of a sphere with radius r.

$S = 4\pi r^2$; S stands for the surface area of a sphere with radius r.

But π occurs in all fields of mathematics—and in the most unexpected places. Here are some examples:

$$e^{i\pi} = -1.$$

This truly remarkable fact follows from $e^{iz} = \cos z + i \sin z$.

$$n! \approx (\sqrt{2\pi n})n^n e^{-n}$$

This is Stirling's formula, which was mentioned in the exercises of Section 1.3. It is a good approximation of $n!$ as n gets large. Notice that this formula relates three interesting numbers: $n!, e,$ and π.

$$f(x) = e^{-x^2}/\sqrt{2\pi}$$

This is the definition for the normal distribution, that bell-shaped curve we see in statistical data.

$$F(n)/n \approx 6/\pi^2$$

Here $F(n)$ stands for the number of square-free numbers $\leq n$. This approximation becomes very good as n gets large. A square-free number is a number made up of primes raised to the first power. For example, if $n = 10$, then the square-free numbers ≤ 10 are 2, 3, 5, 6, 7, and 10. There are six of them; and $6/10$ is close to $6/\pi^2$.

As pervasive and fundamental a number as π is, it is nearly intractable from a numerical standpoint. As we have stated, it was not proved to be irrational until the mid-1700s, and it was not found to be transcendental until the late 1800s. But it has been recognized and studied for as long as mathematicians have lived. In the Old Testament, I Kings 7:23 implies that $\pi = 3$. The Babylonians around 2000 B.C. thought π to be either 3 or $3\frac{1}{8}$. Around 1500 B.C. in the *Rhind Papyrus*, $\pi = 256/81 \approx 3.16049$. Archimedes, around 200 B.C., approximated π using a 96-sided regular polygon. He found $3\frac{10}{71} < \pi < 3\frac{1}{7}$. Ptolemy, the great astronomer, about 400 years later approximated π with $377/120$. This is correct to four places. In the third century A.D. the Chinese geometer Liu Hui approximated π using a 192-sided regular polygon. His estimate was 3.1416. This estimate was also recorded in the sixth century a.d.by the Hindu astonomer Aryabhata, in the *Aryabhitiya* Verse II 28: "Add 4 to 100, multiply by 8, and add 62000. The result is approximately the circumference of a circle of which the diameter is 20000." In the fifth century A.D. the Chinese mathematician Zu Chongzhi found the approximation of $355/113$, which is correct to six places.

Let us begin with a method for approximating π that captures the spirit of both Archimedes and Liu Hui. This approximation involves inscribing regular $(k \times 2^n)$-gons in a unit circle. You can approximate either the area or the circumference of the circle using larger and larger n. We shall approximate the circumference of by finding the perimeter of a regular 2^n-gon. Figure 4.10 depicts a unit circle with center at O. The side of a polygon is depicted by PQ. The unit segment OR bisects the $\angle QOP$ and is perpendicular to PQ. The point S is the intersection of OR and PQ. Let x denote the length of PQ. Let h denote the length of OS. Let y denote the length of PR.

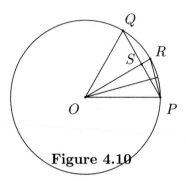

Figure 4.10

Lemma 4.4.27 *The length $y = \sqrt{2 - \sqrt{4 - x^2}}$.*

Proof The Pythagorean theorem gives us

$$h^2 + (x/2)^2 = 1; \quad y^2 = (1 - (h)^2 + (x/2)^2.$$

It follows that

$$y^2 = (1 - (h)^2 + 1 - h^2 = 2 - 2h = 2 - 2\sqrt{1 - (x/2)^2} = 2 - \sqrt{4 - x^2}.$$

So $y = \sqrt{2 - \sqrt{4 - x^2}}$. $\qquad\qquad\qquad\qquad\qquad\qquad\qquad\qquad\square$

Theorem 4.4.28 *The perimeter of a regular 2^n-gon inscribed in a unit circle is*

$$2^n \times \sqrt{2 - \sqrt{2 + \sqrt{2 + \cdots + \sqrt{2}}}}$$

where there are $n - 1$ twos under the square root signs.

Proof We proceed by induction on n for the following statement:

$\mathcal{P}(n)$: The length of the side of a 2^n-gon inscribed in a unit circle is

$$\sqrt{2 - \sqrt{2 + \sqrt{2 + \cdots + \sqrt{2}}}}.$$

$\mathcal{P}(2)$ is true because the side of a square inscribed in a unit circle is of length $\sqrt{2}$.

Suppose $\mathcal{P}(n)$ is true. So $\sqrt{2 - \sqrt{2 + \sqrt{2 + \cdots + \sqrt{2}}}}$ is the length of the side of a regular 2^n-gon inscribed in a unit circle where there are $n - 1$ twos in the expression. Now consider $\mathcal{P}(n+1)$: Lemma 4.4.29

says that the length of the side of a regular polygon with twice the number of sides is $\sqrt{2 - \sqrt{4 - x^2}}$, where

$$x = \sqrt{2 - \sqrt{2 + \sqrt{2 + \cdots + \sqrt{2}}}} \text{ with } n-1 \text{ twos.}$$

But this expression is $\sqrt{2 - \sqrt{2 + \sqrt{2 + \cdots + \sqrt{2}}}}$ with n twos. In order to get the perimeter we simply multiply the length of the side by 2^n. □

This theorem shows what we suspected about the sequence from Example 4.3.2 (d): that $2^n \times \sqrt{2 - \sqrt{2 + \sqrt{2 + \cdots + \sqrt{2}}}}$ converges to π where this expression has n twos under the square roots.

Corollary 4.4.29 *The sequence*

$$\{s_n\} = 2^n \times \sqrt{2 - \sqrt{2 + \sqrt{2 + \cdots + \sqrt{2}}}} \text{ converges to } \pi \text{ where the}$$

expression has n twos under the square roots.

Proof This follows from Theorem 4.4.30, noting that the expression

$$2^{n+1} \times \sqrt{2 - \sqrt{2 + \sqrt{2 + \cdots + \sqrt{2}}}}$$

when divided by 2 gives the measure for the circumference of a semi-circle of radius 1, which is π. □

Today we can approximate π with an inexpensive calculator to several places. Many calculators show it as 3.14159265359. This is accurate to 11 places. In Section 3.3 we looked at its continued fraction expansion. It begins $[3; 7, 15, 1, 292, \ldots]$.

	3	7	15	1	292	
p_k	0	1	3	22	333	355	103993 ...
q_k	1	0	1	7	106	113	33102 ...

We see that

$$|355/113 - \pi| < 1/(113)(33102) \approx .000000267,$$

so this convergent is a very good approximation. Here is the continued fraction expansion a bit further:

$$[3;\ 7,\ 15,\ 1,\ 292,\ 1,\ 1,\ 1,\ 2,\ 1,\ 3,\ 1,\ 14,\ 2,\ 1,\ 1,\ 2,\ 2,\ 2,\ 2,\ 1,\ 84,\ 2,\ldots].$$

Unfortunately, the continued fraction expansion does not show a pattern. As we have mentioned, it does show a dramatic jump in size
of entries with 292 in the fourth place. This is an early symptom of
erratic behavior in the rate of convergence of the continued fraction
to π. This, in turn, is an indication of what we already know; π is
not algebraic.

With the aid of formulas, it is possible to calculate π to many
places. Around 1600, it was calculated to 35 decimals and around
1700 it was up to 100 decimal places. When π was shown to be
irrational in 1761, the search for more digits could no longer be driven
by the search for a cycling of the digits. Its irrationality meant that
this could not happen. But the lure of π to some mathematicians
is inescapable, and more accuracy was calculated. In 1853, William
Shanks calculated π to 707 places. It was pretty rough going past
1000 digits until the age of computers. For example, in 1949, π was
known to 2037 places and it took 70 hours of calculation to arrive
at this. In 1961, 100,000 places were found by computer in 9 hours.
And computers have gotten much faster. In 1975, the millionth place
was found. In 1989, the billionth digit was found.

Now that it is possible to find π to such enormous accuracy, it is
possible to analyze trends in the occurrence of digits. Yet they appear
to be perfectly random. In 1988 a statistical analysis of the first
29,360,000 digits of π was conducted. The most frequent digit was 4,
which appeared 2,938,787 times, while the least frequent digits was
7, which occurred 2,934,083 times. While, in a random sequence, we
would expect that each digit would occur about 2,936,000 times, this
variation is not at all unreasonable. With 29,360,000 random digits
the chances that there would be a string of nine straight instances
of the same number is 29.36%. Indeed there is one such string:
Nine consecutive 7s occur. As we continue to probe deeper into the
infinity of the expansion, we could argue that there will be strings
of hundreds of the same digit. Indeed, we could argue that any
sequence you would ever want would eventually show up—just as we
might argue that, given a typing monkey and a word processor and
an infinite amount of time, the monkey would eventually type the

Bible word for word (thus indicating that $\pi = 3$). It might take a while, though.

Here is a listing of the digits up to the first 0, which, surprisingly, does not occur until the thirty-second decimal place:

$$\pi = 3.14159265358979323846264338327950\ldots.$$

It's only right that this famous number, π, that has no pattern to its decimal expansion and no pattern to its continued fraction expansion can be built up through infinite additions and infinite multiplications with some of the most beautiful and intriguing patterns in all of mathematics. We conclude the book with some of these magnificent formulas.

The first formula for π was found by Francois Viete (1540−1603) the father of modern algebra. It is one of many strange and fascinating equalities.

1. $2/\pi = \sqrt{1/2} \times \sqrt{1/2 + 1/2\sqrt{1/2}} \times$

$$\sqrt{1/2 + 1/2\sqrt{1/2 + 1/2\sqrt{1/2}}} \times \cdots$$

In 1699, π was calculated to 71 decimal places using the formula

2. $\pi = 2\sqrt{3}(1 - 1/(3 \times 3) + 1/(3^2 \times 5) - 1/(3^3 \times 7) + 1/(3^4 \times 9) - \cdots).$

With the invention of calculus in the 1600s, several formulas were invented. They were all approximations of infinite processes, such as infinite series or infinite products. Here are some of the more beautiful formulas involving expressions with π.

3. $\pi/4 = 1 - 1/3 + 1/5 - 1/7 + 1/9 - \cdots$

4. $\pi\sqrt{2}/4 = 1 + 1/3 - 1/5 - 1/7 + 1/9 + 1/11 - 1/13 - 1/15 + \cdots$

5. $\pi^2/6 = 1 + 1/2^2 + 1/3^2 + 1/4^2 + \cdots$

6. $\pi^2/8 = 1 + 1/3^2 + 1/5^2 + 1/7^2 + \cdots$

7. $\pi^2/12 = 1 - 1/2^2 + 1/3^2 - 1/4^2 + \cdots$

8. $(\pi - 3)/4 = 1/(2 \times 3 \times 4) - 1/(4 \times 5 \times 6) - 1/(6 \times 7 \times 8) - \cdots$

9. $\pi^2/6 = 2^2/(2^2 - 1) \times 3^2/(3^2 - 1) \times 5^2/(5^2 - 1) \times \cdots \times p^2/(p^2 - 1) \times \cdots,$
where p is a prime.

EXERCISES

1. Show that the following sets, S, are countable by displaying a 1-1 function from S into $\mathbb{N}$.

    (a) S is the set of odd numbers (both negative and positive).

    (b) S is the set of integer lattice points.

    (c) S is the set of rational lattice points.

2. Using the functions R from Example 4.4.3 and g from Theorem 4.4.4, find $g(x)$ for the following fractions:

    (a) $x = 3/7$

    (b) $x = -11/16$

    (c) $x = 105/13$

    (d) Is there a natural number y for which there is no x such that $g(x) = y$? Explain.

3. Using functions I and g from Theorem 4.4.8, find $g(x)$ for the following algebraic numbers.

    (a) The two solutions of $x^2 + 7x + 4 = 0$

    (b) The smallest real solution for $x^5 + 8x^4 - 3x^2 + 13 = 0$

    (c) The largest solution to $x^3 - 3x^2 + x + 2 = 0$

    (d) The third smallest real solution for $x^7 - 3x^6 - 5x^5 + 3x^3 - 19x - 6 = 0$

    (e) Find natural numbers y such that there is no x for which $g(x) = y$.

4. Given that $0.1234567891011\ldots$ is transcendental, what can you say about

    (a) $17.181920212223\ldots$?

    (b) any number that begins with a natural number n and, following the decimal point, has a decimal expansion consisting of the string of successive digits of the natural numbers that follow n? Give a reason for your answer.

5. Let $y = \log_{10} x$. Show that if x is not a power of 10, then y cannot be rational.

6. Prove that if m and n are natural numbers, then $\sqrt{m}^{\sqrt{n}}$ is transcendental.

7. Show that if α is an acute angle of a Pythagorean triangle, then $\cos \alpha$ is transcendental.

8. Complete the proof of Theorem 4.4.22.

    (a) If $x \neq 0$ is an algebraic number, then e^x is transcendental.

    (b) If $x \neq 1$ is an algebraic number, then $\ln(x)$ is transcendental.

9. Using Theorem 4.4.23, prove that π is transcendental.

10. Show that the following numbers are transcendental:

    (a) $\sqrt{i}^{i}$

    (b) $i^{\sqrt{i}}$

    (c) $e^{\sqrt{\pi}}$

11. Find the ratio of the number of square-free numbers $\leq n$ to n, where n is

    (a) 100

    (b) 1000

    (c) How close to $6/\pi^2$ is the ratio becoming?

12. Look for other patterned continued fraction expansions for terms made from e; for example, $\sqrt[n]{e}$.

13. See how accurate the following terms are to π.

    (a) $99^2/(2206\sqrt{2})$

    (b) $(63/25)(17 + 15\sqrt{5})/(7 + 15\sqrt{5})$

    (c) $\sqrt[4]{9^2 + 19^2/22}$

14. If you carry out π to the extent of the continued fraction expansion

    $[3; 7, 15, 1, 292, 1, 1, 1, 2, 1, 3, 1, 14, 2, 1, 1, 2, 2, 2, 2, 1, 84, 2],$

    to how many decimals would it be accurate?

15. Compare the speed of convergence of these three formulas involving π.

    (a) $\pi/4 = 1 - 1/3 + 1/5 - 1/7 + 1/9 - \cdots$
    (b) $\pi/2 = (2 \times 2 \times 4 \times 4 \times 6 \times 6 \times \cdots)/(1 \times 3 \times 3 \times 5 \times 5 \times 7 \times \cdots)$
    (c) $\pi/6 = x + [(1/2)](x^3/3) + [(1 \times 3)/(2 \times 4)](x^5/5) +$
    $[(1 \times 3 \times 5)/(2 \times 4 \times 6)](x^7/7) + \cdots$, where $x = 1/2$

16. (a) How far must each of the nine expressions of π be carried out so that π is found to be 3.14159?

    (b) Which of the formulas is most efficient? That is, which uses the fewest terms to calculate π to five-place accuracy?

    (c) Using your answer to part (b), use the most efficient formula to calculate π to 10-place accuracy. How many terms does it take?

17. Program a loop like this:

    $Y = A$

    $A = (A + B)/2$

    $B = \sqrt{BY}$

    $C = C - X(A - Y)^2$

    $X = 2X$

    PRINT $(A + B)^2/4C$

    Let the initial values be $A = X = 1$, $B = 1/\sqrt{2}$, $C = 1/4$.

    (a) What do you find?

    (b) How many loops does it take to get 4-place accuracy? 8-place accuracy? 20-place accuracy? How can you tell?

18. (a) Find the length of the side of a regular n-gon where $n = 3 \times 2^k$.

    (b) Find the perimeter of a regular n-gon where $n = 3 \times 2^k$.

    (c) Approximate π using your answer to (b). In particular, how many sides of a regular n-gon are necessary to get 6-decimal-place accuracy? 10-decimal-place accuracy?

19. (a) Find the area of a regular n-gon where $n = 2^k$.

    (b) Find the area of a regular n-gon where $n = 3 \times 2^k$.

    (c) Approximate π using your answer to (a).

    (d) Approximate π using your answer to (b). How many sides of a regular n-gon are necessary to get 6-decimal-place accuracy? 10-decimal-place accuracy?

20. Compare the continued fraction expansion of e to the series expansion, keeping these questions in mind.

    (a) To how many terms must you take each expansion to get 3-decimal accuracy? 6-decimal accuracy? 9-decimal accuracy? 12-decimal accuracy?

    (b) How big are the denominators of your fractions in each case?

    (c) To how many terms must you take each expansion to match the 21-place accuracy of the decimal expansion given?

21. For the following numbers, find the fraction with the smallest denominator that most closely approximates the number to 9-place accuracy.

    (a) $\cos^{-1}(4/5)$

    (b) $\pi + e$

    (c) $\pi \times e$

    (d) π^e

    (e) π^π

    (f) Letting R stand for each of the numbers in these problems and p/q represent your answer, compare $|R - p/q|$ and k/q^2 for various values of k. Does this bear out Theorem 4.4.14?

22. Compare the continued fraction expansion and the series expansion for e^{π}. Find the first term in each sequence that is accurate to nine decimal places. What is that term in the continued fraction expansion? What is that term in the series expansion?

23. Show that, given q and a nonsquare natural number k, there is no rational number p/q such that $|\sqrt{k} - p/q| < 1/Kq^2$, where $K = 2([k]+1)$. The notation $[k]$ stands for the greatest integer $\leq k$.

   (a) Show that, given q, there is no rational number p/q such that $|\sqrt[3]{2} - p/q| < 1/6q^3$.

   (b) Find a number m such that, given q, there is no rational number p/q such that $|\sqrt[4]{2} - p/q| < 1/mq^4$.

   (c) Find a number m such that, given q, there is no rational number p/q such that $|\sqrt[5]{2} - p/q| < 1/mq^5$.

24. In Theorem 4.4.14 (ii) experiment with constants $k > \sqrt{5}$. Find q such that $|\phi - p/q| > 1/kq^2$ for all p, where k is

   (a) .4471

   (b) .4472

   (c) .44721

   (d) .447213

25. Prove that z is transcendental, where

$$z = 1/2 + 1/2^{2!} + 1/2^{3!} + 1/2^{4!} + \cdots$$

26. Theorem 4.4.16 indicates that a sequence of fractions converges in a tame fashion to an algebraic number. The theorem does not apply to transcendental numbers. Examine known transcendental numbers, such as those you can generate from Gelfond's heorem (4.4.18) or those listed in Example 4.4.24. Explain what you see in these continued fraction expansions that shows that convergence that is anything but tame. Explain what you mean. Be as precise as you can.

Chapter 5

Mathematical Projects

In the first four chapters we studied mathematics that has been in the making for hundreds, and in some cases thousands, of years. But it was other people's mathematics. Now it is your turn. This chapter contains 21 mathematical projects that are meant for you, the student. These projects are explorations that are meant to stretch you beyond the exercises in the book. They are based upon the themes and subjects that have been presented, subjects such as primes, factoring, integral domains, modular arithmetic, Farey sequences, the Pythagorean theorem, Diophantine equations, Fibonacci numbers, regular polygons, decimal expansions, continued fractions, and constructible numbers.

It is hoped that these projects will give you a taste of what mathematical research is like. Research in mathematics is unlike research in other fields. There is no guarantee of discovering anything at all; an exploration can go on for months and, in the end, turn up nothing. These mini-research problems are meant to preserve the excitement of the search and, at the same time, reduce the risk of failure. Each project contains several questions that lead you along a well-defined trail. While this guide helps lead to results, it is hoped that there still is enough intrigue for you to experience the thrill of discovery. And each project ends with an invitation for you to pursue questions of your own making. It is here that real frontiers are encountered and discovery becomes highly personal.

The projects vary in familiarity and in difficulty. Some projects are easy to understand but difficult to approach. Some require sophistication to begin but are rather straightforward in their solution.

Some have no known solution; some are well known problems. We will not say which is which. But even if you were to know which description fit which project, it would be impossible to guess how the exploration would proceed. Different students of equal insight can look at the same project in entirely different ways. A single project can take one student, or group of students, hundreds of hours over the course of a semester, while another student, or group, might find a different approach that would take just a fraction of the time. But these various possibilities and perceptions make for exciting research. Naturally it is hoped that every project can offer at least a full month of exciting exploration but, as we say, nothing is guaranteed in mathematical research.

These projects are not meant to be isolated excursions that stay housed within the mind and memory of the explorer. Each exploration should be accompanied by a diary that is a full account of the journey along with a description of all findings. This is your research paper; you should share it (this is especially true if the project is being done in a mathematics class). You should give a formal oral presentation of your exploration as well as a careful write-up of your discoveries.

Good luck, and have a great trip.

5.1   Rings of Factors

This project involves factors and multiples of a sequence of numbers. The sequence of numbers $a_1, a_2, a_3, \ldots, a_k, a_{k+1}, \ldots, a_n$ is a **ring of factors** if $a_k \mid a_{k+1}$ or $a_{k+1} \mid a_k$ for $1 \le k < n$ and $a_n \mid a_1$ or $a_1 \mid a_n$. In other words, for every pair of adjacent members in the sequence one is a multiple of the other. For example, 2, 10, 5, 1, 8, 4 is a ring of factors, its length is 6. The figure shows the sequence arranged in a circle which makes sense since we insist that the last term of the sequence be paired with the first term in the divisibility relationship.

Our exploration involves building rings; rings as long as possible, on a finite set of numbers. A **maximal ring** will use as many of the numbers from the set as possible. The ring just given is not maximal on the set of numbers $n \le 10$. Here is a longer one: 8, 2, 6, 3, 9, 1, 4. It has length 7. We are unaware of any formula that will give you an exact answer to these problems. If you find one, that would be a real achievement. What we ask you to do is develop a strategy that

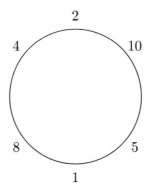

can be used to produce maximal rings.

We shall denote the length of a maximal ring on the set of numbers $n \le N$ by $M(N)$.

1. (a) Find a maximal ring within the set of numbers ≤ 25. What is $M(25)$?

   (b) How many different rings of length $M(25)$ can you find? Write them down.

2. (a) Find a maximal ring within the set of numbers ≤ 50. What is $M(50)$?

   (b) How many different rings of length $M(50)$ can you find? Write them down.

3. (a) Find a maximal ring within the set of numbers ≤ 100. What is $M(100)$?

   (b) How many different rings of length $M(100)$ can you find? Write them down.

4. In each of your answers above, what fraction of all available numbers did you use; that is, what is $M(N)/N$?

5. Talk about the strategy you used to find maximal rings. Characterize those numbers that could not be included in these rings.

6. Using your strategy, find, as accurately as you can, the length of the maximal ring you can build with the universe of natural numbers ≤ 1000. Explain your reasoning carefully.

7. Find the maximum $M(n)/n$ for

   (a) $n \leq 25$

   (b) $25 < n \leq 50$

   (c) $50 < n \leq 100$

8. Make up problems of your own and solve them; for example, here is one: Find a maximal ring where the universe consists of the odd numbers < 100. How does your answer here compare to your answer in question above? Answer the same question for the even numbers ≤ 100.

5.2   Sums of Consecutive Numbers

This project examines the different ways of expressing a number as the sum of consecutive integers. We discussed in Chapter 1 how representing numbers as sums or as products of other numbers has been a time-honored activity of mathematicians of all centuries and all cultures.

Let us look at a couple of examples.

$$12 = 3 + 4 + 5$$
$$15 = 7 + 8; \quad 15 = 4 + 5 + 6; \quad 15 = 1 + 2 + 3 + 4 + 5$$

We see that 12 can be expressed as the sum of consecutive integers in just one way while 15 can be represented in three different ways. Let us introduce a definition and notation to make the conversation easier.

Definition 5.2.1 *The* **consecutive index** *of a number n is the number of ways that n can be written as the sum of consecutive integers.*

Notation: The consecutive index of n will be denoted by $c(n)$.

1. Gather data. For example, for $2 \leq n \leq 50$ list the sequences of consecutive numbers whose sum is n. Note the beginning number of each sequence and the length of the sequence. Find $c(n)$ for each n.

2. Describe the nature of numbers whose consecutive index is

    (a) 0

    (b) 1

    (c) 2

    (d) 3

3. For all natural numbers $\leq m$, which number(s) has (have) the maximum consecutive index where $m =$

    (a) 100

    (b) 1000?

    (c) 10000?

4. Find a formula for $c(n)$ for all natural numbers, n.

5. List the lengths of the sequences of consecutive integers that make up $c(n)$. For each length tell where the starting point of the sequence begins. Be as general as you can.

6. Find the smallest number n such that $c(n) \geq$

    (a) 100

    (b) 1,000

    (c) 10,000

7. Allow 0 and the negative integers in the consecutive sums and define an "extended composite index." Examine questions 1 - 6 using the extended composite index; in particular, observe how your formula for the extended composite index relates to your answer in question 3.

8. Try to justify your answers, especially your formulas in questions 3 and 6.

5.3   Measuring Abundance

This projects quantifies the "compositeness" of a number. In Section
1.4 we introduced different kinds of numbers defined by the abun-
dance of their factors. There were perfect, multiply perfect, abundant, deficient, and weird numbers. All were defined by the sums of
their aliquot factors; that is, the sums of those factors other than the
number itself. In this project we will include the number itself as a
factor in our count. It makes the formulas easier to handle.

Definition 5.3.1 *The* **composite index** *of n is the sum of the factors of n divided by n.*

Notation: We denote the composite index of n by $C(n)$.

1. For numbers ≤ 1000,

   (a) which has the largest composite index? What is that index?

   (b) which odd number has the largest composite index? What
   is that index?

2. (a) Do the same as question 1 for numbers $\leq 10,000$.

   (b) Do the same as in 1 for numbers $\leq 1,000,000$.

3. Comment on the truth or falsity of the following:

   (a) There are numbers with arbitrarily small composite indices > 1.

   (b) If your answer to (a) is true, then, in your discussion,
   display a number with composite index less than 1.001.

   (c) There are numbers with arbitrarily large composite indices.

   (d) If your answer to (c) is true, display a number with composite index greater than:

      i. 10
      ii. 100.

4. Compare the following composite indices.

(a) $C(n)$ and $C(2n)$ if n is odd

(b) $C(n)$ and $C(3n)$ if $3 \nmid n$

(c) $C(n)$ and $C(5n)$ if $5 \nmid n$

(d) Make a general statement about $C(n)$ and $C(pn)$ for primes p if $p \nmid n$.

5. Compare the following composite indices.

(a) $C(n)$ and $C(2n)$ if n has one 2 in its prime form

(b) $C(n)$ and $C(2n)$ if n has two 2s in its prime form

(c) $C(n)$ and $C(2n)$ if n has three 2s in its prime form

(d) Make a general statement about $C(n)$ and $C(2n)$ if n has k 2s in its prime form.

6. Find a formula for $C(n)$ for any natural number, n.

7. If you have not done exercises 12, 13, and 14 in Section 1.4, find those multiply perfect numbers. Add to the list of multiply perfect numbers in Section 1.4 by finding at least one more of index 3 and one more one of 4. Discuss the strategy you used to find these numbers. Describe any patterns or tendencies that you observe in the prime forms of these numbers.

8. Find the average composite index for

(a) the numbers ≤ 1000

(b) all numbers. *Hint:* There is a series that converges to the answer. See if you can find the answer—it has π in it.

(c) Tell how important the computer was to your exploration.

5.4 Inside the Fibonacci Numbers

This project examines the prime form of Fibonacci numbers. The Fibonacci sequence was introduced in Section 1.4. It was noted that many surprising and beautiful relationships hold among the terms of the sequence. The exercises suggested some of these. We hope that these give an indication of why these numbers hold endless interest to both lay people and mathematicians. As with the relationships suggested in the exercises, there are many elegant results concerning the prime forms of Fibonacci numbers. We shall denote the sequence like this: $F_1, F_2, F_3, \ldots, F_n, \ldots$. So, for example, $F_1 = 1$, $F_5 = 5$, $F_{11} = 89$, $F_{15} = 610$, and so on.

1. Answer the following: Where, in the Fibonacci sequence, do the following types of numbers occur?

   (a) Even numbers

   (b) Multiples of 3

   (c) Multiples of 4

   (d) Multiples of 5

2. Test the validity of the following and substantiate your answers.

   (a) If $m \mid F_n$, then $m \mid F_{kn}$.

   (b) If $m \mid n$, then $F_m \mid F_n$.

   (c) If $F_m \mid F_n$, then $m \mid n$.

3. Test the validity of the following and substantiate your answers.

   (a) If p is prime, then F_p is prime.

   (b) If F_p is prime, then p is prime.

4. Find the smallest Fibonacci number F_n such that $p \mid F_n$ where p is

   (a) 2

   (b) 3

   (c) 5

   (d) 7

   (e) 11

   (f) 13

   (g) 17

   (h) 19

   (i) 23

   (j) 29

5. Let F_n be the smallest Fibonacci number such that, for prime p, $p \mid F_n$. Spell out, in detail, the relationship between n and p.

6. Find the smallest Fibonacci number F_n such that

    (a) $4 \mid F_n$

    (b) $8 \mid F_n$

    (c) $16 \mid F_n$

    (d) $9 \mid F_n$

    (e) $27 \mid F_n$

    (f) $81 \mid F_n$

7. Suppose that p is a prime and F_n is the smallest number that p divides. What can you say about the smallest Fibonacci number that p^k divides?

8. Find the smallest Fibonacci number F_n such that

    (a) $6 \mid F_n$

    (b) $10 \mid F_n$

    (c) $15 \mid F_n$

    (d) $35 \mid F_n$

    (e) $30 \mid F_n$

9. (a) Suppose that p is a prime and F_n is the smallest number that p divides. Suppose also that q is a prime and F_m is the smallest number that q divides. What can you say about the smallest Fibonacci number that pq divides?

    (b) Extend your observations to three primes p, q, and r.

    (c) Extend your observations to more than three primes.

10. Find the smallest n such that the Fibonacci number F_n contains, in its prime form,

    (a) all the primes < 100

    (b) all numbers ≤ 100

    (c) $100!$

11. Find the prime decomposition of F_n, where n is

  (a) 120
  (b) 240
  (c) 600
  (d) 1200

12. Try to justify your findings in this exploration.

13. Make up your own Fibonacci-like sequence and study it. For example, look at the sequence that begins $2, 1$, and continues like Fibonacci: $2, 1, 3, 4, 7, 11, \ldots$. Does it display familiar features. How does it relate to the Fibonacci sequence itself.

5.5   Pictures at an Iteration

This project models unary operations on numbers modulo n. A unary operation is an operation that applies to a single number; some examples are: minus, doubling, tripling, squaring, and cubing. The pictures of these operations are very suggestive of patterns and theorems. Let us see what we mean with an example: Let f (mod13 stand for the operation of squaring on the numbers mod 13. So f (mod 13): $n \mapsto n^2$ (mod 13). Here is how the elements are mapped:

$0 \mapsto 0; \; 1 \mapsto 1; \; 2 \mapsto 4 \mapsto 3 \mapsto 9 \mapsto 3; \;\; 5 \mapsto 12 \mapsto 1; \;\; 6 \mapsto 10 \mapsto 9;$
$7 \mapsto 10; \;\; 8 \mapsto 12; \;\; 11 \mapsto 4.$

We can summarize the actions of f (mod 13) with the following diagram showing the paths that various elements take mod 13.

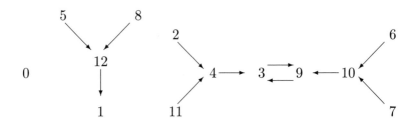

For the exercises we will use the notation f (mod n) for the operation of squaring on the numbers mod n.

1. Show pictures of

   (a) f (mod 5)

   (b) f (mod 7)

   (c) f (mod 11)

   (d) f (mod 17)

   (e) f (mod 19)

2. Talk about what you see in your pictures.

   (a) Describe, in words, the symmetries in your pattern.

   (b) List all the number phenomena you can find. For example, what can you say about the numbers at the ends of branches on your diagrams?

   (c) Try to explain why these phenomena exist.

3. Draw diagrams for other primes to see whether you can find common geometric looks. Can you predict, from the prime, which will have similar patterns? Explain.

4. Draw diagrams for some composite numbers; for example, f (mod 10), f (mod 15), f (mod 16). Name ways in which these pictures differ from those in question 1. Explain what happened and why.

An **isomorphism** between two unary systems (S, f) and (T, g) consists of a one-to-one mapping F from S onto T such that $F(f(s)) = g(F(s))$. For example, let $S = \{1, 2, 3, 4\}$, f denote squaring (mod 5), (so $(S, f) = f$ [mod 5) without the 0 term]. Let $T = \{0, 1, 2, 3\}$ and g denote doubling (mod 4). If we let $F : 1 \mapsto 0$, $2 \mapsto 1$, $3 \mapsto 3$, $4 \mapsto 2$ we have an isomorphism between the two systems. Checking out the condition that $F(f(s)) = g(F(s))$ for $s = 2$, we get $F(f(s)) = F(f(2)) = F(4) = 2$; and $g(F(s)) = g(F(2) = g(1) = 2$. It is true for $s = 2$. A good way of checking for an isomorphism is to draw pictures and see whether they have the same look.

   5. Draw pictures of f and g for sets S and T in the example above. Verify that they are the same. Find another mapping F that constitutes an isomorphism between S and T.

Let us look at the function of doubling; call the function g. So $g : n \mapsto 2n \pmod{m}$.

6. For the systems in question 1, omit 0 from the set and search for isomorphisms between each of these systems and a new system $g \pmod{n}$. The system is (T, g), where $T = \{0, 1, \ldots, n - 1\}$ and g is the doubling operation $\pmod{m}$.

7. Make up a theorem concerning your observations. Explain why your theorem is true.

8. Try to do question 5 with the systems you have diagrammed in 4. What problems arise?

9. Make up problems of your own using this theme. For example, examine the pictures of $h : n \mapsto n^3 \pmod{p}$. Do these pictures look like the pictures of $f : n \mapsto n^2 \pmod{p}$? Do they look like pictures of $j : n \mapsto 3n \pmod{m}$?

5.6   Eenie Meenie Miney Mo

This project involves modular arithmetic in a flexible setting. The idea is to select, from thousands of entries, a winning entry. The entries are numbered and arranged in a circle. Then they are systematically eliminated as the judges progress around the circle. Let's make it real. We want to attend the Final Four NCAA basketball championship. The hundreds of thousands of entries are arranged in a circle and numbered in order, clockwise, beginning with 1. Then the counting begins: "eenie, meenie, miney, mo, catch a tiger by the toe, if he hollers, let him go, eenie, meenie, miney, mo." The last "moe" is the twentieth person. This entry remains in the running, the first 19 are eliminated. The count then continues, excluding 19 more and leaving in the fortieth entry. And so it goes until all but one entry remains. The NCAA figures this process is so complicated that no one can figure it out. Yet we intend to solve this problem. We will then ask the NCAA to allow us to place our own entry in the circle. Since we will know exactly where it should go, we will win a seat at the Final Four.

Let us start with a simpler problem—let us count by 2s rather than by 20s. There are two ways to proceed; we can, beginning with 1, successively include and exclude or we can successively exclude and include. Here is how the two processes would work with 10 entries. First let us successively include, exclude, include, exclude, and so on. We denote the process like this. The parentheses indicate exclusion.

$$1, (2), 3, (4), 5, (6), 7, (8), 9, (10), \quad 1, (3), 5, (7), 9, \quad (1), 5, (9)$$

Stop. Our winner is number 5.

The three circles in the figure represent the three cycles that we toured the circle in order to eliminate all but one number.

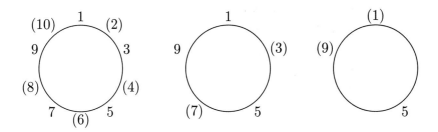

If we begin by first excluding, then including we have the following results:

$$(1), 2, (3), 4, (5), 6, (7), 8, (9), 10, \quad (2), 4, (6), 8, (10), \quad 4, (8)$$

Stop. Our winner is number 4. It is worthwhile to notice some things as we complete the circles. For example, in the exclude-include process, we went around the circle completely three times and began a fourth trip. The survivors that we encountered as we began each cycle were $2, 4, 4$, and finally 4 again. The forms of the numbers that survived the cycles were, respectively, $2k+2$ for the first cycle, $4k+4$ for the second cycle, $8k + 4$ for the third cycle, and $16k + 4$ for the fourth cycle. Of course, there were very few numbers in the latter cycles.

1. Suppose we have N entries and we pick a winner by excluding every other person. That is, we proceed as above; include 1, exclude 2, and so on; that is, $1, (2), 3, (4), 5, (6) \ldots$. Answer the following questions:

   (a) What can you say about the winning number for a circle with $N + 1$ entries as compared to the winner of a circle with N entries? Can you justify your observation?

    (b) What is the number of the winning entry if there were 1000 entries?

    (c) What is the winning number if there were 1 million entries?

    (d) What is the winning number if there are N entries?

    (e) How many cycles are necessary before a winning entry is selected?

2. Suppose that we exclude every third person beginning with 1. That is, we proceed like this: $1, 2, (3), 4, 5, (6), \ldots$. Answer the same questions as in question 1.

3. Suppose that we exclude every fourth person beginning with 1. That is, we proceed like this: $1, 2, 3, (4), 5, 6, 7, (8), \ldots$. Answer the same questions as in question 1.

4. Try to generalize to processes that exclude every fifth person, every sixth person, and so on.

5. Suppose we have 1 million entries and we pick a winner by including (rather than excluding) every other person. That is, $(1), 2, (3), 4, (5), 6, \ldots$. Answer the following questions:

    (a) What form do the surviving numbers take in each successive cycle?

    (b) What is the number of the winning entry?

    (c) How do your answers relate to your answer in question 1(c)?

    (d) How many cycles are necessary before a winning entry is selected?

6. Describe how you would go about answering the questions in 5 if we had a circle with N entries.

7. Suppose we have 1 million entries and that we include every third person. That is, $(1), (2), 3, (4), (5), 6, \ldots$.

    (a) Answer the same questions as in question 5(a), (b), and (d).

    (b) Does your answer relate to the answer of question 2(c)? If so, how?

    (c) Attempt to generalize to N entries as you did in question 6.

8. Try to generalize your results to the processes of including every fourth person, every fifth person, and so on.

9. Using "eenie, meenie, miney, mo. . ." count in 20s, excluding the first 19, including number 20, and so on. Suppose that there are $19, 380, 117$ entries.

    (a) Where should I place my entry so that I win the best seat in the house at the tournament?

    (b) How many times must you go around the circle before finding a winner?

10. Comment on your use of the computer in this project. Do you believe there is theoretical work to be done here? Where might you justify your findings? Be specific.

11. Make up your own problems on this theme. For example, if four tickets were available, where would the second-, the third-, and the fourth-place winners be located?

5.7 Factoring with the Pollard ρ Method

In this project we explore a process for factoring numbers introduced in Section 2.4, the Pollard ρ method. It involves sequences modulo N for various numbers N. Since the numbers tend to get big, this project should be carried out with a computer. The Greek letter rho is used because the shape of this letter is like the sequence $\{u_n\}$; you begin with a string of numbers, like the tail of ρ, and then move into cyclic phase, like the head of ρ. One thing that we should keep in mind, when using algorithmic routines like the Pollard ρ method and the Euclidean algorithm, is the length of the process. While computers are able to perform incredible feats of calculation, they do have their limits. And the natural numbers have no limits; they can just get bigger and bigger. So even running the most powerful computers 24 hours a day, 7 days a week, 52 weeks a year for several years may not get a factorization of a large number. So, for mathematicians who do this sort of work, it is vital for them to get a handle on how many steps a particular process may take. This is a mathematical question, and there are answers. So, as you factor the numbers, pay attention to the number of steps it takes to complete a factorization and compare it to the size of the number being factored. Notice that factorization is not complete until the number in question has been decomposed into its prime form. Finding a factor, whether it be prime or composite, is just the first step toward complete factorization.

1. Experiment with numbers $N = p \times q$, where p and q are less than 100. What can you say about

   (a) The length of the nonrepetitive part of the Pollard sequence?

   (b) The period of repetition of the Pollard sequence? How does it compare to N?

   (c) The period of the sequence $\{d_n\}$? How does it compare to the period of $\{u_n\}$?

   (d) The step at which the first prime, p, is found? How does it compare to the size of p?

   (e) Does the smaller of the two primes always come first?

   (f) Is it possible that more than one factor appears in the sequence $\{d_n\}$?

2. Experiment with numbers N that are products of more than two primes. Can you find examples where the first factor is not a prime? Do a similar analysis to that in question 1.

3. Try to prove Theorem 2.4.13. Let $\{u_n\}$ be the Pollard sequence (mod N) and let $\{d_n\} = \{\gcd(|u_{2n} - u_n|, N)\}$. If N is composite, then there will eventually be an n such that $d_n > 1$ and this term will be a factor of N.

4. Using the Pollard ρ method, factor the following numbers.

   (a) $N = 36,287$

   (b) $N = 199,934,971$

   (c) $N = 269,997,421$

   (d) $N = 1,992,009,497$

(e) $N = 4,294,967,297$

(f) $N = 4,295,098,369$

(g) $N = 30,796,045,833$

5. Take a few large numbers (for example, numbers with 20 or more digits) and factor them.

6. Finish the following statements as precisely as you can.

(a) For a composite number $\leq N$, the period of the Pollard sequence is, at most, ...

(b) If the factor q of N first occurs at or before d_k, then k is, at most, ...

7. If you have not already encountered such examples, try to find examples where

(a) The first $d_n > 1$ is composite

(b) The first $d_n = N$

(c) If the first $d_n = N$, how might you proceed? (Look at question 8 for ideas.)

8. Experiment with different "Pollard-like" sequences; for example, $u_0 = c$ and $c \neq 1$ or $u_{n+1} = u_n^2 + c$ where $c \neq 1$. Do these changes yield significantly different results from the traditional Pollard ρ procedure? Explain.

5.8 Charting the Integral Universe

This project uses the Pythagorean theorem and Farey sequences on the lattice of integer points. Our universe is the integral lattice, the points are stars and our earth is a point also, the origin $(0,0)$. Our questions will involve questions from three different time periods: ancient, modern, and contemporary.

In ancient times, the naked eye and rudimentary telescopes were used to see the stars. Scanning the universe back then revealed just a few of the many stars; for example, there is a star at $(2,1)$ but the stars located at $(4,2)$, $(6,3)$, and, in general, $(2k,k)$ are hidden from view. Measurement of distance to some stars was exact but to other stars it was only a rough approximation. For example, the distance to $(3, 4)$ is 5; the distance to $(2, 1)$ is not easy to measure exactly. The stars located at distances that are whole numbers away from $(0, 0)$, like the star located at $(3, 4)$, are called **special stars** because they were the ones that could originally be measured. The other stars are called **regular stars**.

In modern times two major advances took place in the technology of telescopes; (1) telescopes were made that could "see" hidden stars, and (2) telescopes were made that could measure, with great accuracy, the distances of regular stars from the earth. This revolutionized astronomy and, in particular, gave life to an old theory that the stars were arranged in circular orbits. It was known in ancient times that there were eight stars that were a distance of 5 units away; $(\pm 3, \pm 4)$, and $(\pm 4, \pm 3)$. It was suspected there were more, and indeed there were. These four stars were hidden: $(0, \pm 5)$, $(\pm 5, 0)$. And for stars that were 25 units away all that could be seen were these four: $(\pm 7, \pm 24)$ and $(\pm 24, \pm 7)$. It was a surprise to learn that there were 12 hidden stars 25 units away: $(0, \pm 25)$, $(\pm 25, 0)$, $(\pm 15, \pm 20)$, and $(\pm 20, \pm 15)$. So, all in all, there are 20 stars that filled out this orbit. And the long suspected orbit that was about 8 units away proved to be a true orbit: $(\pm 1, \pm 8)$, $(\pm 8, \pm 1)$, $(\pm 4, \pm 7)$, and $(\pm 7, \pm 4)$. The orbit is almost exactly 8.06226 (actually $\sqrt{65}$) units away. And the suspected orbit that was about $11\frac{1}{2}$ units away was shown to be a true orbit as well: $(\pm 3, \pm 11)$, $(\pm 11, \pm 3)$, $(\pm 7, \pm 9)$, and $(\pm 9, \pm 7)$. The orbit is almost exactly 11.40175 (actually $\sqrt{130}$) units away.

In contemporary times powerful telescopes have been built. They will be used to test the long-standing theory that there are orbits that contain arbitrarily many stars; both special stars and regular stars. One new telescope has just been built that can look at a universe that is bigger by a factor of 10 than the old one.

You will undoubtedly use a computer to generate data for this problem. Just remember that they did not have powerful computers in ancient times and the computers in modern times are not anywhere near as powerful as the ones in contemporary times. So we encourage you to use mathematics along with the computer. It would be great to employ mathematics that might have been appropriate to the spirit of the times.

1. This question involves the ancient times. For centuries the telescopes could see out on the universe (x, y), where $-100 \leq x \leq 100$ and $-100 \leq y \leq 100$.

   (a) How many stars can the telescope see?

   (b) How many special stars can it see?

   (c) What is (are) the farthest special star(s) it can see? How far away is it? (are they?)

   (d) What is (are) the farthest regular star(s) it can see? Approximately how far away is it? (are they?)

   (e) Which distance is most common among special stars? How many stars are at that distance?

   These are questions from the modern era.

2. With the new telescopes, answer the following. Assume that the range is the same as in question 1; that is, (x, y) is such that $-100 \leq x \leq 100$ and $-100 \leq y \leq 100$.

   (a) True or false: All stars that are hidden behind special stars are also special stars. Explain.

   (b) True or false: All stars that are hidden behind regular stars are also regular stars. Explain.

   (c) What orbit(s) has (have) the most special stars? How far away is the orbit? How many stars does it (do they) have?

   (d) What orbit(s) has (have) the most regular stars? How far away is the orbit? How many stars does it (do they) have?

   These questions are from the contemporary era.

3. Test the theory with the new telescope that is 10 times as powerful; that is, it has a range (x, y) such that $-1000 \leq x \leq 1000$ and $-1000 \leq y \leq 1000$. Search for an orbit that has at least 100 stars.

   (a) Is there an orbit you can see with the new telescope that contains at least 100 special stars? If so, how close is it? If not, how powerful must a telescope be to find such an orbit?

(b) What orbit(s) has (have) the most special stars? How near is the closest such orbit? How many stars does it (do they) have?

(c) Is there an orbit you can see with the new telescope that contains at least 100 regular stars? If so, how close is it? If not, how powerful must a telescope be to find such an orbit?

(d) What orbit(s) has (have) the most regular stars? How close is the closest such orbit? How many stars does it (do they) have?

4. Find a formula for the number of

   (a) special stars in an orbit. Distinguish between hidden and visible special stars.

   (b) regular stars in an orbit. Distinguish between hidden and visible regular stars.

5. Using your formula, find

   (a) the closest orbit that contains at least a million special stars. How close is the orbit? How many stars does it contain?

   (b) the closest orbit that contains at least a million regular stars. How close is the orbit? How many stars does it contain?

6. (a) With the new telescope, how many stars are visible? How many are hidden?

   (b) If there were a telescope that could see stars in the universe

   $$-1,000,000 \le x \le 1,000,000$$

   $$-1,000,000 \le y \le 1,000,000$$

   approximately how many would visible? How many would be hidden?

   (c) What is the limiting percentage of visible stars in the universe as we develop telescopes that will reach farther and farther out?

7. Make up a problem of your own and examine it. For example, consider the universe of stars in three dimensions: (x, y, z). Compare your findings with the two-dimensional case.

5.9 Triangles on the Integral Lattice

In this project we study triangles that can be drawn on the lattice of integers. You will be able to generate lots of data with a computer and you should. But you can do a great deal with just pencil, paper, and a calculator, and you should do this too. This way you can develop strategies for solving the questions and be in a position to offer explanations. We will be especially interested in triangles that have integral length sides and integral area. These triangles are called **Heron triangles**. The name is for the mathematician Heron, of Alexandria, who lived in the first century A.D. He is credited with the ingenious formula for finding the area of a triangle when its sides are known:

Heron's Theorem $A = \sqrt{s(s - a)(s - b)(s - c)}$, *where a, b, and c represent the length of the sides of the triangle and s stands for the half perimeter; that is,* $s = \frac{1}{2}(a + b + c)$.

An example of a Heron triangle is the Pythagorean triangle with sides 3, 4, 5. The area of this triangle is 6, an integer. The triangle with sides 13, 14, 15 is also a Heron triangle because

$$A = \sqrt{21(21 - 13)(21 - 14)(21 - 15)} = \sqrt{7056} = 84.$$

We will be interested in which Heron triangles can be drawn on an integral lattice. We know that the slanted sides of a triangle on the integral lattice have length $\sqrt{n}$, where n can be expressed as the sum of two squares. In Section 2.3 we learned what kind of numbers these can be. Of course, the sides of the triangle that are horizontal and vertical may be of any integral length. The area of triangles on integral lattices may be found by Heron's theorem but also may be found by Pick's theorem, (see the exercises in Section 3.3). Pick's theorem applies only to polygons on integral lattices.

1. For triangles with the following length sides, find the area and the perimeter. Also find which ones can be drawn on the integral lattice.

(a) 5, 5, 6

(b) 13, 14, 15

(c) 12, 13, 20

(d) 13, 17, 26

(e) $\sqrt{2}, \sqrt{5}, \sqrt{13}$

(f) $\sqrt{8}, \sqrt{13}, \sqrt{17}$

(g) $\sqrt{8}, \sqrt{13}, \sqrt{41}$

(h) 13, 13, 10

(i) 13, 20, 21

(j) 3, 25, 26

(k) 15, 17, 29

(l) 10, 13, 20

2. List all the Heron triangles that can be drawn on the lattice plane with perimeter ≤ 50 that have one slanted side. Do these triangles have a name?

3. List all the Heron triangles that can be drawn on lattice plane sides with perimeter ≤ 50 that have two slanted sides. Can you build these triangles with the aid of Pythagorean triangles? If, so, explain how it is done.

4. List all the Heron triangles that can be drawn on lattice plane sides with perimeter ≤ 100 that have three slanted sides. Can you build these triangles with the aid of Pythagorean triangles? If, so, explain how it is done.

5. Comment on the truth of the following. Provide reasons for your observations.

   (a) Every Pythagorean triangle is a Heron triangle.

   (b) The length of at least one side of a Heron triangle is even.

   (c) The length of at least one side of a Heron triangle is a multiple of 3.

   (d) The length of at least one side of a Heron triangle is a multiple of 5.

   (e) The area of a Heron triangle is a multiple of 6.

(f) A Heron triangle has at least one integral length altitude.

(g) Every Heron triangle can be drawn on the lattice plane of integers.

A triangle whose perimeter and area are both the same integer is called a **Moon triangle**. Notice that the definition does not require that the length of the sides be integers. An example of a Moon triangle is the Pythagorean triangle 6, 8, 10. The perimeter and area are both 24 (of course the units are different; one is measured in linear units, the other in square units).

6. (a) Prove that a triangle whose inscribed circle has radius 2 has a perimeter and an area that are the same number.

   (b) Find the points of tangency of the triangle 6, 8, 10 with the inscribed circle of radius 2; that is, $x^2 + y^2 = 4$.

7. (a) Prove that there is only one equilateral triangle with area equal to perimeter and it is not a Moon triangle.

   (b) Prove that there is only one isosceles right triangle with area equal to perimeter and it is not a Moon triangle.

8. (a) List all Moon triangles with perimeter (and area) ≤ 100. Distinguish among the Pythagorean triangles, the Heron triangles that are not Pythagorean, and the non-Heron triangles. Also note any triangles that are isosceles.

   (b) Explain your strategy for finding this list (beyond any computer generation).

   (c) Are there Moon triangles with perimeter ≥ 100? If so, list them.

   (d) Is there a Moon triangle of greatest perimeter? If so, what is it? If not, why not?

9. Make up other problems and try to solve them. For example, examine all the Heron triangles that have a common length altitude; say, altitude 48.

5.10   The Gaussian Integers

In this project we examine a system that is very much like the system of integers. In Chapter 4 we learned that the integers form an abstract system called an integral domain. In an integral domain numbers can be added subtracted, multiplied, and divided, but division of one number by another did not always come out evenly; that is, often a remainder was left. This is what made divisibility a very interesting topic. Divisibility was the prime topic (excuse the pun) for our study of the natural numbers. The division theorem and the fundamental theorem of arithmetic were the theoretical benchmarks for our study. The integral domain we examine here is the Gaussian integers. We shall find that, in many ways, the Gaussian integers are like the familiar integers, but they offer some very interesting challenges as well.

Definition 5.10.1 *A number of the form $a + bi$, where a and b are integers, is called a* **Gaussian integer***.*

Addition, subtraction, multiplication, and division are carried out as in complex arithmetic. Let us recall how multiplication and division are done. Letting $x = a + bi$ and $y = c + di$,

$$xy = (a + bi)(c + di) = (ac - bd) + (ad + bc)i$$

$$x/y = \frac{a + bi}{c + di} = \frac{(a + bi)(c - di)}{(c + di)(c - di)} = \frac{(ac + bd)}{c^2 + d^2} + \frac{(ac - bd)}{c^2 + d^2} i.$$

Here are some definitions and notation that will be useful.

Definition 5.10.2 *The* **conjugate** *of the Gaussian integer $a + bi$, is $a - bi$.*

Notation: The conjugate of z will be denoted by $\overline{z}$.

Definition 5.10.3 *The* **magnitude** *of z is $z \times \overline{z}$. So if $z = a + bi$, then its magnitude is $a^2 + b^2$.*

Notation: The magnitude of z is denoted by $M(z)$.

Definition 5.10.4 *If $M(z) = 1$, then z is called a **unit**. If $z = ez'$, where e is a unit, then z and z' are **associates**. A Gaussian integer z is called a **Gaussian prime** if the only divisors of z are units or associates of z. Two primes are **distinct** if one is not an associate of the other.*

Let's do some dividing, some prime searching, and some factoring in the Gaussian integers.

1. Find a quotient q, and remainder r, such that $M(r) < M(w)$ for the following divisors w and dividends z:

   (a) $w = 2 + i$, $z = 7 - i$

   (b) $w = 3 - i$, $z = 5 + 6i$

   (c) $w = 2 + 3i$, $z = 8 - 3i$

   (d) Describe a good general strategy for long division in the Gaussian integers.

2. For Gaussian integers, state the analogue of the division theorem for Integers. Do you believe it is true? Explain.

3. Find the gcd of the following Gaussian integers z and w.

   (a) $z = -2 + 3i$, $w = 5 - i$

   (b) $z = 5$, $w = 3 + 4i$

4. (a) State the Gaussian analogue of Theorem 1.1.11. Do you believe it is true? Explain.

   (b) Write $\gcd(z, w)$ in 3 (a) and (b) in the form of $az + bw$.

5. (a) List the distinct Gaussian primes, z, where $M(z) < 100$, in order of their magnitude.

   (b) What can you say about the numbers that are magnitudes of Gaussian primes?

6. Factor the following Gaussian integers z into Gaussian primes.

   (a) $6 - 7i$

   (b) 13

   (c) $4 + i$

(d) $16i$

(e) $4 - 3i$

(f) 12

(g) $-9 - 5i$

(h) $11 + 13i$

(i) $12 + 21i$

(j) $9 - 32i$

7. State a Gaussian analogue for the fundamental theorem of arithmetic. Do you believe it is true? Explain.

8. Comment on the following statements. If a statement is true explain, why; if false, explain why and, if possible, add conditions to the hypothesis so that it becomes true.

(a) $M(zw) = M(z)M(w)$.

(b) If $z \mid w$, then $M(z) \mid M(w)$ and $M(w)/M(z) = M(w/z)$.

(c) If z is a Gaussian prime, then $M(z)$ is a prime number.

(d) If $M(z)$ is a prime natural number, then z is a Gaussian prime.

(e) If x is a Gaussian prime and $M(x) \mid M(z)$, then either x or its conjugate $\overline{x}$ divides z.

(f) If x is a Gaussian prime and if either x or $\overline{x}$ divides z, then $M(x) \mid M(z)$.

The Gaussian integers are the natural setting for much of the analysis of Pythagorean triples that was done in Section 2.3. The fact that the sum of two squares $a^2 + b^2$ can be factored in this domain is one of the keys. In the domain of Gaussian integers $a^2 + b^2 = (a + bi)(a - bi)$. Another key property of this domain is that it obeys the analogue of the fundamental theorem. A domains that has this property is called a **unique factorization domain** (UFD). The Gaussian integers are a UFD. Here are some theorems that were introduced in Section 2.3 that can be proved elegantly in the domain of Gaussian integers.

Theorem 1. *If m and n are expressible as the sum of two nonzero square integers, then mn is also expressible as the sum of two square.*

(Compare with Theorem 2.3.10.) *Furthermore, if* $m \mid n$, *then* n/m *is also expressible as the sum of two squares.*

Theorem 2. *Given any two distinct natural numbers a and b, then (x, y, z) is a Pythagorean triple where $x = a^2 - b^2$, $y = 2ab$, and $z = a^2 + b^2$. Furthermore, if (x, y, z) is a primitive Pythagorean triple, then such an a and b exist.* (Compare with Theorem 2.3.16.)

Theorem 3. *Every prime number of the form $4k+1$ can be written as the sum of squares of two nonzero integers.* (Compare with Theorem 2.3.9, the Euler-Fermat theorem.)

9. Prove Theorem 1.

10. Prove Theorem 2.

11. Prove Theorem 3.

12. Look at other integral domains, such as numbers of the form $a + b\sqrt{2}$, and analyze them as we have done for the Gaussian integers. This could be a whole project in itself.

5.11   Writing Fractions the Egyptian Way

This project examines how we can write any fraction as the sum of distinct unit fractions. A **unit fraction** is a fraction with numerator 1. Writing fractions as the sum of distinct unit fractions dates back to ancient Egypt. The oldest mathematical document in existence, the *Rhind Papyrus* (2000 - 1788 B.C.) shows these representations of fractions as sums of unit fractions:

$$2/7 = 1/4 + 1/28, \ 2/11 = 1/6 + 1/66, \ 2/97 = 1/56 + 1/679 + 1/776.$$

Apparently, with the exception of 2/3, the Egyptians dealt with fractions as halves, thirds, fourths, and so on, and all other fractions were built from these. Of course, the Egyptians could have expressed 2/97 as $1/97 + 1/97$, but repeating the denominator was not allowed. The equality showing 2/97 as the sum of three unit fractions shows that Egyptians had considerable facility with arithmetic. It turns out that the fraction 2/97 can be written as the sum of just two distinct unit fractions: $2/97 = 1/49 + 1/4753$, and they might have expressed it this way if they had the ability to write fractions as small as 1/4753.

But with a limit of size on the denominator the ability to write arbitrary fractions as the sum of unit fractions is quite a challenge. We will take this challenge into mathematical directions that probably didn't occur to the Egyptian mathematicians.

Let us consider the example of $9/20$. It can be expressed in many different ways. Here are a few with three summands:

$$9/20 = 1/3 + 1/15 + 1/20 = 1/4 + 1/6 + 1/30 = 1/3 + 1/12 + 1/30$$
$$= 1/3 + 1/10 + 1/60 = 1/5 + 1/6 + 1/12 = 1/3 + 1/9 + 1/180.$$

Also it can be written with two summands: $9/20 = 1/4 + 1/5$. Here are three methods of generating these representations.

Method 1

Subtract the largest unit fraction less than $9/20$ and work with the difference. The largest such fraction is $1/3$, so we have $9/20 - 1/3 = 7/60$. Now the largest unit fraction less than $7/60$ is $1/9$, so we take $7/60 - 1/9 = 1/180$. Since the difference is a unit fraction, we are finished. Had it not been we would have applied the principle again. Thus the representation is $9/20 = 1/3 + 1/9 + 1/180$. This method uses a well-defined algorithm; that is, at each step we know how to proceed: Simply take the largest unit fraction less than the remainder. The following methods do not employ well-defined algorithms.

Method 2

Proceed as in Method 1 except successively subtract off unit fractions that need not be the largest unit fractions. For example, $9/20 - 1/6 = 17/60$; $17/60 - 1/5 = 1/12$. So $9/20 = 1/6 + 1/5 + 1/12$. This method can generate many different answers.

Method 3

Write $9/20$ in various nonreduced forms. For example, $9/20 = 18/40 = 27/60$. Now find ways of writing the numerator as the sum of numbers that are also factors of the denominator. For example: looking at the version $18/40$, we see that $18 = 10 + 5 + 4 + 1$. Note that 10, 5, 4, and 1 are all factors of 40. So $9/20 = 1/4 + 1/8 + 1/10 + 1/40$.

Here is another alternative: Consider $27/60$. Note that $27 = 20 + 6 + 1 = 20 + 4 + 3 = 15 + 12$. These yield the following sums: $1/3 + 1/10 + 1/60$; $1/3 + 1/15 + 1/20$; and $1/4 + 1/5$.

As we see from the example of $9/20$, fractions can be written as the sums of distinct unit fractions in a variety of ways. It would be difficult to say there is a "best" way, but let us make some definitions with this in mind. A representation for a/b will be called **algorithmic** if it is obtained by Method 1. A representation will be called **frugal** if it employs the fewest number of unit fractions. A representation will be called **economical** if its largest denominator is the smallest of the largest denominators for all representations. The algorithmic representation will be denoted by $A(a/b)$. A frugal representation will be denoted by $F(a/b)$; it need not be unique. The economical representation will be denoted by $E(a/b)$. We shall assume that a/b and all representations of it are in reduced form. So, for 9/20 we have $A(9/20) = 1/3 + 1/9 + 1/180$, $F(9/20) = E(9/20) = 1/4 + 1/5$.

1. Find the algorithmic representation for the following fractions.

   (a) 4/41

   (b) 21/68

   (c) 7/11

   (d) 12/17

   (e) 35/37

   (f) 47/72

2. Prove or disprove the following statement: The reduced proper fraction a/b can be written in the algorithmic representation as the sum of at most a different unit fractions.

3. (a) What can you say about $A(a/b)$ if $a/b = 2/n$? Prove your statements.

   (b) What can you say about $A(a/b)$ if $a/b = 3/n$? Prove your statements.

   (c) What can you say about $A(a/b)$ if $a/b = 4/n$? Prove your statements.

4. Find $A(a/b)$, $F(a/b)$, and $E(a/b)$ for the following a/b.

   (a) 7/12

   (b) 8/15

   (c) 11/49

   (d) 43/105

   (e) 151/273

   (f) 275/504

5. (a) Find the fraction of smallest denominator, d, such that
       $F(n/d)$ needs four summands. What is the fraction? If
       there are more than one with the same denominator, give
       the one with the smallest numerator.

   (b) Find the fraction of smallest denominator, d, such that
       $F(n/d)$ needs five summands. What is the fraction? If
       there are more than one with the same denominator, give
       the one with the smallest numerator.

6. This problem concerns writing unit fractions as sums of two
   unit fractions; for example, 1/6 can be so represented in four
   different ways: $1/6 = 1/7 + 1/42 = 1/8 + 1/24 = 1/9 + 1/18 = 1/10 + 1/15$.

   (a) Find all the different ways that $1/n$ can be expressed as
       sum of two unit fractions where n

       i. 7

       ii. 8

       iii. 10

       iv. 12

   (b) Find a formula for the general case $1/n$.

   (c) For what denominator(s) n for $n \leq 1000$ does the unit
       fraction $1/n$ allow the most ways of being written as the
       sum of two different unit fractions? How many ways is it?

7. Make up your own problem on this subject; for example, if
   negative as well as positive unit fractions are allowed, answer
   question 3.

5.12   Building Polygons with Dots

In this project we build polygons with dots. We will then search
for sets of dots that can be arranged in more than one polygon.
The number of dots required to build a polygon is often called a

polygonal number. The term should not be confused with the term "polygon number" that is introduced in Section 4.2. The figure shows how we can generate formulas for triangular and square numbers.

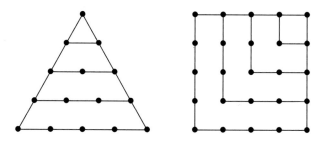

Notice that the successive triangles are formed by counting the dots from the top. We have 1 dot, then $1+2$ dots, then $1+2+3$ dots, $1+2+3+4$ dots, and finally $1+2+3+4+5$ dots. These numbers are, respectively, 1, 3, 6, 10, and 15. We studied triangular numbers in Section 1.4 and we showed that the general formula for the number of dots in a triangle with n rows of dots is $n(n+1)/2$. Similarly, for squares, we can count the successively larger squares by moving from the dot at the upper right down and proceeding clockwise about the square. We have 1 dot, then $1+3$, then $1+3+5$, $1+3+5+7$ and finally $1+3+5+7+9$ dots in the large square. These numbers are, respectively, 1, 4, 9, 16, and 25. The formula for a number of dots in a square with n dots on a side is, clearly, n^2. Next we move to pentagonal numbers. By analogy it appears as if pentagonal numbers would be generated by counting 1, $1+4$, $1+4+7$, and so on. The figure below shows that this is the case for the numbers 1, 5, and 12.

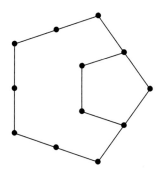

Listing the first few triangular numbers, $1, 3, 6, 10, 15, 21, 28, 36$, we find that two of these numbers are perfect squares: $1 = (1 \times 2)/1 = 1^2$ and $36 = (8 \times 9)/2 = 6^2$. There are lots more. In listing the first few pentagonal numbers, $1, 5, 12, 22, 35, 51, 70, 92, 117, 145, 176, 210$, we find two triangular numbers, 1 and $210 = (20 \times 21)/2$. There are lots more.

Finding polygonal numbers of one type that are also of another type involves solving Diophantine equations. For example, triangular numbers that are square involves solving the Diophantine equation: $n^2 = m(m + 1)/2$. This Diophantine equation can be transformed into another Diophantine equation that we have had practice solving: $x^2 - ky^2 = l$. We encourage the explorer of this project to spin out as many answers as possible with a computer, although it would take a Herculean effort for a computer to answer all of the questions below.

1. Triangular numbers that are square.

   (a) Find the Diophantine equation $x^2 - ky^2 = l$ that models the equation $n^2 = m(m + 1)/2$.

   (b) Find 20 triangular numbers that square numbers.

2. Pentagonal numbers that are square.

   (a) Find a formula for the mth pentagonal number. Call it $p(m)$.

   (b) Build a Diophantine equation $x^2 - ky^2 = l$ that models the equation $n^2 = p(m)$.

   (c) Can this Diophantine equation be solved? If so, find 10 pentagonal numbers that are square.

3. Hexagonal (six-sided) numbers that are square.

   (a) Draw a hexagonal number that has more than 25 points.

   (b) Find a formula for the mth hexagonal number. Call it $h(m)$.

   (c) Build a Diophantine equation $x^2 - ky^2 = l$ that models the equation $n^2 = h(m)$.

   (d) Can this Diophantine equation be solved? If so, find 10 hexagonal numbers that are square.

4. Try question 3 with heptagonal (seven-sided) numbers that are square.

5. Try question 3 with octagonal (eight-sided) numbers that are square.

6. Try question 3 with nonagonal (nine-sided) numbers that are square.

7. Look for pentagonal numbers that are triangular. Are there any? Do the same for hexagonal, heptagonal, octagonal, and nonagonal numbers that are triangular numbers.

8. Can you find any number, besides 1, that is a polygonal number for three different polygons? You may go beyond nine-sided figures if you wish.

9. You can build a square number out of two different triangular numbers. Show this. Also see if you can build other polygonal numbers out of triangular numbers. Find neat formulas that do this.

10. Create other questions of this type; for example, can you form a triangular number from two squares of the of the same size? That is, $2n^2 = m(m+1)/2$. Can you do the same for other polygonal numbers from 2 squares? 3 squares? etc. Try different size squares if you like.

5.13   The Decimal Universe of Fractions, I

This project and the next explore, in detail, the decimal expansion of fractions; in particular, the expansions of proper fractions n/q, where q has a periodic decimal expansion. In Part I we will examine the numbers that make up the individual decimal expansions of fractions n/q. In Part II we will examine the many different types of decimal expansions.

Recall that $1/7 = 0.\overline{142857}$. The expansion of $1/7$ has period 6. Recall also that $2/7 = 0.\overline{285714}$, $3/7 = 0.\overline{428571}$, $4/7 = 0.\overline{571428}$, $5/7 = 0.\overline{714285}$, and $6/7 = 0.\overline{857142}$. We may represent this with the digits 1, 4, 2, 8, 5, and 7 arranged in a circular fashion in a clockwise direction. As the decimal is moved around clockwise, the expansions of fractions 1/7, 3/7, 2/7, 6/7, 4/7, and 5/7 are displayed. The decimal is placed before the 1 indicating 1/7 in the figure.

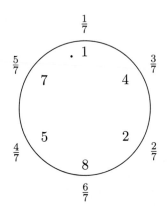

Recall that $1/13 = 0.\overline{076923}$ and $2/13 = 0.\overline{153846}$. The period of each of these expansions is also 6. We represent this with two circles.

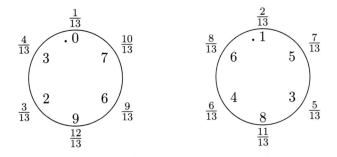

Notice that as we move the decimal point about the circle on the left we see 1/13, 10/13, 9/13, 12/13, 3/13, and 4/13. Around the circle on the right we see 2/13, 7/13, 5/13, 11/13, 6/13, and 8/13. We learned in Section 3.1 that if p is a prime other than 2 or 5, n/p has a decimal expansion of size k, where $k \mid p - 1$. So we may say that each p has associated with it a fraction, n/p, whose decimal expansions lie around circles; there are $(p - 1)/k$ circles and each circle has k digits and k associated fractions around it. We shall use the following terminology: We shall call these circles the **rings** for p. The number of digits in the circle will be called the **size of the ring**. In these examples, 7 has one ring of size 6 and 13 has two rings of size 6. If you like you may think of the prime, p, as being at the center of a universe of rings that circle around it. The same terminology can apply to numbers q, where n/q is periodic; that is, numbers q that are not divisible by 2 or 5.

1. Look at the ring for 7, pictured above. Suppose the decimal is placed next to digit a and denote the associated fraction by $n_a/7$. If you move the decimal clockwise around the circle $180°$, it ends up placed next to digit b. Let the associated fraction be $n_b/7$. What can you say about the following sums?

   (a) $a + b$

   (b) $n_a + n_b$

   (c) Does the same thing hold true for the rings associated with 13? Does the fact that there are two rings for 13 and only one ring for 7 make any difference?

2. Suppose that the decimal point is moved around the circle $120°$ at a time. So three digits are involved: the original digit a, the digit next to the decimal point after the $120°$ rotation (call it b), and the digit next to the decimal point after the $240°$ rotation (call it c). And there are three associated fractions as well: n_a/p, n_b/p, n_c/p.

   State the questions that are analogous to those in question 1 (a), (b), and (c) and answer them.

3. Suppose that the decimal point is moved about the circle $60°$ at a time until you have gone all the way around. So there are six digits and six associated fractions involved.

   State and answer the questions that are analogous to those in question 1 (a), (b) and (c).

4. (a) Draw the rings associated with the primes 11, 17, 19, and 31.

   (b) Given your experience with questions 1, 2, and 3, ask and answer the same questions and state your observations about the rings associated with 11, 17, 19, and 31. Explain carefully how your questions may have to be modified and how your answers differ from those in 1, 2, and 3.

(c) State your findings in as general conjecture as you can.

5. Prove your conjecture of question 4(c). You should recall Corollary 3.1.14; it says, in our language, if k is the size of a ring associated with p, then $k \mid p - 1$ and $10^k \equiv 1 \pmod{p}$. The following statement is also true and it may be helpful. If you use it, then be sure to prove it.

   Lemma *As you move the decimal clockwise around one digit, the numerators of the associated fractions n_a/p and n_b/p are related as follows: $n_b \equiv 10n_a \pmod{p}$.*

6. (a) Draw the rings associated with the numbers q: 21, 77, 91, 407.

   (b) Discuss the number of rings that are associated with each q.

   (c) Given your experience with questions 1, 2, 3, and 4, ask and answer the same questions and state your observations about the rings associated with 21, 77, 91, and 407. Explain carefully how your questions and answers may differ from those for prime numbers p.

7. Given your experience with questions 1, 2, 3, and 4, ask and answer the same questions and state your observations about the rings associated with 27, 81, 49, 121, and 169. Explain carefully how your questions and answers may differ from those for prime numbers p and composites pq.

8. State and prove a theorem about the sum of numerators of the fractions in a ring. If it matters whether the ring is for a prime, a composite, or a power of a prime, take that into account in your theorem.

9. Notice that in the ring for 7 the digits represented are 1, 2, 4, 5, 7, and 8. For the rings of 13 the digits represented are 0, 1, 2, 3 (twice), 4, 5, 6 (twice), 7, 8, and 9.

   (a) Check the frequency of the digits in the rings for several primes p. What do you find?

   (b) Check the frequency of the digits in the rings for several composites q. Do you find a similar results to the primes? Explain.

5.14   The Decimal Universe of Fractions, II

This project studies the number of rings and the sizes of the rings
in the universe. This requires that you collect lots of data. To get
a feel for some of the numbers asked, it would be good to know the
numbers of rings and their sizes for all primes well past 1000. You
need not know the digits in the decimal expansions, however, for this
project.

1. (a) If p and q are distinct primes, what can you say about
       the number of rings and their sizes for the number pq?
       Extend your observations to the product of three primes
       and beyond.

   (b) For prime p, what can you say about the number of rings
       and their sizes for the number p^2, p^3, and, in general, p^n?
       If the behavior differs substantially between primes, tell
       about it.

2. Which prime < 1000 has

   (a) the largest ring?

   (b) the most rings?

3. Answer question 2 for composite numbers.

4. What is the smallest prime that has

   (a) at least 100 rings?

   (b) a ring of size at least 1000?

5. Answer question 4 for composite numbers.

6. (a) Find the smallest prime that has n rings for each n, where
       $n = 1, 2, 3, 4, 5, 6, 7, 8, 9$, and 10.

   (b) Find the smallest prime that has rings of size r for each
       r, where $r = 1, 2, 3, 4, 5, 6, 7, 8, 9$, and 10.

7. Find all the numbers q that have rings of size 20. Which q has
   the least number of such rings? Which has the most?

8. Find the smallest q that has rings of size exactly

   (a) 100

(b) $2 \times 3 \times 5 \times 7 \times 11 \times 13$

9. Using a computer, collect data on the number of rings and their sizes for primes up to at least $10,000$. Then answer the following statistical survey. What percent of primes have

   (a) a single ring?

   (b) an even number of rings?

   (c) rings of an even size?

   (d) Make any general observations you wish based on the data you have collected.

5.15   The Making of a Star

This project examines the famous five point star and regular pentagon. First we construct these figures with straightedge and compass; then we examine the numbers that represent the lengths and areas.

    The following diagram shows the pentagram, a figure that has been attributed mystical properties since ancient times. One reason for this is its close connection with the golden mean, which, in turn, is related to the famous Fibonacci sequence of numbers. The figure shows the two pentagrams, $ABCDE$ and $UVWXY$, along with the five-pointed star, $ACEBDA$. Let r denote the radius of the circumscribed circle and s the inscribed circle of $UVWXY$; let R denote the radius of the circumscribed circle; and let S denote the inscribed circle of $ABCDE$.

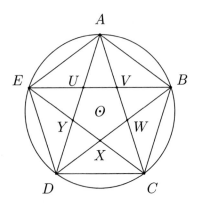

1. Construct the figure of the pentagrams with straightedge and compass.

2. Suppose that UV is of length 1. Write the following constructible lengths in terms of square roots.

   (a) EU

   (b) EV

   (c) EB

   (d) R

   (e) r

   (f) s

   (g) S

3. Find the perimeters of

   (a) $ABCDE$

   (b) $UVWXY$

   (c) $AVBWCXDYEU$

4. (a) Find the angle measures for triangles AUV, AUE, AOE, and ACE.

   (b) Find the cosines of those angles. Express them as constructible numbers.

5. Find the areas of

   (a) $ABCDE$

   (b) $UVWXY$

   (c) $AVBWCXDYEU$

6. Express the following ratios as constructible numbers.

   (a) R/r

   (b) R/S

   (c) P/p, where P is the perimeter of $ABCDE$ and p is the perimeter of $UVWXY$

   (d) P/Q, where P is as above and Q is the perimeter of $AVBWCXDYEU$

   (e) M/m, where M is the area of $ABCDE$ and m is the area
       of $UVWXY$

   (f) M/N, where N is the area of $AVBWCXDYEU$

7. Make a five-sided pyramid with base $UVWXY$ by folding up
   the triangles of the star (that is, UAV, VBW, WCX, XDY,
   and YEU) and creating the sides. Express the following as
   constructible numbers.

   (a) The height, h, of this pyramid

   (b) The ratio h/r

8. The golden mean is represented geometrically in Section 4.2.
   Numerically it is the positive solution to the polynomial equation $x^2 - x - 1 = 0$. This number is labeled ϕ. In reviewing
   your answers to questions 2, 3, 4, 5, 6, and 7, express as many
   of them as possible in terms of ϕ.

9. Make up your own problems and pursue them; for example, find
   other relationships among the measurements of the pentagram,
   or pursue the same questions for the regular hexagon and the
   star of David, a regular septagon, or a regular dodecagon.

5.16   Making Your Own Real Numbers

In this project you will create your own real numbers. (OK, the real
numbers already exist, but they are still your creations.) As we have
seen, there are several different ways of presenting real numbers. In
Chapter 4 real numbers are written as decimal expansions, continued
fraction expansions, infinite series, infinite products, and Cauchy sequences. Here we offer a recipe for another way of presenting real
numbers. Let's see how it works. Choose a sequence, any sequence,
of natural numbers a_1, a_2, $a_3, \ldots, a_n, \ldots$ and form the decimal expansion $0.a_1a_2a_3 \ldots a_n \ldots$. This is a standard expansion if all terms,
a_i, are single digits. If a_i has more than one digit, then a natural
adjustment is necessary; simply carry the digits over to the previous
digits just as you do in ordinary addition.

   Here is an example. If you chose the arithmetic sequence: $1, 4, 7,$
$\ldots, 3n + 2, \ldots$, then the real number, r, that is associated with the
sequence looks like this:

$0.147\,(10)\,(13)\,(16)\,(19)\,(22)\,(25)\,(28)\,(31)\ldots =$

$$
\begin{aligned}
&0.147 \\
+\ &.0010 \\
+\ &.00013 \\
+\ &.000016 \\
+\ &.0000019 \\
+\ &.00000022 \\
+\ &.000000025 \\
+\ &.0000000028 \\
+\ &.00000000031\ldots
\end{aligned}
$$

$= 0.1481481481\ldots.$

It looks like the sequence is converging to $0.\overline{148} = 4/27$. In fact, it is and we can prove it by the following argument. Letting r stand for $0.147(10)(13)(16)(19)(22)\ldots$, we have

$9r = 10r - r$

$\quad = (1.47(10)(13)(16)(19)(22)\ldots) - (0.147(10)(13)(16)(19)(22)\ldots)$

$\quad = 1.33333333\ldots = 4/3.$

So $r = 4/27$.

Notation: The real number associated with the sequence $\{a_n\}$ will be denoted by $r(a_n)$.

Let's recall some examples of sequences that have appeared in this book.

1. $1, 3, 5, 7, 9, \ldots, 2n - 1, \ldots$
2. $1, 2, 4, 8, 16, \ldots, 2^{n-1}, \ldots$
3. $1, 4, 9, 16, 25, \ldots, n^2, \ldots$
4. $1, 3, 6, 10, 15, \ldots, n(n+1)/2, \ldots$
5. $1, 8, 27, 64, 125, \ldots, n^3, \ldots$
6. $1, 1, 2, 3, 5, 8, 13, \ldots$
7. $1, 1, 3, 7, 17, 41, 99, \ldots$
8. $1, 1, 3, 4, 11, 15, 41, 56, \ldots$

The above sequences are specific examples of the following generalized versions:

1. Arithmetic sequences: These sequences are of the form

$$a, a + d, a + 2d, \ldots, a + nd, \ldots.$$

2. Geometric sequences: These sequences are of the form

$$a, ar, \ldots, ar^{n-1}, \ldots.$$

3–5. Figurate sequences: These sequences are made up of numbers that are used to form geometric figures: 3, square numbers; 4, triangular numbers; 5, cubic numbers.

6. Fibonacci sequence: After the first two numbers $1, 1$, the sequence proceeds like this: $a_{n+2} = a_{n+1} + a_n$.

7, 8. These sequences are formed from quotient tables. Recall the tables for $\sqrt{2}$ and $\sqrt{3}$; 7. begins with $1, 1$, and proceeds $a_{n+2} = 2a_{n+1} + a_n$ while 8. begins with $1, 1$ and proceeds: $a_{n+2} = q_n a_{n+1} + a_n$, where the q_i are $1, 2, 1, 2, \ldots$.

1. Write out, to 10-decimal accuracy, the real number $r(a_n)$ associated with the following geometric sequences $\{a_n\}$. See if you can find $r(a_n)$ exactly.

   (a) $1, 2, 4, 8, 16, \ldots$

   (b) $2, 6, 18, 54, 162, \ldots$

   (c) Make some general observations about the real number $r(a_n)$ associated with a geometric sequence $a, ak, \ldots ak^{n-1}$.

2. Write out, to 10-decimal accuracy, $r(a_n)$ for the following arithmetic sequences $\{a_n\}$. See if you can find $r(a_n)$ exactly.

   (a) $1, 2, 3, 4, 5, 6, \ldots$

   (b) $1, 3, 5, 7, 9, 11, \ldots$

   (c) $2, 4, 6, 8, 10, 12, \ldots$

   (d) Make some general observations about $r(a_n)$ associated with an arithmetic sequence $a, a + 2a, \ldots, a + nd, \ldots$.

3. Write out, to 10-decimal accuracy, $r(a_n)$ for the following figurate sequences. See if you can find $r(a_n)$ exactly. Are you still getting fractions as your answers?

   (a) $1, 3, 6, 10, 15, \ldots$

   (b) $1, 4, 9, 16, 25, \ldots$

   (c) $1, 8, 27, 64, 125, \ldots$

(d) 1, 4, 10, 20, 35,... These are tetrahedral numbers: 1, 1 + 3, 1 + 3 + 6, 1 + 3 + 6 + 10, ...

(e) Make any general observations that you can about the real numbers associated with these sequences.

4. Write out, to 10-decimal accuracy, $r(a_n)$ for the following Fibonacci and Fibonacci-like sequences. See if you can find $r(a_n)$ exactly.

(a) 1, 1, 2, 3, 5, 8, 13,...

(b) 2, 1, 3, 4, 7, 11, 18,...

(c) 1, 2, 4, 7, 12, 20, 34,... These are sums of the preceding Fibonacci numbers; that is, $1 = 1$, $2 = 1 + 1$, $4 = 1 + 1 + 2$, $7 = 1 + 1 + 2 + 3$, ...

(d) 1, 1, 1, 3, 5, 9, 17, 31, 57,... These numbers come from summing the three previous numbers, beginning with 1, 1, 1.

5. Write out, to 10-decimal accuracy, $r(a_n)$ for the following quotient table sequences. See if you can find $r(a_n)$ exactly.

(a) 1, 1, 3, 7, 17, 41, 99,... Beginning with 1, 1, proceed with $a_{n+2} = 2a_{n+1} + a_n$.

(b) 1, 1, 4, 13, 43, 142,... Beginning with 1, 1, proceed with $a_{n+2} = 3a_{n+1} + a_n$.

(c) 1, 1, 5, 21, 89, 377,... Beginning with 1, 1, proceed with $a_{n+2} = 4a_{n+1} + a_n$.

(d) Make general observations about these quotient table type of sequences.

6. Write out the first 10 terms of the sequences defined as follows: beginning with 1, 1, proceed with $a_{n+2} = q_n a_{n+1} + a_n$, where the q_i are given. Then write out, to 10 decimal accuracy, $r(a_n)$ for these quotient table type of sequences. See if you can find $r(a_n)$ exactly.

(a) Let q_i be 2, 1, 2, 1, ...

(b) Let q_i be 1, 1, 2, 1, 1, 2....

(c) Let q_i be 3, 2, 3, 2, 3, 2, ...

7. Make some general observations about $r(a_n)$ for the following types of sequences:

    (a) Quadratic sequences; that is, numbers of the form $an^2 + bn + c$; $n = 1, 2, \ldots$.

    (b) Cubic sequences; that is, numbers of the form $an^3 + bn^2 + cn + d$; $n = 1, 2, \ldots$.

    (c) Polynomial sequences; that is, numbers of the form $a_k n^k + a_{k-1} n^{k-1} + \cdots a_1 n + a_0$

    (d) Generalized Fibonacci sequences:

        i. $m, n, m + n, m + 2n, 2m + 3n, \ldots$ Beginning with first two terms m, n the sequence proceeds like this: $a_{n+2} = a_{n+1} + a_n$.

        ii. $j, k, l, j + k + l, j + 2k + 2l, 2j + 3k + 4l, \ldots$ The sequence proceeds like this: $a_{n+3} = a_{n+2} + a_{n+1} + a_n$.

    (e) The sequence $s_1, s_2, s_3, \ldots, s_n$ of partial sums; that is, $s_i = a_1 + a_2 + a_3 + \cdots + a_i$ where the real number associated with $a_1, a_2, a_3, \ldots, a_n$ is already known.

8. Create your own real numbers with imaginative sequences of your own dreams. How do the real numbers you make up differ from those we have asked about?

5.17 Building 1 the Egyptian Way

In this project we attempt to add distinct unit fractions together to make 1. For example, $1/2 + 1/3 + 1/6 = 1$. Also, $1/2 + 1/4 + 1/9 + 1/12 + 1/18 = 1$. The latter sum is more appealing to us because we will be interested in stringing together lots of unit fractions. Of course, if we don't restrict the size of the denominator, it is possible to build 1 with very many very small unit fractions. In fact, we can build 1 from an infinite number of numbers where the denominators grow without bound. The geometric series shows this: $1 = 1/2 + 1/4 + 1/8 + 1/16 + \cdots + 1/2^n + \cdots$. So we intend to limit the size of the denominators and pose the question as to how long a string of different unit fractions we can arrange to add to 1. Here are two ways of proceeding.

Method 1 We may choose a fixed denominator and list its factors and attempt to add them to equal 1. Let us choose the denominator

6. Its factors are $1, 2, 3,$ and 6. We may choose 1, 2, and 3 and add them and get 6 exactly. So $1 = 1/6 + 2/6 + 3/6 = 1/6 + 1/3 + 1/2$. Of course, this works nicely because 6 is a perfect number. Let us choose 36, a number that is relatively rich with factors. Its aliquot factors are $1, 2, 3, 4, 6, 9, 12,$ and 18. It turns out that 36 is a normal number, so we may find a subset of aliquot factors that adds to 36. In fact, we can do this in more than one way: $6 + 12 + 18 = 36$, $1 + 2 + 3 + 12 + 18 = 36$, $2 + 3 + 4 + 6 + 9 + 12 = 36$. We shall choose the longest string of summands. This gives us $36/36 = 2/36 + 3/36 + 4/36 + 6/36 + 9/36 + 12/36 = 1/18 + 1/12 + 1/9 + 1/6 + 1/4 + 1/3$. Here we have demonstrated a string of six unit fractions whose denominators are ≤ 18 that add to 1. This string is one unit fraction longer than the string we cited previously.

Method 2 A second method of approach is to create long strings of unit fractions from shorter ones. Here is how this works: $1 = 1/2 + 1/2$. But we know that $1/2 = 1/3 + 1/6$, so we may substitute that term in and get $1 = 1/2 + 1/3 + 1/6$. Continuing this substitution process, we may again substitute for $1/2$ and write $1 = 1/3 + 1/6 + 1/3 + 1/6$. Now we must change this because we have duplication of both the $1/3$ and the $1/6$. Noting that $1/3 = 1/4 + 1/12$, we obtain $1 = 1/3 + 1/6 + 1/4 + 1/12 + 1/6$. We still have duplication of the $1/6$, but we know that $1/6 = 1/9 + 1/18$. Making this substitution, we get $1 = 1/3 + 1/6 + 1/4 + 1/12 + 1/9 + 1/18$. This is the same series we found with Method 1.

We should tell you that this project, like Project 5.1, has no known answer. It is not clear that either of the two strategies is the best; in fact, there is a string longer than six unit fractions with denominators ≤ 18 that adds to 1. The good thing about this project is that the exploration takes place in uncharted waters; anything you come up with will likely be new. The bad thing about the project is that the exploration takes place in uncharted waters and you may not come up with much. But that is the risk of doing mathematics research. One thing is almost certain; if you think you have found a maximal string, you probably haven't. There is likely one that is longer.

1. (a) Find the longest string of unit fractions that add to 1 whose numerators are ≤ 18.

   (b) What percentage of the available denominators, d, where

$2 \leq d \leq 18$, did you use?

(c) Can you find different strings of unit fractions that accomplish this goal?

(d) What common denominator could you have used to generate your string?

(e) Did you generate your string by splitting up single unit fractions into two unit fractions? How?

2. Find the longest string of unit fractions that add to 1 whose numerators are ≤ 25. Answer questions (b), (c), (d), and (e) of question 1.

3. (a) Find the longest string of unit fractions that add to 1 whose numerators are ≤ 50.

   (b) Can you find different strings with the same maximal length? What is your maximal length?

   (c) What denominators must be left out of all your strings? What numbers seem to be included in your several different maximal strings?

4. (a) Find the longest string of unit fractions that add to 1 whose numerators are ≤ 100.

   (b) Can you find different strings with the same maximal length? What is your maximal length?

   (c) What denominators must be left out of all your strings? What numbers seem to be included in your several different maximal strings?

5. Describe your method for finding the unit fractions above. Did you create a way different from the suggested ones? Did you use a computer? Tell your strategy.

6. (a) Do you see any relationship between the length of your string of fractions and the cap on denominators given? What might it be?

   (b) How long would you guess a maximal string would be that added to 1 with denominators ≤ 200? Try to verify your guess.

(c) Find a number $10 < N \leq 100$ such that, given N as the cap on the denominators, the maximal string of unit fractions $1/d_i$ $(d_i \leq N)$ that sum to 1 uses the greatest percentage of denominators available.

5.18   Continued Fraction Expansions of $\sqrt{N}$

In this project we examine the infinite continued fraction expansions of the square roots of nonperfect squares. We know, from Section 4.3, that the irrational numbers that do have periodic expansions are roots of quadratic equations; that is, numbers of the form $q + r\sqrt{n}$ for some rationals q and r and natural number, n. If these numbers are, in fact, square roots, then the patterns are more apparent. This is good news. But it turns out to be better than just good news. There are patterns everywhere. They are almost endless.

In the questions that follow we shall use our notation from Section 4.3; so N, n, k, and a_i will refer to this notation: $\sqrt{N} = [n; \overline{a_1, a_2, \ldots, a_k}]$.

1. Gather data; for example, find the continued fraction expansions of $\sqrt{N}$ for $N \leq 100$, where N is not a perfect square.

2. What can you say about

   (a) the number k?
   (b) the relationship between n and a_k?
   (c) the numbers a_j, where $2 \leq j < k$?

3. Generally comment about any symmetry you see in the expansions.

4. What can you say about N if

   (a) $k = 1$?
   (b) $k = 2$, and $a_1 = 1$?
   (c) $k = 2$ and $a_1 = 2$?
   (d) $k = 2$, and $a_1 = n$?

5. What can you say about the continued fraction expansions of $\sqrt{N}$ if N is of the form

(a) $n^2 - 2$?

(b) $n^2 + n + 1$?

(c) $n^2 + k$? where $k \mid n$? Can you relax the condition that $k \mid n$?

6. Finish these thoughts:

   (a) If n appears in the periodic expansion, it is placed ...

   (b) If $n - 1$ appears in the periodic expansion, it is placed ...

   (c) With the exception of $2n$, the largest number in an expansion is ...

7. For a given n, how many expansions begin their period with a

   (a) 1?

   (b) 2?

   (c) 3?

   (d) n?

8. Consider all those numbers n where $n \leq 1000$; that is, all numbers $\sqrt{N}$ where N is not a perfect square and $N \leq 1,000,000$. Which n has the most different expansions?

   (a) of period 2? How many does it have?

   (b) of period 4 with $a_1 = 1$? How many does it have?

   (c) of period 6 with $a_1 = a_2 = 1$? How many does it have?

9. Pursue other explorations. Here are some suggestions. Find the square root that has the longest consecutive sequence of 1s. Find the square root that has the longest period. Investigate the continued fraction expansions of square roots of fractions. Can you find a relationship between the expansions of $\sqrt{N/k}$ and $\sqrt{N}$? Look at patterns other than the typical square root pattern (for example, $[n; \overline{n}]$), and $[n; \overline{1}]$, and see what kinds of numbers you get.

5.19   A Special Kind of Triangle

In this project we examine triangles that have one interior angle twice the size of another. It turns out that these triangles lead to a nice use of Diophantine equations and also give us a look at the angles associated with regular pentagons and heptagons. Everyone knows about the famous 3, 4, 5 triangle. It is a Pythagorean triangle; that is, a right triangle with integer length sides. But the 4, 5, 6 triangle is interesting too; it is one of the special kinds of triangles we are studying here. It has the property that it has integer length sides and one of its interior angles is twice another. Let's check it out.

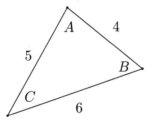

The figure shows the triangle with sides $a = 6$, $b = 5$, $c = 4$. These sides are opposite their respective angles A, B, C. The law of cosines says that $a^2 = b^2 + c^2 - 2bc \cos A$. In this case we have

$36 = 16 + 25 - 40 \cos A$. So $\cos A = 1/8$ and $A \approx 82.819244°$.

Similarly,

$16 = 25 + 36 - 60 \cos C$. So $\cos C = 3/4$ and $C \approx 41.409622°$.

Incidentally, $B \approx 55.771134°$.

Here is a general figure with the property that one of its angles is twice the size of another. Here $\angle B$ is twice the size of $\angle A$.

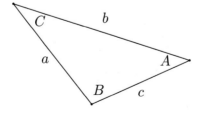

Figure 5.1

The law of sines tells us that $a/\sin A = b/\sin B = c/\sin C$. In this case we have $a/\sin A = b/\sin 2A$. Using the laws of sines, cosines, and elementary trigonometric identities, we can form a Diophantine equation in a, b, and c. Finding a relationship between b/a and c/a leads to an elementary equation in these two variables. This project proceeds as we did in Section 3.4, solving for a, b, and c for integers. Using these ideas, we can expand our investigation and develop triangles that don't necessarily have integer sides but contain angles of size $2\pi/5$ and $2\pi/7$.

1. Referring to Figure 5.1, find a Diophantine equation in a, b, and c.

2. Letting $x = b/a$ and $y = c/a$, find an equation linking x and y. Graph this equation.

3. Let $P = (0, -1)$, $Q = (q, 0)$, and R be the point of intersection of the line PQ with the curve on your graph in question 2. This point is $(x, y) = (b/a, c/a)$. What are the limitations on q so that the segments of length $a, b,$ and c

   (a) form a triangle?

   (b) form an acute triangle?

   (c) form an obtuse triangle?

   (d) form a triangle with integer length sides?

4. Find all the triangles with integer sides with perimeter ≤ 100 where one interior angle is twice another. Write down all the sides and all the angles.

5. Find the triangle(s) with perimeter ≤ 1000 that are close to isosceles as possible.

6. (a) Find the number(s) q such that the triangle $x, y, 1$ is isosceles.

   (b) Where have you seen the number(s) q before? Explain.

   (c) Is q rational? constructible? arithmetic?

   (d) Letting α and 2α represent interior angles of a triangle, is $\cos \alpha$ rational? constructible? arithmetic?

7. Find a q such that the interior angles are α, 2α, and 4α. Write a polynomial in x, one of whose roots can be used to build a triangle with these angles and with sides x, y, 1. Find the size of the angles.

(a) Is q rational? Constructible? arithmetic?

(b) Is $\cos\alpha$ rational? constructible? arithmetic?

(c) What can you say about the interior angles of a regular septagon?

(d) What can you say about the seventh polygon number, P_7?

8. Pursue other questions; for example, what can you say about the areas of these triangles? What can you say about triangles where one interior angle is triple another?

5.20 Polygon Numbers

In this project we analyze regular polygons; in particular, the lengths of the sides and the cosines of the central angles. A regular polygon has n equal sides and n equal angles. The central angle is of size $360°/n$. We found lots of information about regular polygons in Section 4.2. For example, we learned that it is possible to construct regular p-gons for $p = 3$, 5, 17, 65, 257, and 65537. We saw that $\cos(72°) = (-1 + \sqrt{5})/4$. It is impossible to construct a regular 18-gon because $\cos(20°)$ is not a constructible number. However, $\cos(20°)$ is arithmetic; it is a solution to the cubic $8x^3 - 6x - 1$. And we learned that all polygon numbers are algebraic.

Since the regular 17-gon, the heptadecagon, is included in the following problems, here is one way it can be constructed. This was the procedure of English mathematician John Lowry in 1819.

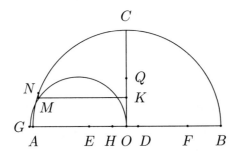

On the radius OC of a semicircle, find the midpoint Q, and on the diameter AB, perpendicular to this radius, lay off from the center O the distance $OD = 1/8$ of the radius. Lay off DF and DE each equal to DQ, and EG and FH equal, respectively, to EQ and FQ. Lay off OK, the mean proportional between OH and OQ; that is, $OH : OK = OK : OQ$. Through K draw KM parallel to AB, meeting the semicircle described on OG in M. Draw MN parallel to OC, cutting the circle about O in N. The arc AN is $1/17$ of the circumference.

In the following problems we restrict our examination to regular n-gons, where $3 \leq n \leq 18$, that are inscribed in unit circles.

1. Explain how to construct each of the constructible regular polygons with at most 18 sides. Try to explain why the constructions really do work.

2. Find $\cos \theta$, where θ is the interior angle for each of the constructible polygons. Express your answer in terms of square roots.

3. Find P_n for each of the constructible regular n-gons. Express your answer in terms of square roots.

4. Find a polynomial of smallest degree that has P_n as a root.

5. (a) Find $\cos \theta$, where θ is the interior angle for each of the polygons where P_n is arithmetic but not constructible.

   (b) Express your answers in terms of roots.

   (c) Find a polynomial of smallest degree that has $\cos \theta$ as a root.

6. (a) Find P_n for each of the regular n-gons whose sides are arithmetic but not constructible.

   (b) Express your answers in terms of roots.

   (c) Find a polynomial of smallest degree that contains the root, P_n.

7. (a) Find $\cos \theta$, where θ is the interior angle for each of the regular polygons not covered in the questions above.

   (b) Find the irreducible polynomials that each satisfies.

8. (a) Find P_n for each of the regular n-gons not covered in the previous questions.

   (b) Find the irreducible polynomials that these numbers satisfy.

   (c) Express these numbers to 10 decimal accuracy.

9. Ask more questions about regular polygons; for example, what are the perimeters of the polygons inscribed in a unit circle? What are the perimeters of the polygons circumscribed on a unit circle? What can you say about the areas of each of these polygons?

5.21   Continued Fraction Expansions

This project deals with the entries in the continued fraction representations of real numbers. In particular, we are interested in the frequency and the size of the entries. So this is a statistical study that requires gathering lots of data.

It has been said that, when performing the Euclidean algorithm on two numbers n and m, you really need only do simple subtraction most of the time. This means that, most often, the quotient that you obtain in the division process is 1. In turn, this means that the continued fraction expansion of the n/m has lots of 1s. If this is true, then it is natural to ask what the frequency of 1s is. Does this preponderance of 1s carry over to expansions of irrational numbers. Is the preponderance of the same magnitude for irrational numbers as it is for rational numbers?

In Section 4.4 we alluded to the fact that algebraic numbers have "tamer" continued fraction expansions than transcendental numbers have. While we may not call it "wild", we know that the expansions of π and e show exciting behavior; π has a dramatic jump in an early entry in its expansion and e and linear combinations using e have unbounded entries. As for algebraic numbers, we know that roots of linear polynomials (that is, fractions) have terminating expansions. And roots of quadratic polynomials have repeating expansions. A natural question is whether roots of lower degree polynomials have, in some sense, tamer entries than roots of higher degree polynomials. Generally speaking, is there a way of describing the tameness of the expansions of various types of algebraic numbers? And is there a way of describing the "wild" characteristics of transcendental expansions?

1. (a) Find the continued fraction expansions for at least 1000 "randomly" chosen proper fractions a/b.

   (b) For each of the fractions count the number of instances of each of the numbers in the representations. Then calculate the frequency of the numbers 1, 2, 3, 4, 5, and the frequency of the numbers ≥ 10.

2. (a) Choose at least 1000 irrational numbers and find their continued fraction expansions, $[a_0; a_1, \ldots, a_m, \ldots,]$ for a fixed $m \geq 10$. Try to mix up your choice of irrational numbers. For example, don't take them all to be constructible, arithmetic, or algebraic.

   (b) For each of the fractions count the number of instances of each of the numbers, $a_1, \ldots, a_m$ in the representations. Then calculate the frequency of the numbers 1, 2, 3, 4, 5 and the frequency of the numbers ≥ 10.

   (c) Compare your findings here to your findings for rational numbers and irrational numbers.

3. Lumping together all your data from questions 1 and 2, calculate frequencies of the numbers 1, 2, 3, 4, 5 and the frequency of numbers ≥ 10. Can you find a formula that could fit your results?

4. (a) Gather lots of data on continued fraction expansions of algebraic numbers. Make sure you get your data from roots of polynomials of various degrees, both small and large. To get going you can start with numbers of the form $\sqrt[n]{k}$. Carry your expansions out to at least 10 places; the more places, the better. The entries that concern us are $a_1, \ldots, a_m$ for as large an m as you can manage.

   (b) Consider the magnitudes of the entries; in particular the largest of the entries. Is there evidence that roots of higher-degree polynomials have larger maximum entries or a greater frequency of larger entries? Be as specific as you can.

5. (a) Gather data on the continued fraction expansions of transcendental numbers as you did for algebraic numbers.

   (b) Consider the magnitudes of the entries; in particular, the largest of the entries. Compare the frequency and maximum sizes of the entries with those of algebraic numbers. What do you find? Be as specific as you can.

6. Make up a definition of "tame" and "wild" continued fraction expansions of real numbers and give examples from the expansions of real numbers you have gathered in this project. Make appropriate conjectures and try to justify them.

7. Analyze all of your data in other ways. For example, take your results from question 1 and compare the size of a and b in your fraction a/b with the length of the continued fraction expansion of a/b. Make up a theorem that begins like this: Given a and b, the Euclidean algorithm will take, on average, k steps to find $\gcd(a, b)$ where k is ....

8. Comment on the limitations of your computer as it tries to generate continued fraction expansions. Do you believe that you can reasonably speak to the issues of "tame" and "wild" with the limitations you have cited?

Further Reading

The following books can serve as an introduction to further study. This list is a *very* small sample of the available books that touch on subjects covered in this book.

Beiler, Albert, *Recreations in the Theory of Mathematics*, 2nd ed., Dover Publications, New York, 1966

Bell, E. T., *Men of Mathematics*, Simon Schuster, New York, 1937

Boyer, Carl and Merzbach, Uta, *Introduction A History of Mathematics*, John Wiley Sons, New York, 1989

Burn, R. P., *A Pathway into Number Theory*, Cambridge University Press, Cambridge, 1992

Courant, R. and Robbins H., *What Is Mathematics?*, Oxford University Press, New York, 1941

David, Donald, *The Nature and Power of Mathematics*, Princeton University Press, Princeton, N.J., 1993

Hardy, G. H., *Ramanujan*, Cambridge University Press, Cambridge, 1940

Huntley, H. E., *The Divine Proportion*, Dover Publications, New York, 1970

Khinchin, A. Ya, *Continued Fractions*, University of Chicago Press, 1964

Levecque, W. J., *Elementary Theory of Numbers* , Addison Wesley, Reading, Mass., 1962

Niven, Ivan, *Irrational Numbers*, Carus Mathematical Monographs, Number 11, The Mathematical Association of America, Washington D.C., 1956

Niven, Ivan, *Numbers: Rational and Irrational*, Random House, New Haven, Conn., 1961.

Ogilvy, C. Stanley and Anderson, John T., *Excursions in Number Theory*, Dover Publications, New York, 1988

Richards, Stephen, *A Number for your Thoughts*, S. P. Richards, New Providence, N.J., 1984

Roberts, Joe, *Lure of the Integers*, The Mathematical Association of America, Washington D.C., 1992

Silverman, Joseph, *A Friendly Introduction to Number Theory*, Prentice Hall, Upper Saddle River, N.J., 1997

van der Waerden, B. L., *A History of Algebra*, Springer Verlag, New York, 1985

Wells, David, *The Penguin Dictionary of Curious and Interesting Numbers*, Penguin Books, London, 1986

Wisner, Robert, *A Panorama of Numbers*, Scott Foresman, Glenview, Ill., 1970

Index